FAX: Facsimile Technology and Systems

Third Edition

For a listing of recent titles in the *Artech House Telecommunications Library*, turn to the back of this book.

FAX: Facsimile Technology and Systems

Third Edition

Kenneth McConnell
Dennis Bodson
Stephen Urban

Artech House
Boston • London

Library of Congress Cataloging-in-Publication Data
McConnell, Kenneth R.
FAX : facsimile technology and systems / Kenneth R. McConnell, Dennis Bodson, Stephen Urban. — 3rd ed.
p. cm. — (Artech House telecommunications library)
Includes bibliographical references and index.
ISBN 0-89006-944-1 (alk. paper)
1. Facsimile transmission. I. Bodson, Dennis. II. Urban, Stephen. III. Title IV. Series.
TK6710.M33 1999
621.382'35—dc21 99-23716
CIP

British Library Cataloguing in Publication Data
McConnell, Kenneth R.
FAX : facsimile technology and systems. — 3rd ed. — (Artech House telecommunications library)
1. Facsimile transmission
I. Title II. Bodson, Dennis III. Urban, Stephen
621.3'82'35

ISBN 0-89006-944-1

Cover design by Lynda Fishbourne

685 Canton Street
Norwood, MA 02062

International Standard Book Number: 0-89006-944-1
Library of Congress Catalog Card Number: 99-23716

10 9 8 7 6 5 4 3 2 1

To our wives,
Jane, Priscilla, and Rita

Contents

Preface

Now that fax has risen from a century of neglect to its rightful place in telecommunications, is there anything new to say about it? The answer is yes, and the updated information will be found in this third edition of *FAX*.

With fax on local area networks (LANs) and the Internet, 64-Kbps digital fax on the integrated services digital network (ISDN), higher resolution and higher speed fax modems, fax standards for color, fax capability included in new computers and personal digital assistants (PDAs), simultaneous voice and fax, and higher speed handshake, there have been many changes in fax technology, which are reflected and discussed in this new edition of *FAX*.

From the primitive methods of sending messages over distance to the sophisticated communication systems of the modern office, fax has emerged from obscurity to being an indispensable tool for business, government, professional, and education personnel.

Because readers of previous editions have expressed an interest in this history, we have expanded the section on the evolution of fax, recording the eventual rise of fax over telegraph as a preferred method for sending messages. Today, fax copies have replaced telegrams for fast communication of important messages. Postal Telegraph vanished years ago, and Western Union now sends money orders rather than telegrams.

The term *fac-simile* came into being on November 10, 1885, when it appeared in a patent application. The hyphen was later removed, and then the term was shortened to *fax*. In this book, we use *facsimile* and *fax* because both are still in use. The term *telefax* is still used in Europe. Additional word evolutions, including some that seem contradictory, have crept into the fax language. The recent *pel density* is unrelated to density in the optical sense. Likewise,

/25.4 mm ("per 25.4 mm") rather than */in* ("per inch") is the result of an international compromise. *Pel*, *pixel*, and *baud* are often used incorrectly. As language evolves, what was incorrect may become commonly used. In this edition of *FAX*, we use alternative terms as they are commonly used by those in the field. More details are given in the text and the glossary.

The third edition of *FAX* provides extensive coverage of Group 3 standards, which are ever evolving. However, while some of the material in this book covers standards work that does not yet have final approval, that could change at any time. We strongly recommend that the actual standards printed be checked before proceeding with any work based on them. We have used reasonable care to ensure the accuracy of the material presented here, but we cannot be responsible for undesirable consequences that may result from use of the information contained herein.

The focus of the third edition of *FAX* is on the more than 100 million Group 3 fax machines and computers with fax capability. Starting with the basics of fax, this new edition proceeds to show the evolution of fax from a mechanical contraption to a specialized computer. Also included is an explanation of what is inside fax machines (scanners, printers, modems, architecture), protocols, image compression, communication channels for fax, color fax, fax test charts and testing procedures, standards organizations, and fax standards. Further coverage reports on standards in process for Group 3 with higher speed modems, improved protocols, simultaneous voice/fax, and fax on the Internet.

The information in this handbook is intended for all who are interested in fax: users, engineers, designers, imaging specialists, consultants, computer programmers, company executives, communication planners, government regulators, marketing personnel, distributors, dealers, retailers. Communication history buffs may be particularly interested in Chapter 2.

Chapter 1 is written in nontechnical language to assist beginners who are considering the purchase of a fax machine or who want to improve their method of using fax. Technical or semitechnical language is used in many parts of the book. Those who are familiar with fax may want to skip the nontechnical parts. For any readers who might be unfamiliar with fax terminology, there is a comprehensive glossary.

Some information is provided about the special fax machine designs required for applications such as transmission of news photos, weather maps and photos from satellites, newspaper printing masters, and fingerprints. The number of these specialized fax machines is very small (less than 0.1% of the number in office use).

Acknowledgments

Particular recognition and thanks go to Gregory Bain, George Constantinou, Jim Dahmen, David Duehren, Bob Easson, Stephen Perschau, James Rafferty, Dick Schaphorst, Steven Karty, John Shonnard, and George Stamps for providing material and reviewing text as the manuscript was being prepared.

The authors also wish to thank the following people for their helpful information and advice: Bill von Alven, Ajay Batheja, John Bingham, Philip Bogosian, Bob Boykin, Paul Brobst, Karl Clough, Lester Davis, Joseph Decuir, Bruce DeGrasse, Michel Didier, Roger Free, Eugene Gavenman, George Giddings, Glenn Griffith, Ralph Grant, Herb Israel, Chuck Jacobson, Bob Krallinger, Granger Kelley, Ken Kretchmer, John Long, Gary Lucas, Tim McCullough, Lloyd McIntyre, John Munch, Jr., Toby Nixon, Janet Orndorff, Alan Pugh, Ed Reilley, Bob Robinson, Dave Shaler, Herman Silbiger, Neil Starkey, Hiroshi Tanaka, Bob Trachy, Bill Tyrrel, Virginius Vaughan, Bill Webb, Don Weber, Pierre-Andre Wenger, Jim Wilcox, Yasuhiro Yamazaki, Aoki Yoshimori, Charles Zeigler, and Mohamed Zennaki.

Kenneth McConnell gives special thanks to the late Austin G. Cooley, the facsimile pioneer and his mentor. Information from Cooley's writings as well as knowledge gleaned from working for many years under his leadership are embedded in Mr. McConnell's chapters. Kenneth McConnell also thanks his wife, Priscilla (who started training at the Signal Corps Laboratories to become a facsimile engineer just a few months after he took his first facsimile engineering job there), for her intensive help with this book over a period of three years, gathering new material, editing all his work, and even some writing. The expanded coverage of the evolution of fax and the earlier communication methods that it replaced would not have been possible without her assistance. He also is grateful for her suggestions to make the manuscript more easily understandable and, he hopes, more interesting.

1

Fax Basics

1.1 What Fax Does

Fax makes remote copies of documents anywhere in the world at speeds almost as fast as making copies on an office copier and at a cost often less than that of postage. The fax machine is somewhat like two office copiers electrically connected by a telephone line. The following description of what goes on inside the fax machine will help the reader to understand subsequent chapters.

Compare talking on a telephone with faxing a printed page to a distant point. Words spoken into a telephone are converted by its microphone into electrical signals that travel over the telephone line to the receiving telephone, where they are changed back to sound by that set's earphone. For fax, the imaging portion of the sending "office copier" produces electrical signals that represent the page being sent within the same tonal (frequency) range as voice. Fax signals pass over the telephone line connection the same way as voice. At the receiving end, the signals are changed back into an image and printed by the other half of the receiving "office copier." A telephone line can connect two fax machines anywhere in the world.

1.2 How Facsimile Works

The sending portion of the fax machine must read, or *scan*, the page being sent and produce electrical signals that represent it. This task can be compared to a person taking two to four minutes to read a page into a telephone. Reading starts at the top left side and a line of characters is viewed as the eye scans across the page.

For sending by fax, the reading is done electronically, also starting at the top left corner of a page. Fax reads 2 to 10 pages a minute, much faster than most people read. Fax does not recognize the printed letters but reads the small black dots that form each character.

To visualize this process, imagine the page being sent as printed on very fine graph paper with 200 squares per inch. Fax starts by reading only the row of squares across the tops of the printed characters. Each square is either a black dot or a white dot. Starting at the left end of the row, the black dots and the white dots are read in sequence (1,728 dots in all). The process then repeats for the next row. One full line of text takes 10 to 20 additional rows (*scanning lines*). A whole page comprises 2,200 scanning lines (at fine resolution).

Next, imagine that the recording paper at the fax receiver has a pattern of graph paper squares that corresponds to the transmitter pattern of squares. The receiver effectively "sees" the transmitter squares and copies the black ones into their corresponding positions. After all the scanning lines have been read and sent to the fax receiver, the black squares needed to reproduce the original image are filled in, forming the characters and lines of the page sent. Thus, a fax transmission is converted into a fax copy.

The technique to bring this process about uses an electronic eye to convert the page image into electrical signals, with a strong pulse for each white dot (square) and a weak pulse for each black dot (square). All the adjacent white dots in the scan line up to the start of black dots are counted and then converted to a coded number. A short digital number can be the code for a long string of dots to achieve faster transmission. The same process is used to code adjacent black dots, producing alternating white and black codes. Tones are produced (by a modem) to send this information over a telephone line. The modem then takes a group of 16 successive digital numbers and sends only one symbol to the fax receiver (for even faster transmission). The symbols follow each other 2,400 times a second as a signal,[1] which goes through the regular dial-up telephone system in a manner similar to voice.[2] Listening on a telephone, one would hear a hissing sound similar to that produced by an AM radio tuned between stations. At the fax receiver, the transmitter processes are reversed and each digital number is transformed into a string of dots. After the dots reach across the entire page, the line is printed. Each square across the

1. Phase/amplitude modulation.
2. To decode the signal as the same digital number formed at the transmitter, the modem must synchronize on the received symbol rate.

recording page has its own writing element coupled to the corresponding memory element so that each black dot is printed in the correct position.

1.3 Fax as a Tool for Home or Office

The ability to send exact copies to any other location easily, rapidly, and at a reasonable cost has made fax equipment and personal computer facsimile (PC-fax) a necessary part of most businesses, small and large. Hospitals, law offices, real estate firms, architectural designers, lobbying groups, political organizations, marinas, delicatessens, computer software sales, and technical support departments all depend heavily on fax. Now that most personal computers (PCs) have built-in fax capability, people at home are discovering it as another useful way to communicate with the outside world.

For documents already in page form, sending them by a Group 3 fax machine usually is cheaper, faster, and far simpler than sending them by PC-fax or express mail services. Fax copies can be made automatically anywhere in the world by merely placing the pages into the fax machine hopper and pressing a button. Many machines can store pages in memory after "reading" them from the hopper. The pages feed through rapidly and the originals can be put away during transmission. Such fax automatic features free the sender from attending the operation.

For text and image files generated in a computer or residing in computer memory, built-in PC-fax capability makes it easy to send them directly to another PC-fax or a Group 3 fax machine. The file is sent as a fax from the print menu by selecting "fax" instead of a printer. The pages are sent automatically without being printed. This method produces better fax copy than sending from a fax machine, because spots read in a fax scanning operation do not exactly match the edges of text or graphics, causing ragged edges in the fax copy. The PC-fax character generator skips the scanning step and aligns text precisely with the printing squares at the fax receiver, making the received copy much sharper. Graphic images are also sharper.

The regular voice telephone network is used to connect the sending and receiving fax machines, but talking on a fax call is unusual. All operations may be automatic, including dialing, sending the pages, and disconnecting (hanging up) after all pages are received at the distant location. The facsimile machine first dials the distant fax machine listed for a single button pressed (a telephone is not needed). The receiving fax machine answers the call, and then the fax machines send short messages back and forth (*handshaking*) to select the highest speed possible; no action is required by an operator at either the sending end or the receiving end. Fax copies can be sent anywhere another fax machine is

plugged into a telephone jack. No conversion of signals is needed to cross international borders.

Understanding the pros and cons of different setups helps determine the kind of equipment to install. Whether a fax machine or a PC-fax would be better depends on how it will be used. Although connection to its own telephone line is preferred, sharing a line for voice calls or with a computer is possible. Fax machines that record on plain paper are preferable, but they still cost more than those that record on thermal paper. The lowest priced Group 3 machines now being sold in the United States cost less than $200 (only 1.5% of the price of the earliest Group 3 machines in 1980).

1.4 Mail Delivery Changes in the 1990s

Sending documents via fax mail, e-mail, or "snail mail" (U.S. Postal Service) is a choice now widely available (unfortunately, to junk mailers as well). Many package delivery systems also handle mail. In the few cases where fax cannot be used because originals are required, the older systems, such as the USPS, FedEx, and UPS, are the obvious means of delivery. But when speed is important, fax transmits in seconds. Although e-mail should go in seconds, it may be delayed by Internet traffic or problems with the Internet service provider (ISP). When the recipient needs hard copy in hand immediately, the fax machine is preferable to PC-fax, which usually requires some processing at the receiving end. When the quality of the received copy is more important, PC-fax is better. As for e-mail, converting it into print or graphics is more complicated than producing documents from fax-received copy.

Undoubtedly, mail systems, as well as other ways of communicating, will continue to change as we move into the millennium.

2

From Drum Beats to Fax Beeps: Telecommunication Beginnings

Drum beats sounding in the African air, smoke plumes rising over ancient plains, shouts from hilltop to hilltop, wails from Australian aborigines' tubes, clanging Chinese "tam tams"—this was "telecommunicating" in the ancient world, eons before telefax, telegraph, telephoto, and their telecousins: telephone, Telex™, Teletype™, television. By using fire and sound waves for messaging in the years before electricity was available, our ancestors showed great ingenuity.

In this section, it is our intent to trace early telecommunication as the root of facsimile to show how the earliest forms of communication led to the telegraph and fax and how they are related. Reasonable care was used in determining the sequence and accuracy of sources, but occasionally we found some sources to have conflicting information.

"Send more arms!" or "Persians coming by sea!" might have been the meaning in a spiral of smoke or flash of fire seen by an ancient Greek or Roman observer. From mountain top to mountain top or tower to tower, the messages went out across the miles. Greek literature tells of signals from fire posts that announced the end of the Trojan War. The Bible, the Talmud, and Roman histories mention fire beacons as means to transmit warnings and news. The Persians established communication by a line of shouting men. Victory or defeat in military operations often depended on which side could communicate faster.

Even in the twentieth century with all the methods of rapid communication, the news media must wait for white smoke to rise out of Vatican chimneys before they can report the election of a new Pope. In the fourth century B.C., runners were commonly used to deliver news over distances. Modern day marathon races commemorate the 26-mile run from Marathon to Athens in 490 B.C. after a small Greek army was victorious over a much larger Persian army. The famous victory message was "Rejoice, we conquer!" but the more important part of the message contained a warning of imminent attack from the Persian fleet. When the Persians sailed into Athens and saw that they were expected, they turned back without attacking. With the discovery that a certain kind of pigeon could be trained to fly between specific destinations came another way to send messages over distance. From as early as 2350 B.C. right up to the twentieth century, these little birds have performed an airborne delivery service. Roman magistrates were reported to have used homing pigeons to fly in with "late for dinner" or "cancel appointment" messages. Genghis Khan used a pigeon relay system in the twelfth century A.D., and the British Air Force used them in 1918 to drop messages into French and Belgian territory occupied by the Germans. As recently as 1981, Lockheed engineers used them to transmit microfilm negatives of drawings to a location 25 miles away. With their pigeon-feed salaries and traffic-free routes, the birds proved to be a fast and cost-effective system.

Although the term *telegraph* is universally associated with Samuel Morse and his revolutionary form of electrical communication, telegraphy (from the Greek "communicating over distance") had been in use 20 centuries before electricity was harnessed. One might think that the pre-electrical communication was not a telegraph, but it was called a telegraph and the methods were basically the same. Signaled information (telegraphy) had to be manually copied as it was received, whether from optical telegraphs or from electrical telegraphs. An electrical telegraph had been demonstrated 70 years earlier than the Morse 1844 patent, and the first fax design was a recording telegraph.

2.1 Ancient Telegraphy

There are many ancient Greek accounts of the telegraph and its origins. Pre-electric telegraph systems in many forms were in use over a widespread area in many ancient empires, including Egypt, Babylon, Greece, and Rome. Curiously, there seems to have been a virtual standstill in new communication ideas from those times until about the ninth century A.D. Then some pioneers began to experiment with clocks, and much later early-sixteenth-century navies used flag signaling. The invention of the telescope in 1608 fed into the imaginations of inventors of those days, and better systems of telegraphy began to appear.

We have chosen one system from around 360 B.C. and one from the late eighteenth century A.D. to describe in more detail.

2.1.1 Water Telegraph

Even before 300 B.C., the water telegraph was used, but no water moved between stations. To envision its principle, imagine that both sending and receiving stations had stop watches and the message sent corresponded to the length of time between a start signal (torch, drum beat, or other) and a stop signal. By selecting the time corresponding to the desired message, one of many predetermined messages could be sent. For "stop watches," the ancients used water timers in the form of matching water containers with plugged holes near the bottom. Before starting to send the message, the containers were filled. As the first signal was sent, the plugs were removed from both the sending and the receiving containers. The second signal was sent when the sending water level reached the desired message. The receiving station then shut off the water flow and observed the water level to know which message was sent. Each message corresponded to a level of water in the containers. A floating vertical pole with marks made it easier to read the message. If four sets of successive signals were sent for four water containers, thousands of messages could be selected. Alternatively, combinations of marks could represent characters and numbers to send any message. By 150 B.C., a net of these devices spanned thousands of miles across the Roman Empire.

2.1.2 Chappe's Optical Telegraph (Semaphore)

In late-eighteenth-century Europe, many scientists were concentrating on "building a better mousetrap" in the communications field, particularly telegraphy. One method for development was the semaphore, whereby messages went out from strategically located posts, often towers, by an arrangement of flags, lights, or mechanical arms visible at a distance. The French system designed by Claude Chappe found favor with the controlling government agencies and was adopted in the last years of the eighteenth century. Those were turbulent times, with many countries warring along their borders. Napoleon Bonaparte was in the midst of the turmoil and recognized that rapid communications would give him an advantage. He expanded the French system throughout Western Europe, undoubtedly facilitating his conquests. One version of Chappe's invention consisted of a horizontal beam with an arm at each end that was set at the proper angle in increments of 45 degrees (eight positions). The horizontal beam was set on one of three vertical positions. An arrangement of ropes and pulleys allowed the semaphore settings to be changed rapidly. By using a set of two successive semaphore patterns, 36,864 words would be available ($8 \cdot 8 \cdot 3 = 192$; $192^2 = 36{,}864$). About 25,000 words and phrases of the French language were assigned a semaphore combination in the code books. Another version had eight angles for the center arm and sent one character per semaphore pattern. With telescopes, the distance between relay posts could be increased. The rate of sending information was comparable to that of early electrical telegraphs, and the Chappe system prevailed in some locations until about 1880.

Our own folklore illustrates the combined use of simple semaphore and horseback courier. In Longfellow's poem, "Paul Revere's Ride," the courier Revere receives the lantern signal "one if by land, two if by sea" and delivers the warning using the fastest means then available—his horse. In those days, sending written messages was limited to physical transportation and delivery, requiring relay teams of runners or horseback riders operating between stations along the way. Telegraph methods before electricity required human effort with teams of flag wavers, shouters, torch signalers, or semaphore operators to copy information and relay it from one station to the next. The advent of electrical telegraphy did not eliminate the need for human effort—the message could be transmitted over electrical channels, but both sending and receiving operators had to perform manual tasks as each character was sent and received. Elimination of manual operations did not occur until facsimile came into use.

2.2 Electrical Telegraphy

Although electricity was discovered many years before the first fax machine design, fax was one of the first applications of electricity. Almost nothing was known about it when the word *electric* was coined by Sir William Gilbert, physician to Queen Elizabeth I and King James I of England. After 17 years of experimenting with electricity and magnetism, he prepared the treatise "The Magnet, Magnetic Bodies, and the Great Magnet the Earth," which covered everything known about electricity and magnetism. Gilbert presented his paper to the Royal Society in 1600 and is credited as the founder of electricity and magnetic sciences. Gilbert's treatise stimulated practical compass designing, but not much was done with static electricity until 1663, when German physicist Otto Van Gorky built a machine that generated a spark. Sir Gilbert had become intrigued by the experiments reported in 600 B.C. by the ancient Greek philosopher Thales, who had observed attractive effects between lodestones and iron. Attractive effects were also reported with a piece of amber. After being rubbed with a cloth, the amber attracted certain light objects. Thales believed that these substances possessed souls.

In 1752, Benjamin Franklin undertook a high-risk experiment by flying a kite during a thunderstorm. He attached a metal spike and one end of a string to the kite and a metal key to the end of the string. Fortunately, Franklin was not electrocuted, which he almost certainly would have been if the kite had been hit by a bolt of lightning. Instead he was able to draw large sparks from the key when electricity flowed down the rain-soaked string. He observed that the electricity had the same properties as the static electricity produced by friction. It took until 1774, 174 years after Gilbert's treatise, before high electrostatic voltages were utilized to power the telegraph demonstrated by George Lesage and Charles Morrison in Geneva, Switzerland. It had 24 separate wires stretched through a doorway between two rooms, from the sending station to the receiving station. When a wire from the friction electrostatic generator touched a selected wire at the sending end, a pith ball at the receiving end moved toward the wire, indicating the letter sent. Apparently, the plan to send over a distance was to put each wire into a separate pipe with insulating beads to center it. An electric telegraph line was established in 1793 between Paris and Lille, France, and in 1796 between Madrid and Aranjuez, Spain, but operational problems, such as losing the static charge because of high humidity, made for erratic performance.

In 1799, low-voltage-sustained electrical current that could be easily controlled was produced when Alessandro Volta of Italy constructed the first chemical-electric cell. It had one terminal on a strip of copper and another terminal on a strip of zinc, separated in an acid solution. Steady current flowed when a wire was connected between the terminals. Further experimenting led to a voltaic pile (later called a *battery*), a sandwich of multiple layers of zinc and copper discs separated by flannel discs soaked in the solution. In 1800, 200 years after Gilbert's treatise on electricity, the Royal Society of England heard a paper on the voltaic battery and watched a demonstration of its capability. The invention of the voltaic cell and battery furnished the essential element for electrical communication, leading to inventions of a practical telegraph and the facsimile machine. In 1820, Christian Oersted of Denmark discovered that current flowing in a wire near a compass caused the needle to deflect. Although this was not the first system of signaling with electricity, it inspired others to develop a practical telegraph powered by current from a low-voltage battery rather than a very high voltage and electrostatic attraction for an indicator. For the first time, it seemed possible to communicate instantaneously from the sending station to the receiving station, no matter how far away. Then a race was on to see who would be first to exploit low-voltage electricity for a practical telegraph. Many inventors came up with telegraph systems.

In 1831, Joseph Henry built and operated a telegraph with batteries over a distance of 1 mile. He later assisted Samuel Morse in his development of telegraph. In 1829, Henry had discovered how to greatly increase the attractive force of an electromagnet by insulating wire and wrapping it around an iron core, allowing a large number of turns to be used [1]. In 1837, Charles Wheatstone and William Cooke developed and installed a five-needle telegraph between two London railway stations. At the receiver, two of either five or six needles on a triangular grid pattern pointed to the same intersection, selecting the letter sent. It was more efficient than the earlier system that required a separate wire for each letter, but it still needed too many wires. Two-needle and single-needle systems were devised later, and the Wheatstone-Cooke single needle system was used for many years, even after the Morse telegraph started.

2.2.1 Morse

Although he was not the first to use a single-wire circuit between the sending and the receiving station, Samuel F. B. Morse is credited with inventing the telegraph, an invention that revolutionized telecommunications. Morse had achieved fame as a sculptor and portrait painter before becoming interested in electricity. A personal tragedy drove him to seek faster ways of communication; while traveling overseas, he did not hear of his wife's illness before her untimely death, and the death notice itself took another four days to arrive.

Morse attended lectures about electricity and observed a demonstration of an electromagnet invented by William Sturgeon. He became interested after seeing that current sent through turns of wire wrapped around an iron bar attracted another iron bar. The idea of an electric telegraph came to him in 1832 on his return from Europe aboard the packet ship *Sully*, when he discussed electricity and magnetism with his fellow passengers. He was fired with enthusiasm for his idea and immediately started working on it. Shortage of money forced him to devote part of his time to a teaching position at New York City College while constructing his telegraph. In 1837, five years after beginning his project, he demonstrated his first telegraph. The recorder was built around a picture frame from one of his canvasses. An electromagnet pulled down a pencil, marking dots and dashes onto a strip of moving paper driven by clock works. A weight on a cord over a pulley lifted the pencil between marks. Each letter of the alphabet had a different key with a distinctive code. To send, the key for the letter selected was drawn through a slot and closed electrical contact springs to generate a group of short and long current pulses from a battery. The receiver printed the pulses as groups of dots and dashes on the moving paper strip. Earlier systems lacked the advantage of his telegraph code or the single-wire telegraph line.[1]

In 1838, Morse demonstrated his apparatus to the U.S. Congress in Washington, D.C., hoping to get a grant to continue his work, but he was refused. With a second attempt in 1843, Congress passed a bill on the last night of the session, appropriating $30,000 for testing the telegraph. The line was strung between the Supreme Court chambers in Washington and the Baltimore & Ohio railroad station in Baltimore. On May 24, 1844, Morse tapped out the famous "What hath God wrought." Success with the operation of this line followed, but when Morse offered his invention to the government, Congress refused, on the recommendation of the postmaster-general, who was "uncertain that the revenues could be made to equal the expenditures." Congress authorized the sale of the telegraph to private interests of the government-financed Washington-Baltimore railroad and gave up the opportunity to develop the telegraph as part of the postal system. The Morse telegraph patent was not issued until 1848.

Morse coded the alphabet efficiently by assigning the shortest codes to the letters used most often. This code concept was probably more important to communications development than the equipment designs of the Morse telegraph itself. Today, a similar concept codes Group 3 fax signals efficiently, with assignment of the shortest code words to the black marks and the white spaces that occur most frequently. The Baudot code, often thought to be the first binary code, was not patented until 1874. By 1856, telegraph operators had learned how to read relay clicks and no longer needed the paper printout. In 1861, the first transcontinental telegraph message was sent from California to Washington, D.C. Western Union and Postal Telegraph dominated the record communication systems for many years but were eventually replaced by Teletype™, TWX™, and Telex™, and those systems have now given way to fax and e-mail.

2.3 Facsimile Starts

At the same time that the Morse telegraph system to transmit documents was developing, another system was emerging. That other system was *facsimile,* now more often called *fax.* Both telegraph and fax had many things in common. Both sent pulses made from switching contacts, interrupting current from a battery to convey information. Both received by marking on paper. Both

1. The Morse picture frame instrument now resides in the Smithsonian Museum, Washington, D.C.

connected sending and receiving units with a single wire, using ground to complete the circuit. With the electrical telegraph, someone at the receiving station had to write down the message, even after the Morse telegraph marked the message on paper as dots and dashes. But when facsimile was invented, the received message was readable *without human processing or translation.*

2.3.1 Bain

On November 27, 1843, a Scottish physicist named Alexander Bain obtained English Patent No. 9,745 for a recording telegraph (facsimile unit) (Figure 2.1). Bain designed an electric pendulum clock that used battery power to keep it running. His system synchronized two clocks, locking their pendulum swings together. The pendulum furnished both driving power for fax scanning or fax

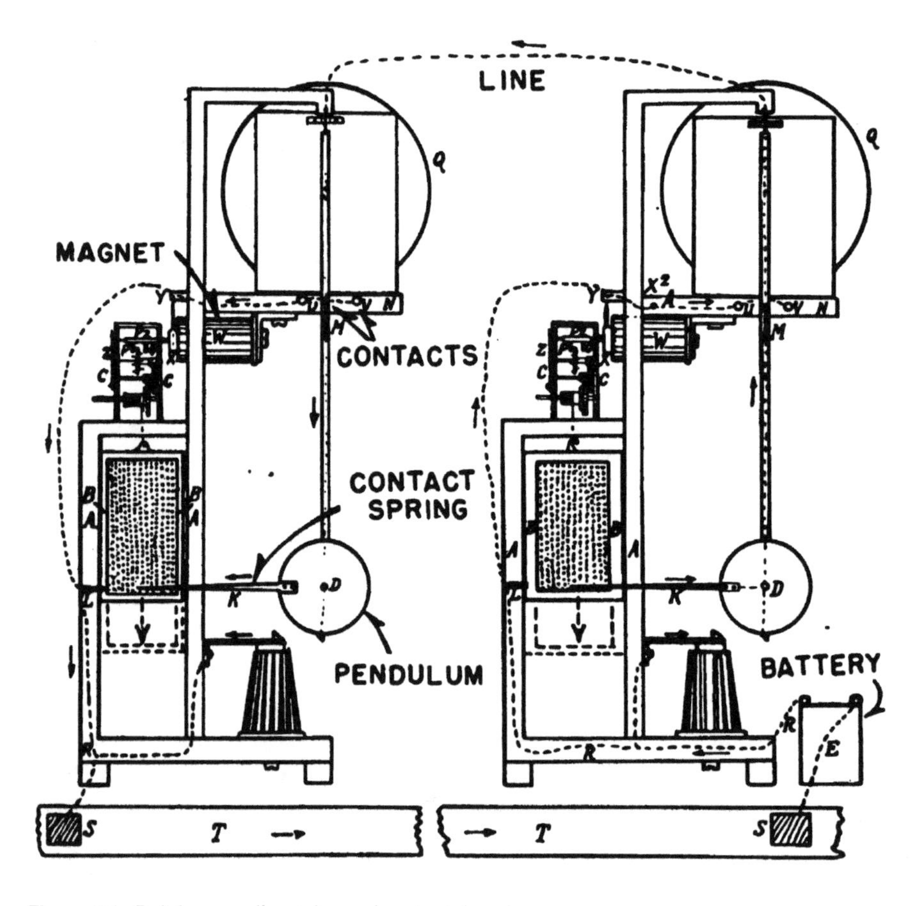

Figure 2.1 Bain's recording telegraph patent drawing.

recording and the timing standard for keeping them in step. To prevent the pendulums from drifting, Bain reset them at the start of each swing.

At the sending end, electrical contact between the stylus (K in Figure 2.1), a spring on the swinging pendulum arm, and the metal printer's type (B) completed a circuit from the battery, sending current through the receiver stylus. At the receiving end, a stylus on its swinging pendulum arm made marks on damp electrolytic paper whenever current flowed. A battery (E) was connected in series with the transmitter stylus (K), a telegraph wire (the dotted line), the receiver stylus (K), and back to the battery (E) through earth ground (T). The transmitter stylus (K) wiped over the raised metal type (B), closing the circuit each time it touched the raised metal typeface. As the receiver stylus wiped over damp electrolytic paper (A), current passing through it made marks corresponding to the pattern traced by the pendulum swing at the transmitter. The printer's type and the recording sheet were in frames held up by cords. An escapement mechanism (C) let out the cords to lower the frames one line height (scanning line) for each swing of the pendulum. When the type was touched by the stylus at the transmitter (corresponding to black), marking started at the fax receiver. Marking continued until black signal was no longer sent from the fax transmitter. A few years later, in time for the London Fair of 1851, Bain had constructed a working model.

2.3.2 Bakewell

A second facsimile machine, constructed by English physicist Frederick Bakewell (1848, English Patent No. 12,352), was shown at the 1851 fair. Bakewell's rotating drum and leadscrew design probably were constructed before Bain's pendulum design. While the Bain scanner used electrical contact between a stylus and a metal typeface, Bakewell substituted a more convenient method by using insulating ink written on a metal surface. His system still would not scan documents written on paper, however. At short distances, a low-voltage battery was adequate to provide the required current through the stylus to mark the chemically treated paper at the receiver. Bakewell's drum was rotated by gravity, with a driving weight and a cord wrapped a few turns around another drum. As the driving weight lowered, the cylinder used for transmitting or receiving was rotated through a gear train (Figure 2.2).

2.3.3 Caselli

In 1865 in France, Giovanni Caselli inaugurated the first commercial facsimile service. The fax transceiver, which he named "Pantelegraph," was patented in September 1861. This ancestor of the modern fax machine was huge, about

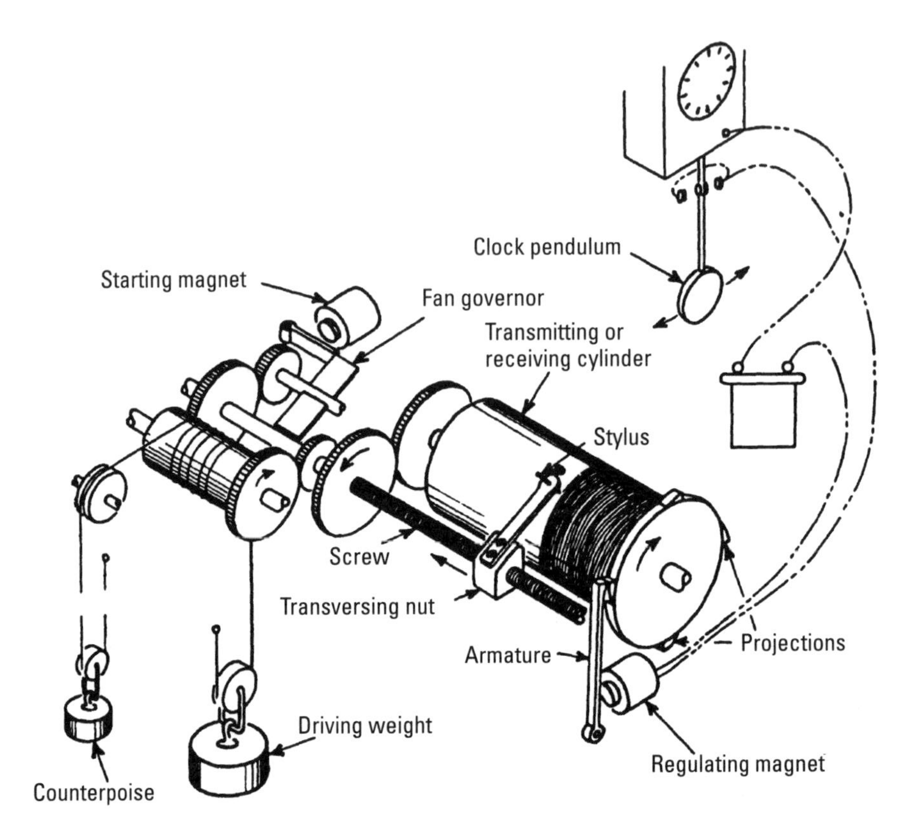

Figure 2.2 Bakewell's rotating cylinder.

7 ft (213 cm) high [2] and with a pendulum almost that long. The electrically driven pendulum needed that length to provide enough power for moving a scanning stylus and a recording stylus through a clever but complex mechanism. In Figure 2.3, the image on the left is an enlargement showing details of the scanning and recording mechanisms from the midsection of the right image.

Oscillating motion from the pendulum swing was connected by a rod (W in Figure 2.3) through a rocker arm (Y) to both the transmitting stylus (V) and the recording stylus. The styli moved back and forth in an arc, touching the outside of the sending and receiving drum segments (T). A leadscrew moved the styli one line per swing. The page to be faxed was drawn with insulating ink on a metal surface (similar to Bakewell's machine).

Caselli, an Italian, formulated ideas for his inventions when he was teaching physics at the University of Florence. Research into telegraphic transmission of images led to his first fax machine design. On November 10, 1855, he

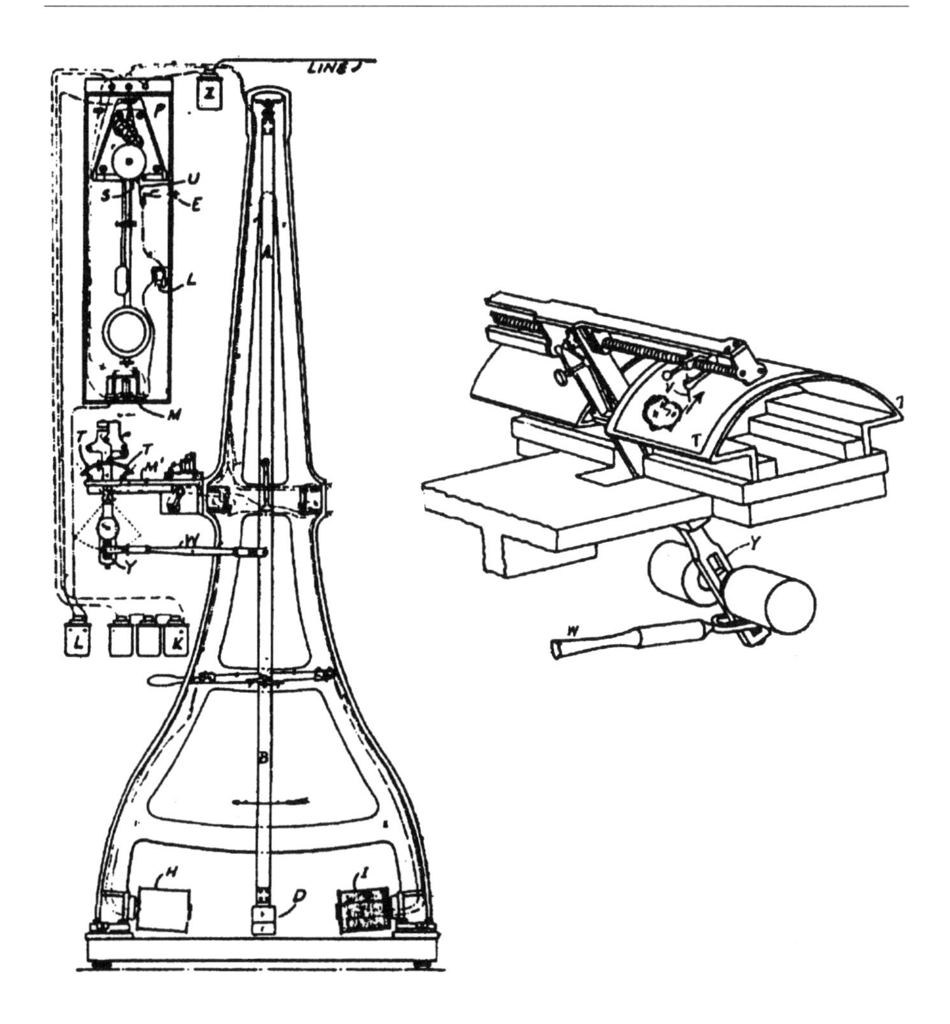

Figure 2.3 First commercial fax, by Caselli.

applied for an English patent for "improvements in transmitting *fac-simile* writings and drawings by means of electric currents" and stated that "the object [is]... to convey with great rapidity the *fac-simile* of handwriting and drawing to any distance whatever." The word "facsimile" was surely used before this time to mean "an exact reproduction or copy," but this is the first record we have found on use of the word for sending copies over distance by electricity. Caselli's 1855 Pantographic Telegraph fax machine was a cross between those by Bakewell (a drum) and by Bain (a pendulum). This first design led to the start of an improved design and probably was instrumental in getting funding for the Pantelegraph.

In Caselli's day, an inventor's success often depended on the sponsorship of a wealthy patron. Such a benefactor for Caselli was the Grand Duke of Tuscany, who paid the expenses and promoted the invention. In 1857, Caselli traveled to Paris, where he assisted Jean Foucault in building his pendulum. Caselli, bolstered by Foucault's recommendation, obtained the assistance of inventor and mechanical engineer Paul Froment. The Pantelegraph design was completed and assembled in Froment's workshop. The machine so excited the Parisian scientific world that a Pantelegraph Society was created to promote its adoption. An important and ardent supporter was Emperor Napoleon III (Bonaparte's nephew), who watched a demonstration on May 10, 1860, in Froment's workshop. The emperor gave access to the telegraph lines Caselli needed to continue his experiments. Using a telegraph line between Paris and Amiens, he conducted a successful intercity test. Favorable French press for the Pantelegraph produced such interest that members of the aristocracy, as well as those from the scientific and commercial worlds, swarmed to Froment's workshop for demonstrations. Crowning those successes, Caselli accepted the invitation of King Victor-Emmanuel of Italy to demonstrate his machine in a series of events at the Florence Exhibition in 1861. Figure 2.4 is a reproduction of copy received over a long-distance telegraph circuit that year.

Two years later, the French governing bodies authorized utilization of a telegraph line between Paris and Marseille. In England, authorization was granted for a four-month trial using a line between London and Liverpool. In 1865, Caselli initiated fax service in France between Paris and Lyon. That circuit was later extended to several other French cities. Figure 2.5 shows a demonstration of Caselli's system for the emperor at the French Bureau of Electric Telegraph in 1866.

When word of the capabilities of the Pantelegraph reached the Chinese emperor, he sent representatives to Paris to assess its usefulness. The scouts were sufficiently impressed by the concept of a device that could transmit the Chinese ideograms to recommend its introduction into Peking. Negotiations were started but never completed.

Although the Pantelegraph had been designed to transmit images, it was perfectly able to send written messages, like today's fax machines. The French Telegraphs Administration, threatened by a superior system, imposed such a prohibitive tariff for handwritten dispatches that Caselli was unable to continue. This shortsightedness (or perversity) on the part of governing bodies limited use of the Pantelegraph to the transmission of trademarks or bank signatures, which other systems could not do. In both France and Italy, governments were beset by competing interests. Unfortunately, the initial enthusiasm for Caselli's machine waned as the Pantelegraph Society failed to grasp the opportunity for market development. The promise of a universal system

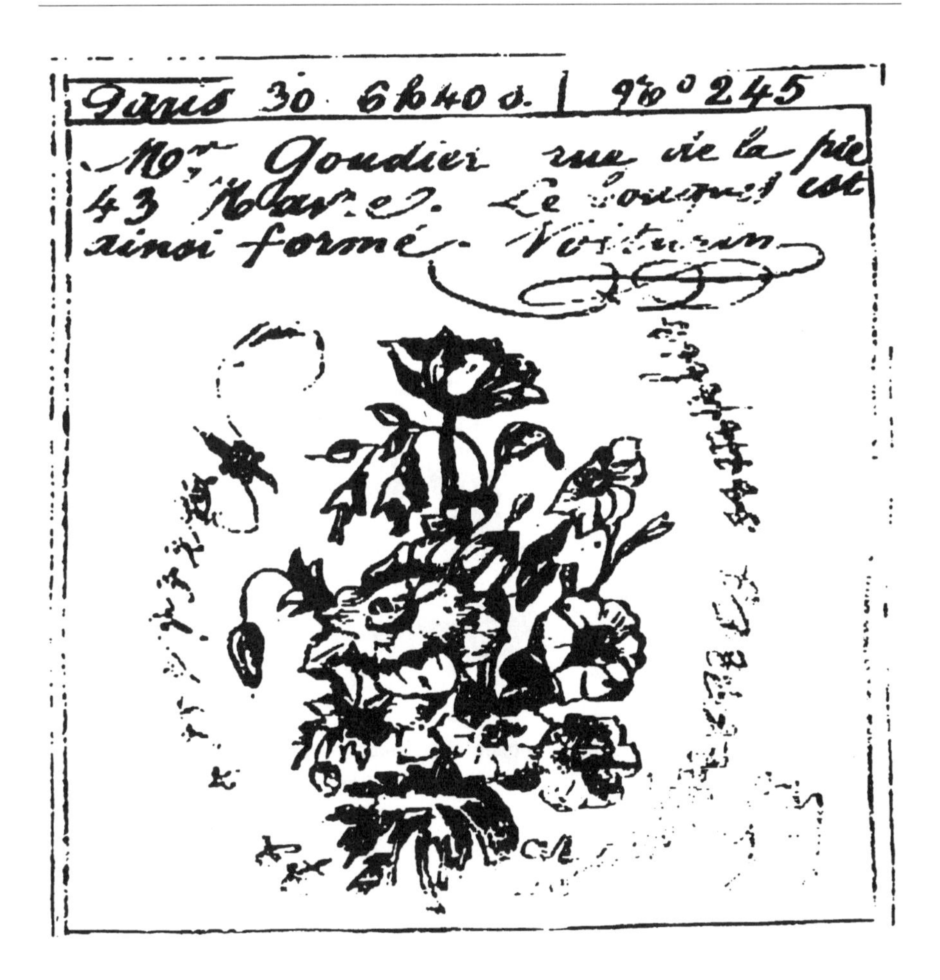

Figure 2.4 Received copy on Pantelegraph, 1861.

available to the common people was not fulfilled, and in 1870 the Pantelegraph was eased into obsolescence. This story has the familiar ring of an imaginative genius unappreciated in his native country, achieving great success elsewhere to public acclaim but ultimately failing due to political maneuvering. The development of fax technology might have accelerated if the Caselli fax network operating over French telegraph lines had not been thwarted.

Pantelegraphs still exist and are on display in a few museums. The newly renovated Musée des Arts et Metiers in Paris has one of them. To celebrate the 100-year anniversary of the Pantelegraph, demonstrations of its capabilities were run in 1961 and again in 1982, when they "operated faultlessly, six hours a day, for several months" (according to Julien Feydy in an article in the

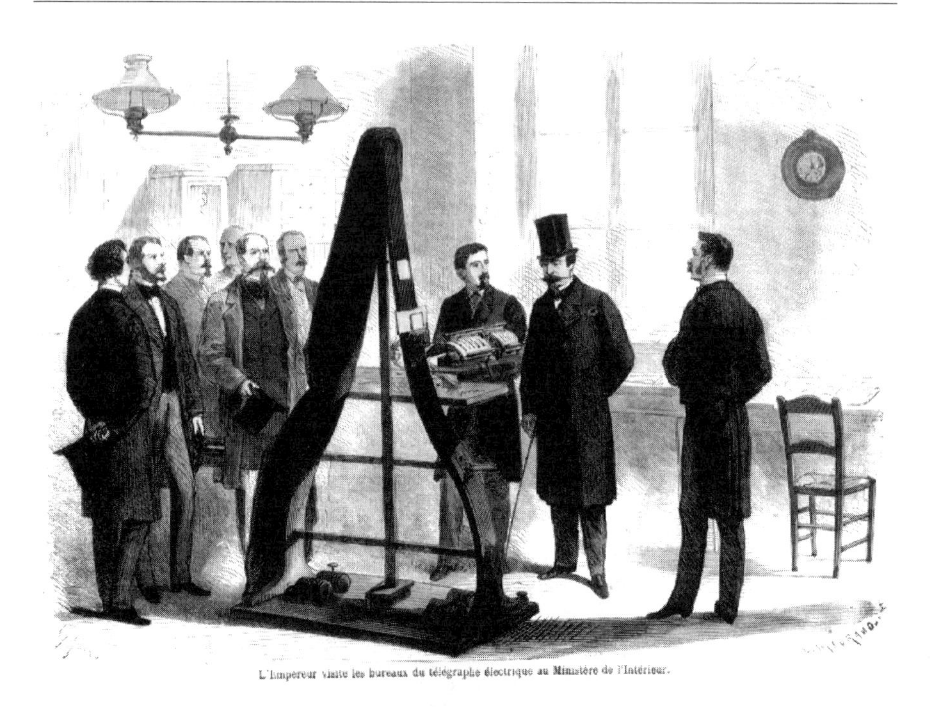

Figure 2.5 Napoleon III views the Pantelegraph. (*Source:* France Telecom Organisme National de Soutien Communication. Reprinted with permission.)

June 1995 issue of *La Revue*, the publication of the Musée). A CD-ROM issued by the museum shows a video of a Pantelegraph in operation, with the pendulum ticking away at about 120 scans per minute.

2.3.4 Photographs by Fax

Another step forward in facsimile development came about in the late 1800s, when transmission of photographs by fax was first demonstrated. The picture sent was on a plate made by the screening process used for printing photographs (similar to today's technique). A fine pattern of dots small enough to be barely noticeable was on the screened plate; the size of the dots determined the shade of gray represented. One method used contact scanning of the metal printing plate, in which raised dots substituted for Bain's metal typefaces, sending current when a dot was touched. In another method, dots on the screened plate were depressions (like rotogravure printing). The depth that the stylus moved into the hole controlled the resistance of a rheostat (variable resistor) connected in series with a battery to control the amount of dc current sent.

2.3.5 Korn

In 1902, Dr. Arthur Korn of Germany demonstrated the first practical fax system for sending photographs, using many new ideas and improvements (Figure 2.6). Up to this point, some form of contact scanning was needed to generate the fax signals, even for photographs. As the distances between stations increased, higher voltages were switched at the scanning stylus, resulting in much sparking as the stylus passed over the insulating ink on the conducting surface of the page being sent. Although relays solved the sparking problem, their very slow response rate severely limited the sending speed, making fax impractical for most applications. Korn's system overcame the contact scanning problem by sending photographs from film negatives using optical scanning, giving "eyes" to the fax machine. Korn's fax recording was also on film. His system had a tuning fork frequency standard in the sending and receiving fax machines, with phonic wheel motors driving both units in synchronism. These improvements established Korn as the leading figure in the facsimile field for many years.[2]

The transmitter lamp (23 in Figure 2.6) was focused by a lens (22) into a small spot of light on the transparent original wrapped around a glass drum (4). Light passed through the original and the drum, was deflected downward by a mirror (29), and was picked up by a selenium cell (2). The synchronous motor (18) turned the drum (4), which moved vertically 0.0127 inch (0.5 mm) per revolution.

In 1906, Korn's newsphoto facsimile equipment with optical scanning and photographic recording on film was put into regular service over telegraph circuits from Munich to Berlin. This was nine years before the telephone repeater was developed, making long-distance telephony practical. The Korn system later expanded to provide news picture service to London, Paris, and Monte Carlo (500 miles). Transmission time was 12 minutes per 5-by-8-inch page at a resolution of 51 lines/inch (2 lines/mm).[3] Korn's receiver had a recording rate of 300 pixels/s, which was later increased to 1,000 pixels/s when telephone lines with amplifiers became available. The first design used a low-pressure gaseous discharge as the light source for recording. Relays were much

---

2. In 1911, Dr. Korn published a book on fax titled *Handbuch der Phototelegraphie und Telautographie* (publisher unknown). A leader in European facsimile from 1900, Korn emigrated to the United States in 1939. He gave a series of lectures on telephotography at Stephens Institute of Technology in 1940 and was employed as a consultant by Times Facsimile Corporation prior to his death in 1945. Much of the material on the history of fax cited in the Times Facsimile *Service Bulletin* came from his book and lecture notes.

3. Because telephone amplifiers did not exist, it is presumed that a telegraph channel was used for the 500-mile network. That rate was very fast for a telegraph channel (75 pulses/s maximum).

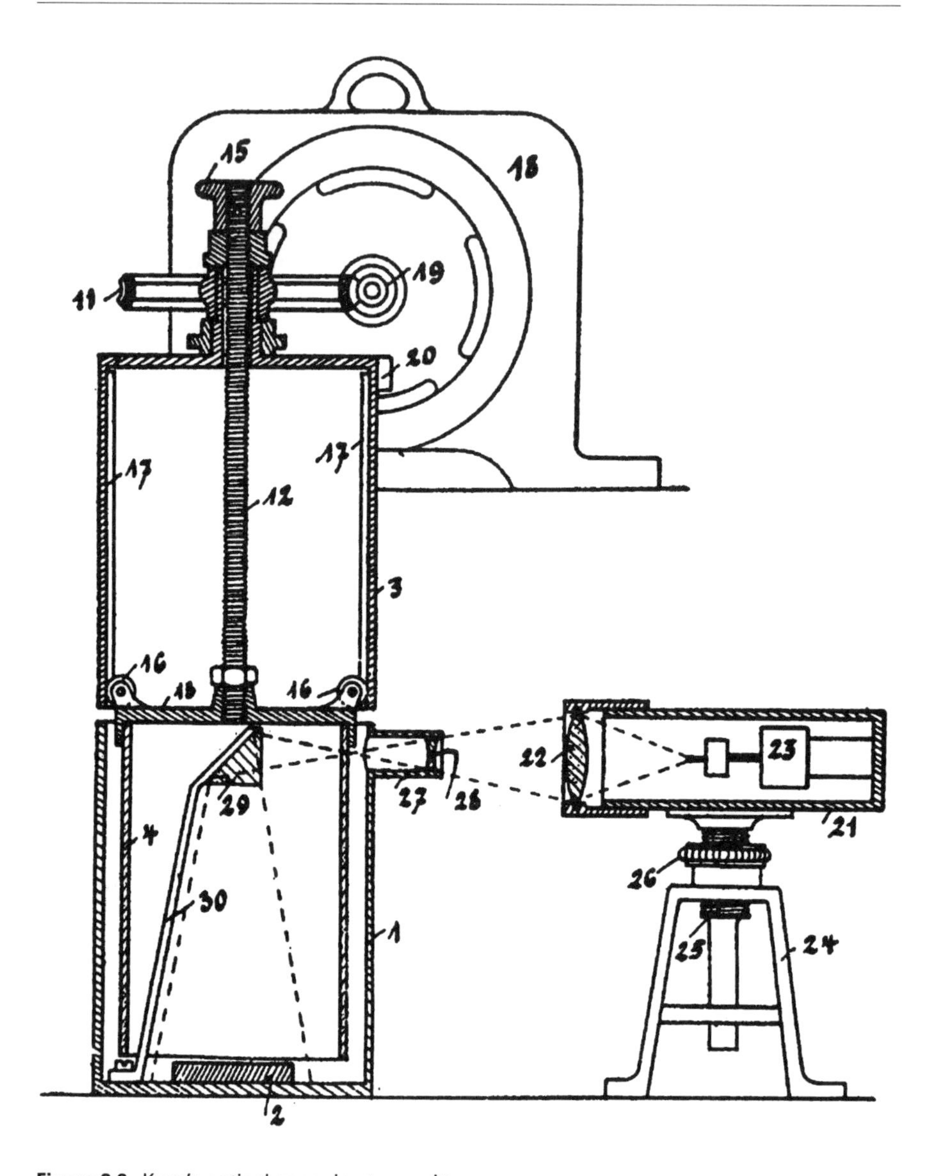

Figure 2.6 Korn's optical scanning transmitter.

too slow to turn on the lamp under control of the received fax signal. Korn's spark gap modulator was marginal on speed. (A similar type of light source became the most popular one for photographic fax recording after the vacuum tube was invented.) Korn then developed an electromagnetic light shutter or string galvanometer (71 in Figure 2.7).

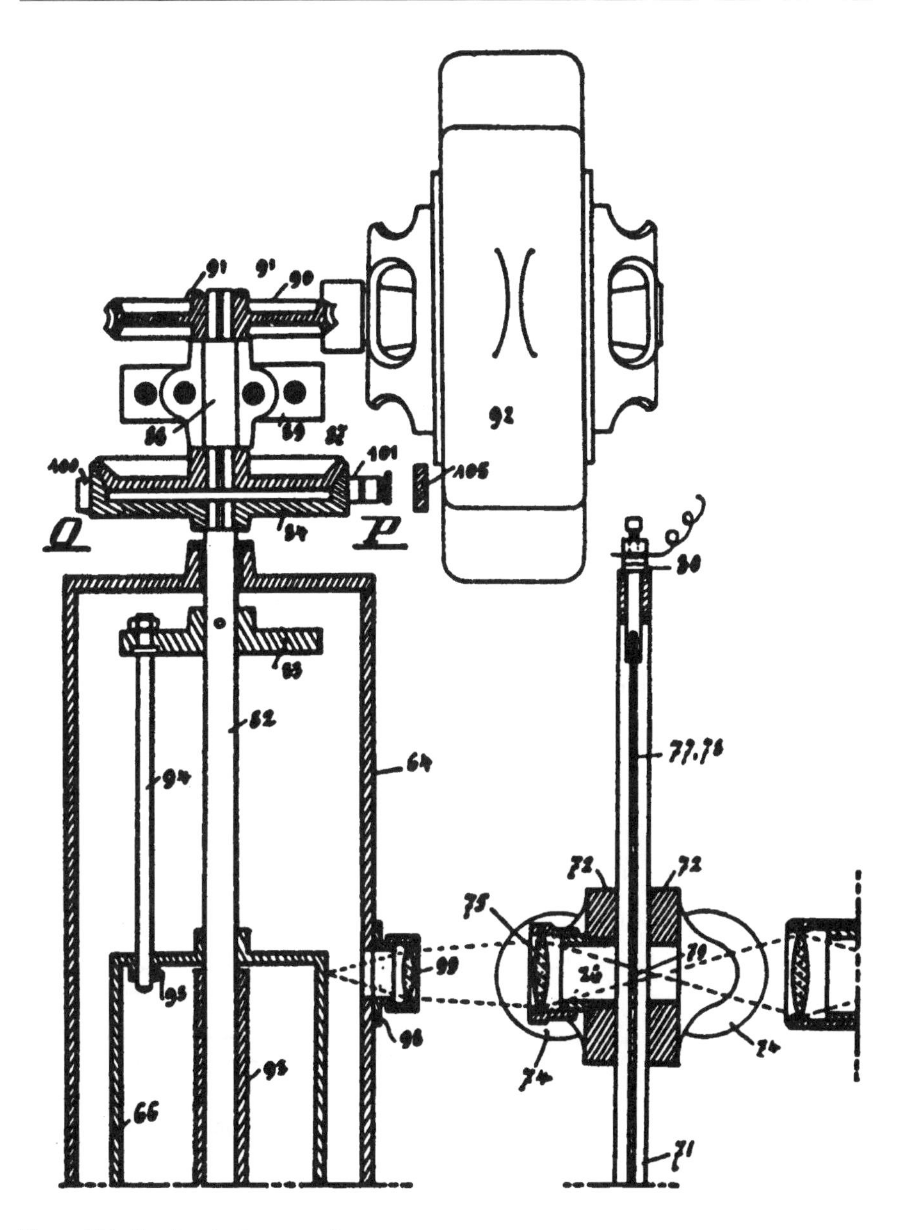

Figure 2.7 Korn's telephoto receiver.

The shutter was a small piece of aluminum foil held by a loop of wire (77–78 in the figure) in a magnetic field. A steady light source was focused onto the shutter. It was closed when there was no signal, to prevent exposure of film on the recording drum. When a fax signal from the scanning of white was received, current passed through a coil, opening the shutter. The light passing through was focused to record on photographic film. The selenium photosensor, with its slow response speed, poor stability, and lack of uniformity in electrical characteristics, still limited performance of the fax scanner. A gaseous-medium photoelectric cell patented in 1908 by two German scientists, Elster and Geitel, overcame these limitations but would not pass the heavy currents needed in existing dc-loop facsimile equipment designs. When the vacuum tube became available, the photoelectric cell greatly improved performance of fax transmitters [3].

2.4 Fax Development Problems

Achieving good fax machine performance took many years, with contributions from thousands of people in fax and other communication arts. Progress was slow because improving performance in one problem area often revealed another problem. That a fax machine could be made to work even before the invention of the telephone is an amazing accomplishment (Figure 2.8).

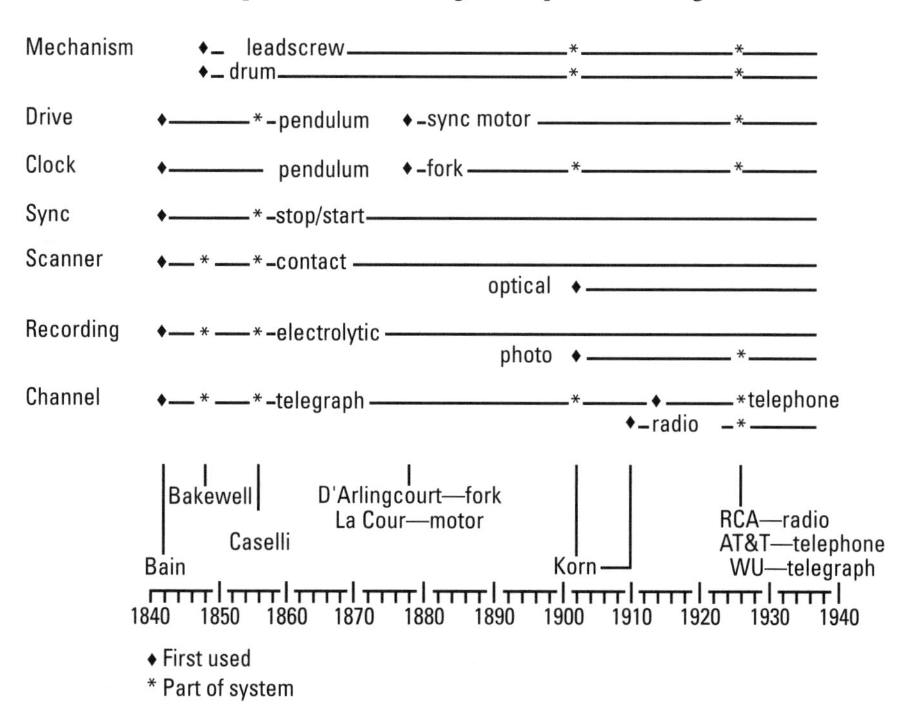

Figure 2.8 Development of facsimile technology, 1840–1936.

The power to move the scanning stylus and page of the early machines came from clock weights, gravity, or a pendulum. No photosensors existed. Cameras, which had just started to be used, took minutes to expose a picture. The communication channel was a telegraph line with no amplifiers or repeating relays.

2.4.1 Synchronization

Synchronization between sending and receiving fax machines was a serious problem. To understand the problem, consider that both the page being sent and the received page have an electronic grid overlay of squares, like graph paper. The square size would be very small, corresponding to the size of each dot used to build up the received image. To make an image of characters and lines sent from the fax transmitter, the receiver must know which squares to color black. If the columns of squares across the page were assigned letters (a, b, c,...), and the rows of squares down the page were assigned numbers (1, 2, 3,...), each black square could be identified at the transmitter and the information sent by a series of signal codes and marked at the fax receiver. The code a1 would signify a black dot in the upper left corner. Codes b1, b2, and b3 would start recording a vertical line near the left edge of the paper. The system just described had no synchronization problem, but there was no way of implementing a grid overlay into a fax system until digital techniques became available.

On earlier fax systems, instead of information on *where* to mark the recording sheet, information on *when* to mark the recording sheet was sent. A scanning spot the size of one electronic grid square was physically moved across the page being sent, or a drum with the page mounted was rotated to produce a similar result. The receiving fax had a similar movement of a spot over a recording sheet. When the scanning spot started to move over a black mark, a black signal was sent to start marking the recording sheet. The mark was made wherever the recording spot was at the time the signal was received. There was no electronic grid for the dot to snap to; it was a real-time system.

Placing the black dots in the proper squares was much more difficult before digital fax machines were developed. Analog fax machines (Group 2 and earlier) had a single scanning spot at the fax transmitter and a single recording spot at the fax receiver. Once the transmitter started sending a page, the receiver spot had to track the scanning spot by starting at exactly the same time and moving at exactly the same speed. Because the fax receiver had no way of locking in for synchronism on the received fax signal, each fax machine required a very high precision clock (frequency standard). Most fax machines before today's Group 3 did not have the writing speed of the fax recorder (printer) locked in synchronism with the scanning speed of the fax transmitter. Instead, the speed of the recording spot across the page was locked to the receiver frequency standard, and the speed of the scanning spot across the page was locked to the transmitter frequency standard. The standards were not locked together but were completely independent. The accuracy requirements for those standards were much more stringent than for a good clock, even though the clock might run for days without needing to be reset (this was before modern crystal-controlled clocks and watches). With fax, consider that the total length of a line for scanning an 8.5-by-11-inch page at 100 lines/inch is $8.5 \cdot 11 \cdot 100 = 9{,}350$ inches. A clock stability of 1 min/day (1 part in 1,440) would give about 6.5 inches of skew[4] in the received copy, making it unacceptable. Accuracy of 1 sec/day (1 part in 86,400) would cause a skew of about one-tenth of an inch per page, but after a week or a month the clock likely would be off at least three times that rate. Skew was quite a problem in network operations, such as those for newsphotos, and the fax machines had to reset their frequency standards often. In the 1940s, the precision of tuning fork frequency standards (but not crystals) became good enough for most applications.

Bain's pendulum was kept in step by a synchronizing (sync) pulse for each swing to avoid skew of the recorded fax image. A latch magnet stopped the pendulum at the end of its swing until the sync pulse was received. Even if the pendulum rates varied slightly, the change of line length would not be noticed, and the left edge of the received copy would be straight. Any variation in the spot speed caused the marking to drift out of the proper position. To limit the error in recording spot position, the spot speed at the receiver was slightly faster than the fax transmitter. At the end of a line, the recording spot movement stopped and waited for the fax transmitter to start the next line. The positional error thus was limited to the amount of drift between the spots in one line. The stop-start synchronism used by Bain in the first fax unit continued to be used in later fax designs for 100 years, and electrolytic recording lasted into the 1960s.

Unless Bakewell's transmitting and receiving cylinders were in exact step (in sync), the received image shape would be extremely distorted, making the system useless. He avoided the synchronization problem altogether for his exhibit in London. Because his fax machine had the scanning and recording cylinders on the same shaft, they rotated in perfect synchronism; however, the fax image was sent only from one side of a machine to the other side of the same machine. In later designs, when Bakewell tried sending between separate fax machines, he had little success in obtaining readable copy. He attempted to

4. At the bottom of the fax recording page, a black dot that should be at the left edge of the page would have drifted over by 6.5 inches. Halfway down the page, the drift would be half as much, or 3.25 inches.

lock the receiver cylinder rotation to a pendulum by correcting the drum position with electromagnetic braking two to four times per drum revolution. That system was too crude to be useful. The rotating drum synchronizing problem was finally solved about 50 years later.

Caselli used Bain's synchronized electric clock system, with an improved method of pendulum synchronization eliminating most errors caused by atmospheric disturbances. Probably both lightning and solar storms produced extraneous currents in the single wire transmission circuit and the ground return path. The pendulum gave way to a tuning fork frequency standard. Although Joseph Henry is reported to have constructed the first electric motor in 1829, and Gustave Froment had made different types of electric motors between 1844 and 1848, those motor designs were not used in fax machines of the era [1]. Rotating drums became practical after Poul LaCour of Copenhagen made his synchronous motor in 1878. The motor consisted of an electromagnetic coil next to the teeth of a gear. The gear was the motor rotor, which turned one tooth per pulse. To drive the motor, LaCour used an arrangement of electromagnets and electrical contacts with the tines of a tuning fork. Constant frequency pulses generated by a battery and the tuning fork contacts operated the motor in step with the tuning fork.

2.4.2 Telegraph and Telephone Lines

When the first fax equipment was made, there were no telephone or radio channels, and telegraph channels were dedicated to a single user, usually with multiple wires strung between two telegraph stations. We find no record of fax being used to send to a remote location until Caselli installed his fax system in France. Although facsimile and the telegraph were evolving at the same time, the Morse telegraph was dominant, and many telegraph lines already had been installed when Caselli needed them. By 1852, there were over 23,000 miles of telegraph lines in the United States alone. Telegraph lines communicated with dc pulses, but there was no low-frequency cutoff, so fax could send very long pulses to represent black areas. Resistance of the telegraph wire and the ground return limited the length of the circuit. Attenuation of the telegraph line itself gradually increased with frequency but probably did not limit the sending speed for fax use. Longer telegraph channels needed relays for repeating the pulses, but they limited the sending rate for fax to a very low speed. Telegraph lines could seldom send more than 30 or 40 (75 maximum) pulses/s, too slow for most fax applications. Fax generated a pulse for each mark scanned, and the rate could be very high compared with the dots and dashes of the telegraph. With the advent of telephones for voice communication, the single-wire channel with ground return carried a different type signal. Voice sound vibrations changed the resistance of a carbon granule transmitter button, causing the dc current passing through it to vary. The current amplitude and frequencies representing those sounds passed through the telephone line to the earphone of the telephone receiver, causing a diaphragm to vibrate and reproduce the original sounds. For fax use, telephone lines could send about 200 pixels/s but could not communicate over the long distances that telegraph lines could.

A major communications milestone occurred in 1915 when vacuum tube repeaters (amplifiers) were added to telephone lines. Shouting on long-distance calls was no longer required. More important for facsimile though was the possibility of sending signals much faster and much farther. Audio transformers in the signal path did not pass the dc fax signals. To send them over a telephone channel, inventors developed a modulator in which the fax signals controlled the amplitude of a carrier frequency signal that would pass through. This was similar to controlling the loudness of a single high note from a musical instrument (see Section 2.4.3).

As the demand for telephone service increased, the older open-wire telephone lines were replaced by cables whose multiple pairs of wires handled more voice circuits. These telephone lines have high shunt-capacitance between the wires, causing the signal loss to increase rapidly with frequency, shortening the required distance between repeater amplifiers. Loading coils added in series with each wire converted the line into a low-pass filter, reducing transmission loss and improving the quality of voice transmission; however, the loading coils introduced most of the envelope-delay distortion. The combination of attenuation of higher frequencies by longer telephone lines and the effect of the amplifier blocking low frequencies from dc to a few hundred hertz shaped the attenuation of a telephone channel into a U-shape curve, which still exists today. Fax signals were then sent using double-sideband amplitude modulation of a carrier frequency of 1,800 Hz because it was centered in the envelope-delay passband of about 600 to 3,000 Hz. At higher speeds, envelope-delay distortion, caused by slightly longer transit time over the telephone line for frequencies near the edge of the voice band, degrades the received picture quality, causing smearing of line edges, ghosts, and other defects, making it unacceptable in many cases. Later, 12 voice channels were sent over a single telephone channel. Frequency multiplexing systems use 4,000-Hz bandpass filters to separate the channels. Although the upper frequency limit for a channel is about 3,600 Hz, the overall end-to-end characteristics on telephone calls can be relied on only to 3,000 Hz. The analog telephone switched network described here was known by various names as it evolved, including "plain old telephone service" (POTS), long distance, and direct distance dialing (DDD). Today it is referred to as the public switched telephone network (PSTN) or the general switched telephone network (GSTN).

2.4.3 Modulation

Fax started about a century before data transmission, when there were no modems available to transport the signals over telephone lines. The first modulator was probably Rosing's light chopper in 1911. A rotating disk with a ring of holes or teeth interrupted the light path in the scanner, generating a constant frequency carrier signal at the photocell. The signal amplitude was controlled by the density of the page being scanned: maximum when white was scanned, lower when gray was scanned, and no signal when black was scanned. A few systems, including the highly successful Western Union Deskfax, continued to use the light chopper modulator for many years.

When scanning a pattern of alternate black and white lines, a light chopper modulator generates four different signal frequencies. The lowest, the same rate of amplitude changes per second as black lines per second, is called baseband or the f_m as shown in Figure 2.9(a).

Baseband signal is undesirable. A telephone line impairment called *envelope delay distortion* causes these lowest frequencies to interfere with the main signals, distorting the received image. The following frequencies are generated:

- f_m (modulating frequency, or baseband);
- f_c (carrier frequency);
- f_l (lower sideband frequency) $= f_c - f_m$;
- f_u (upper sideband frequency) $= f_c + f_m$.

When text or graphics are sent, the significant frequencies are scattered but confined to the baseband range of 0 to f_{mm} plus the sideband range of f_l to f_u, as shown in Figure 2.9(b). f_{mm} is the maximum modulating frequency, generated when the finest lines the fax can send are scanned. Baseband can be removed with a high-pass filter if the sending speed of the fax system is not too high (where the highest baseband frequency, f_{mm}, is below the lower sideband, f_l). Instead of spinning-wheel light choppers, some fax scanners substituted

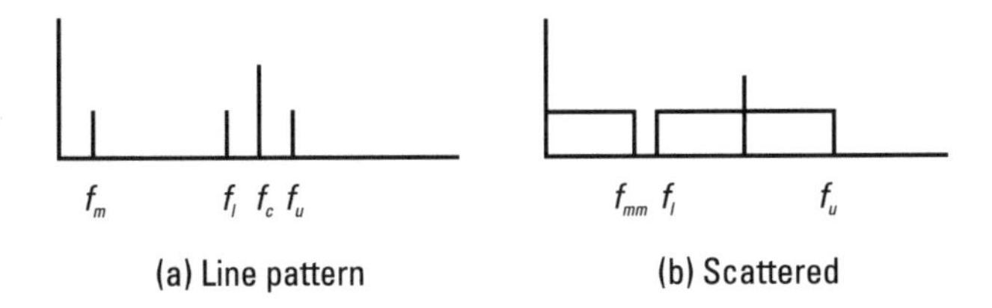

Figure 2.9 Amplitude modulation (AM) spectrum: (a) line pattern and (b) scattered.

a vibrating mirror galvanometer in the light path to alternately pass or block the light beam.

Fax scanners that did not interrupt the light beam usually needed dc amplifiers and vacuum tube modulators, but they were unstable. To avoid dc amplifiers, some designs used a phototube with the carrier applied directly. Others used a photomultiplier, taking advantage of its high sensitivity, high speed, and high-level output signal. Photomultipliers were, however, very sensitive to even small changes in the power supply voltage. Most required manual adjustment of the voltage every time the fax transmitter was turned on and often drifted out of tolerance during the day. One design that avoided modulator instabilities inserted the carrier frequency (1,800 Hz) in a balanced bridge circuit with an RCA 1645 photocell. The carrier was the only voltage applied, and the photocell acted like a light-sensitive resistor in series with a rectifier. The advantage was that the modulator did not drift and was not sensitive to supply voltage; however, baseband was mixed in with the AM-modulated signal. All these modulators imposed fax speed limitations caused by baseband.

Modulators were later devised with an adjustment to balance out baseband. One was a spot projection illumination system developed by Acme Teletronix. A tungsten lamp was focused to the small spot, forming the elemental area on the transmitter drum. A mirror galvanometer in that optical path vibrated at carrier frequency to alternately deflect the light beam between two phototubes. One conducted on the positive portion of the carrier and the other on the negative portion in a push-pull arrangement to produce a sinusoidal carrier and balanced modulation without baseband. A great improvement in performance of modulators that balanced out baseband came about after John Shonnard persuaded RCA to make the 5652, a photocell similar to its 1645 but with a second cathode substituted for its anode. Electrons from the first cathode traveled to the second cathode just as well as the anode it replaced. As the voltage of the carrier frequency swung from positive to negative, the first cathode acted as an anode. A simple twist of the phototube balanced the amount of light between cathodes with equal photocurrents, eliminating the baseband. Once set, no further adjustment was needed.

2.5 After World War I

In this book, certain systems are discussed with some detail. The systems were selected to show how differences in available technology affected the approach needed for development of the facsimile art, leading toward today's digital fax. An attempt to cover all significant contributions to fax progress is beyond our scope.

2.5.1 Cable Fax

In 1920, the Bartlane system was the first new postwar method of transmitting photographs by facsimile. Pictures were sent between London and New York via transatlantic cable. The Bartlane system, which was named after its British developers, H. G. Bartholomew and M. L. MacFarlane, used a digital method of encoding fax signals. The picture to be sent was first printed as a screened image (as used in newspapers). It was then scanned, and the size of each screen dot was assigned a code that an automatic tape perforator punched as a line of holes across a standard paper telegraph tape. The number of holes punched corresponded to the size of the dot on the image, ranging from all holes for large dots to one hole for small dots. The tape was then sent as a telegraph message on the standard Western Union automatic cable equipment and received as a duplicate tape that the receiving equipment converted into a photographic image. In 1921, another cable facsimile system by Eduoard Belin of France sent photographs over the transatlantic cable. Cable fax was used for only a few years before being displaced by newer fax systems.

Higher speed photofacsimile services developed by Western Union, RCA, and AT&T made their appearance in 1924 and were the next step in transmission of pictures (in the United States). Western Union's Telepix sent over its telegraph lines, RCA's Photoradio sent over shortwave (high frequency, or HF) radiotelegraph channels, and AT&T's telephoto system sent over its telephone wire channels. Fax equipment developed by Belin for phone line transmission was popular in Britain and France by 1928, and transmissions were called "Belinos."

2.5.2 Western Union

The Western Union Telepix system received little support and was soon discontinued. The company continued to be active in developing fax equipment for both photo transmission services and for sending telegrams. An improved medium for receiving telegrams and other documents by fax was Teledeltos™, a dry burn-off recording paper developed by Western Union. It worked for photographs, as well as text. The image was immediately visible and no processing was required for long-term storage of images. A number of fax machine designs were made for sending automatically after the customer placed the telegram in a slot and pushed a button. In one fax system, the user wrote messages with a pencil on 3/4-inch-wide Teledeltos™ tape. Contact scanning was used with four styli on a rotating disc. The same styli and paper were used for sending and receiving. An 11-inch strip took about 15 seconds to send.

2.5.3 RCA

RCA engineers were goaded to come up with an improved method of transmitting radio messages as described in a paper by Richard H. Ranger in June 1925.

> Mr. Owen D. Young, Chairman of the Board of Directors of the Radio Corporation of America stated, at a banquet, that he was tired of all the arduous effort behind a twenty-four hour job of sending radio messages by telegraphy from a transmitting operator to a receiving operator who put down the letters one by one at a distant point. Instead of this, the new possibilities of radio should make it feasible to say: "ZIP, and a page of the London Times is in New York City." He added, "Not being an engineer, I am not interested in the details; that is your job." [4]

The Photoradio system used standard radiotelegraph transmitting and receiving facilities. A dot converter keyed the radio frequency (RF) carrier on and off at a constant keying rate. Signals from the fax transmitter controlled the duration of each pulse generated by the dot converter, with the time varying in proportion to the scanner signal amplitude. This method is known as constant frequency variable dot (CFVD). It helped to overcome the radio propagation impairments of fading and multipath distortion. A room full of equipment was used at each city to provide service between New York, San Francisco, London, Buenos Aires, and Berlin. The rotating drum of the fax transceiver was driven at 30 or 60 rpm by a dc motor and alternator on the same shaft with a braking system. An 810-Hz steel tuning fork provided signal to drive the motor amplifier. To maintain the high accuracy required, the fork was inside two concentric ovens, holding temperature variations to 0.01°C. The picture size was 8.5 by 12 inches. The drum speed could be set in 6 steps from 20 to 120 rpm. Lines per inch could be set in 12 steps from 40 to 300.

2.5.4 AT&T

The components of AT&T's telephoto transmitting and receiving system took up an entire room. The equipment sent 4.25-by-6.5-inch (108-by-165-mm) pictures. Resolution was 100 lines/inch with a drum speed of 100 rpm for sending at 1 inch/min. The equipment was built like a lathe with a rotating drum and leadscrew. A dc shunt-wound motor drove the drum directly with a generator operating at 300 Hz, operating in a servo loop. The 300-Hz tuning fork was made of nickel-chromium steel for a small temperature-frequency coefficient. It was operated in an oven held to 50°C, ±0.1°C. The frequency was maintained to an accuracy of a few parts per million by manual adjustment when necessary. Vestigial sideband amplitude modulation was used with a carrier frequency of 2,400 Hz.

The AT&T newspaper telephoto picture transmission service used selected and conditioned private line telephone channels. Facsimile was the only high-speed data transmission service at that time. The solutions used to correct envelope-delay distortion and many other telephone line impairments were helpful many years later. The telephone line performance specifications set for facsimile were almost identical to those for data transmission over a voice channel adopted about 20 years later. Herbert Ives did much of the work on the program and is credited with the telephone line equalization used in telephotography. Envelope-delay distortion limits of no more than ±300 μs were needed for the telephone line between sending and receiving facsimile machines. Newly designed filters added propagation delay in the center of the voice band, making the delay more nearly constant, and amplitude-versus-frequency filters corrected the frequency response over the range required for facsimile signals (1,000 to 2,600 Hz). Selection of lines for conditioning avoided those with electrical disturbances and distortions affecting the fax picture quality. Electrical disturbances include abrupt variations in line loss, envelope-delay distortion, noise, echo, and crosstalk. Each leg of a facsimile telephone circuit had separate filters to compensate for that particular section. The fax lines were made from the regular telephone service lines. Manual connection of the conditioned legs provided a "private line" point-to-point circuit or network in which many newspapers received photos by fax at the same time. Before sending, a key switch on a simplexed dc control circuit set the other telephoto stations on the network to receiving mode. It also locked out the echo suppressers and the stepping of attenuation regulators.

2.6 Early Fax Newsphoto Services

Worldwide distribution of pictures by fax began in the 1930s and 1940s with the use of private line networks and HF radio circuits, with services provided by Associated Press, NEA Acme, Reuters, and INP. All of them recorded the fax images with a full gray-scale range for each pixel, providing good copies of photographs. By contrast, today's Group 3 and Group 4 facsimile units record only black and white dots, producing a lower quality gray scale in spite of the higher resolution.

2.6.1 Associated Press

Associated Press (AP) acquired an improved AT&T telephoto system with a larger picture size of 11 by 17 inches (279 by 432 mm) and initiated its Wirephoto™ service for newspapers on January 1, 1935. The AP system thus became the forerunner of today's private line newsphoto systems. The system required delivery of photographs to one of the fixed locations where an AT&T transmitter was available. As the AP telephoto network developed, a portable facsimile transmitter was needed to cover breaking news. After a competing portable facsimile transmitter had been developed for the *New York Times*, arrangements eventually were made to procure portable units from the Times Facsimile Corporation. The facsimile receivers operated in a darkroom at the newspaper offices.

After the laser became a reality, a laserphoto recorder was developed for AP with the assistance of MIT Professor Schreiber. This fax machine did not require a darkroom and eventually replaced the earlier fax recorders. Today, out-of-town pictures printed in newspapers are sent by facsimile from almost any place in the world. Thousands of newspapers receive pictures at the same time. AP uses a direct-broadcast satellite channel operating at 9.6 Kbps. Resolution is 170 lines/inch for an 8-by-10-inch picture, sent in 2 or 3 minutes. Eight bits are sent per pixel, for 256 gray-scale shades. A helium-neon laser spot is swept across the recording paper by a mirror galvanometer, exposing dry silver photographic paper. The image is fixed and developed by a thermal unit inside the fax machine.

2.6.2 Acme

Acme Newspictures started in 1923 as the news photograph division of NEA Services, a well-known news agency. Later, under L. A. Thompson, Acme Telectronix designed and built its own fax machines and assembled a wire and radio fax network to expedite the delivery of newsphotos. One of its best known units was Trans-ceiver CNP, designed during World War II. For transmitting, the carrier frequency of 1,920 Hz had double sideband amplitude modulation. For receiving, a lightproof box containing the drum with film or photographic enlarging paper wrapped around it, was snapped into place in front of the recording optics. The box was loaded in a darkroom. The recording light source was a crater lamp such as the Westinghouse CR-2 or the Sylvania R1130. The drum was rotated without gear reduction by a 100-rpm multipole motor made by General Electric. Standard resolution was 100 lines/inch, but other resolutions up to 300 lines/inch were available. The frequency source for the motor was a 1,920-Hz bimetallic tuning fork from Riverbank Laboratories.

Model FOA was a 300-line/inch, high-resolution version made for the U.S. Navy. The frequency of its tuning fork and carrier was 2,400 Hz.

Acme Newspictures went through different ownerships, including Fairchild, which was developing a scanner for making screened halftone-plate printing masters, and United Press International (UPI), which continued to produce picture transmission equipment [5]. UPI developed a unique black-white electrostatic recording system to simulate gray scale, in which a pixel has its black-white ratio controlled to represent the desired average density of the pixel. A mid-gray pixel has the left half white and the right half black. Each recorded pixel is first written with a full width charge corresponding to black over a rhombic-shaped area the height of the recording line. A single stylus mounted on a belt moves across the page to apply the recording charge pattern. If the pixel is black, nothing more is done. For half gray, the charge is neutralized over half the spot width; for white, over the entire spot. The lightest gray is a bar that is 1/64th the width of the pixel. At normal viewing distances, it is difficult to tell the difference between the fax recording and a photographic print. Both Reuters and UPI now have laser newsphoto recorders similar to AP's. An electronic picture desk at newspaper offices stores the pictures in memory as they are received from the newsphoto network. The editor can then view them on a cathode ray tube (CRT) screen, select those desired, edit them electronically, and print them for use in the newspaper.

2.6.3 *New York Times*

When AP bought the Wirephoto™ system from AT&T without inclusion of the *New York Times*, it caused concern to publisher Adolph Ochs, who was not told even though he was an AP director. The *New York Daily News* and other large newspapers had contributed money for the acquisition. The *Times* photo service, Wide World Photos, could no longer use the AT&T wirephoto service to compete with fax delivery of AP newsphotos to other papers. Ochs's son-in-law, Arthur Hays Sulzberger, obtained permission to develop a portable fax transmitter so pictures could be sent from any location on a regular telephone call instead of using the expensive Wirephoto™ network between fixed locations. After initial attempts in that pursuit were unsuccessful, Sulzberger contacted Austin G. Cooley, who had been conducting experiments in fax since 1920. As the student in charge of MIT experimental radio station 1XM, Cooley had helped Professor Hainsworth make radio fax transmissions. Cooley's development of the Rayfoto system (see Section 2.11) had gained him a good reputation in fax. Sulzberger and Cooley signed a contract in October 1934 for 60-day delivery of a fax system with a portable transmitter and fixed station recorder that would operate without making an electrical connection to the phone lines.

Cooley and his colleagues began working round the clock to construct the photofax system. Meanwhile, Sulzberger involved Walter Gifford, head of AT&T, in a dialog about telephone use. He asked whether it made any difference what language was used in telephone conversations. Gifford assured him that any conversation other than profane was acceptable. Turkish, gibberish, even fax machine language would not bother AT&T. Western Electric engineers had assured Gifford that picture transmission by fax would not work for more than 100 miles except on the expensive, specially selected and conditioned Wirephoto™ circuits set up by AT&T, so he had no qualms about the implicit go-ahead to Sulzberger [6].

The new fax equipment was delivered in 60 days, after initial back-to-back tests worked perfectly. When acoustic coupling to the phone lines through the telephone caused echoes in the received pictures, an open core magnetic coupling unit was soon made. It clamped on the outside of the box that contained the open core induction coil for the telephone. Leakage flux made it perform like a transformer between the two units. During early tests, the new fax transmitter was in San Francisco when on February 13, 1935, a disastrous event occurred off the coast of Monterey, California. The U.S. Navy dirigible *Macon* had crashed, and newspaper photographers rushed to the scene. The *Times* grabbed the opportunity to test fax transmission with the equipment. Pictures of the survivors appeared on the front page of the *New York Times* the next morning. AT&T executives and AP Wirephoto™ customers were incredulous. Now, sending a picture by fax cost the same as a long-distance telephone call instead of needing the $3,000,000 special Wirephoto™ circuits [7].

In 1935, Cooley was hired to run the newly formed Wide World Wire Photos, later called Times Facsimile Corporation (TFC) to build and operate facsimile equipment for the *New York Times.* Opening the POTS to fax allowed pictures to be sent directly from the location of a news event to *Times* headquarters in New York or for distribution of Wide World Photos from the *Times* to its customers. That greatly increased the operational flexibility and reduced costs. The news photographer made a print of a picture for the facsimile operator, who sent it from a portable facsimile transmitter about the size of a small suitcase (a radical reduction in size from the room full of equipment required by AP). A regular telephone call was made to the *Times* picture darkroom. The portable fax transmitter operated from a 6-volt automobile battery, so even a coin-operated telephone could be used. A dc motor turned the lead-screw and a motor generator provided the high voltage needed by the vacuum tubes. A stable oscillator provided the frequency for sync-motor-amplifier and for the tone that was amplitude modulated to carry the picture signals over the telephone line. A tuning fork with a shutter acted as a strobe for light on the fax transmitter drum, and the oscillator was adjusted until the drum appeared to be stationary. At the receiver, the facsimile carrier furnished the frequency for the sync-motor-amplifier, keeping the receiver drum locked to the transmitter drum. If, however, the telephone circuit was multiplexed to provide more than one voice path on the same pair of telephone wires, the received frequency was translated slightly from the sent frequency, skewing the received image. Even a 0.01% error was unacceptable.

That problem was solved by correction of the frequency error with a drift compensator, invented for this purpose. A motor was used as a transformer with the fax signals passing between the stator and rotor windings. If the motor shaft was not rotated, the output frequency was the same as the input frequency. But when the shaft was rotated slowly with a second dc motor, the fax signal carrier frequency was shifted up or down, adjusting the fax receiver motor speed to be the same as the fax transmitter motor. That corrected the frequency offset introduced by mismatch of carrier frequencies in the telephone system. The recording lamp flashed as a strobe light on the receiving drum for each phasing pulse. A rheostat on the dc motor was adjusted until the drum appeared stationary. The system was abandoned when tuning fork oscillators with accuracy of about 10 parts per million were developed for facsimile use, and each fax unit sync motor then operated from its own standard. Other small-frequency standards with the long-term stability required for facsimile units did not exist. Even good crystals aged a few parts per million per month and the frequency would need resetting about six times a year to keep the older fax machines in step with new ones.

When the *Times* began fax operations in 1935, it was difficult to fabricate the equipment because few of the parts needed were available. All the electronic circuits had to be designed using vacuum tubes, resistors, capacitors, and other components. Some of the audio and power transformers, filter coils, solenoids, and other electromechanical components were designed and made at TFC. Many of the mechanical, electromechanical, and electrical parts required were unique and were pushing the state of the art for precision. Because there was no established source, facsimile manufacturers had to design and make the parts or make arrangements with companies with the background and expertise to develop what was needed. For portable fax units, there was no motor that could be purchased with the rotational stiffness required for high-quality recording. A 1,800-Hz, 1,800-rpm, 1/100-hp synchronous motor was designed and manufactured by TFC. The gap between stator and rotor was only 0.001 inch. Dies for the motor laminations were made in the *Times* experimental machine shop. The stamping, heat treating, machining, assembly, and testing also were done inhouse. Obtaining precision gears, bearings, and machining leadscrews with enough accuracy to prevent picture streaking was a serious problem. Even the optics for some of the facsimile systems required unique designs.

For the *Times* newsphoto equipment, a 1,800-Hz tuning fork was developed by Garret V. Dillenback. A nickel-chrome steel (similar to Invar™) tuning fork with a positive coefficient of frequency with temperature was silver soldered to a carbon steel tuning fork that had a negative coefficient of frequency with temperature. The laminated, bimetallic tuning fork was purchased (Riverbank Laboratories, B. E. Eisenhour patent) with a slightly negative temperature coefficient. The coefficient was tediously adjusted in steps by removing a small amount of steel from the thin carbon steel side with a surface grinder. Heat treating then removed the strains produced by grinding. After a check of frequency versus temperature, a smaller grinding step was taken, and the process repeated.

The *Times* sent a list of news photos available to 30 to 40 customers (mostly in the East and the Midwest) and then faxed the ones selected. Among the customers were the *Atlanta Constitution*, the *Boston Globe*, the *Chicago Daily News*, the *Detroit Free Press*, the *Philadelphia Inquirer*, the *St. Louis Post Dispatch*, and the *Toronto Globe and Mail*. Wide World Photos also sent pictures from bureaus in Washington and Chicago. Not all calls, including many to Boston, furnished good fax pictures. A good telephone line could be selected with a hook click test. A sharp click indicated a good line, but if a "plink" sound was heard, the circuit was rejected. It made no sense to send a fax and develop the received film only to find the line distortions made the received copy useless. It helped to know the circuit numbers of good lines between cities. With the four-channel C carrier, it was best to have one of the two center channels to avoid envelope-delay distortion caused by the telephone company channel filters.

Long-distance telephone operators could be a problem. They had been instructed to disconnect a circuit immediately when the customer was finished or when something went wrong with the call. Because facsimile calls were longer than voice calls, operators often monitored calls to ascertain when they were over. "Are you through?", a common inquiry on lengthy telephone calls, caused streaks in the received picture. On hearing the facsimile signal gibberish, operators often disconnected calls. On pay phone calls, the fax operator had to persuade the long-distance operator not to monitor and promise to pay after the coins ran out, even if it took half an hour. At that time, a wired connection to the telephone line was illegal, so inductive coupling was used. Phone installations had a bell box that contained the transformer to isolate the telephone from the line. The fax operator fitted a large clamp around the box, and a winding on the clamp coupled the fax transmitter signal to the transformer through leakage flux. Later, small, easier-to-use UC and KC magnetic coupling coils that clamped on the telephone receiver were used. Before starting a fax transmission, the coil was rotated for the loudest sound in the telephone receiver.

2.7 World War II Era

In the late 1930s, as the use of fax for newsphotos was growing rapidly in the United States, the news stories out of Europe and Japan were becoming more and more ominous. Pictures alongside news reports of foreign events were attracting greater public attention. "On March 14, 1939, Hitler invaded the remainder of Czechoslovakia despite his previous Munich pledge;... On April 7, Mussolini, desirous of sharing in the spoils, attacked Albania and soon had that country under his control... He [Hitler] demanded of Poland the return of Danzig, as well as numerous concessions along the Polish corridor. To strengthen his demands he mobilized a large army along the Polish border." [8]

2.7.1 The Royal Tour

American readers sought escapism in reading about less troublesome events. In the spring of 1939, when England's King George VI and Queen Elizabeth started a royal tour of Canada, news organizations gave extensive coverage of the five-week train journey. A team of reporters, photographers, and fax operators accompanied the royal entourage as it traveled west from Quebec, with many stops along the way, to Vancouver and then returned through Niagara Falls, Washington, D.C., New York City, Hyde Park (where Franklin Roosevelt entertained them American style with a weenie roast), the Gaspé Peninsula, Newfoundland, Prince Edward Island, and Nova Scotia. Faxed pictures of the royal couple's visits with many government officials and other dignitaries were sent daily to the newspapers. Fax services included Acme, AP, International News Photos, Muirhead, RCA, Reuters, and the *New York Times.* At Halifax, the King and Queen boarded the *White Empress of Britain* for their return to England. On June 17, 1939, the press corps went aboard the Polish ship *Pilsudski* to return to the United States. On the way, the reporters were discussing the trip and sought the captain's opinion about the European problems. "Will there be a war?" asked Web Miller (United Press Berlin chief). The room suddenly became silent. After a short pause, the captain replied, "Yes. On September 1 the German troops will cross the Polish border. Poland will lose, but Poland will fight."

"On the morning of September 1, 1939, the invasion of Poland started, and two days later France and Britain, living up to their promises, went to her assistance. World War II had begun" [9]. The *Pilsudski* became a British troop ship and was sunk by a German submarine on November 26, 1939.

2.7.2 The Plattsburg Maneuvers

In July 1939, the U.S. Army conducted maneuvers in Plattsburg, New York. Among those attending were military attachés from Germany, France, and Japan. Papier mâché tanks and guns gave evidence of the sorry state of U.S. military equipment, but sending pictures by fax gave promise of one area of superiority. As part of the maneuvers, an interesting contest was run by the Army to pit fax against carrier pigeons for sending pictures. The *Times* newsphoto fax equipment used in the test easily outdid the pigeons. Only one or two of the birds escaped being caught by hawks and arrived hours late. The success of fax in this experiment triggered interest by the U.S. Army and a German general who looked over the fax equipment. A later news item that the Germans had shot down a Polish radio tower alerted officials to the folly of such open access to U.S. military information, and maneuvers were abruptly stopped.

2.7.3 Military Fax (Ft. Monmouth)

Around 1940, the U.S. military wanted a small-sized portable fax transceiver. The Signal Corps Laboratories at Ft. Monmouth, New Jersey, received a one-page directive to test available commercial facsimile equipment and buy 250 fax units suitable for military use. It is hard to imagine that such a brief order on a single sheet of paper could launch the military on a facsimile development and production program of such magnitude. A Facsimile Section was established as part of the Wire Branch, and the task was started under the direction of Henry Burkhard. Information on U.S. fax machines was collected, and equipment was purchased from Acme Teltronix, Associated Press, and the *New York Times*, but all the equipment was much too large for military needs. The *Times* offered to build a portable fax transceiver in their small suitcase-sized newsphoto transmitter case and won a contract for $1,100. When the *Times* set up a new corporation devoted to military facsimile, the general manager, Julius Adler, told Austin Cooley, "We want you to do all you can to meet the needs of the military. If you can do it without losing money that will be fine, but if you do lose some, that will be all right too" [10]. In two months, Cooley and his team delivered their "suitcase" according to specifications. The large darkroom-operated fax-receiver functions were squeezed into the small transmitter case along with an added direct recording function for receiving fax images without a darkroom.

The new fax unit could send photographs or messages, receive positives on photographic paper, receive negatives on film, or receive on Teledeltos™, a direct recording paper made by Western Union. Operation was from a 6V automotive battery or from an additional 115V, 60-Hz ac power supply. A few more transceivers were ordered for testing. The fax machines were hand-wired point to point with the discrete components soldered from pins on tube sockets to a tie point or ground. Each technician wired a fax machine from a circuit diagram rather than from production-engineered layout drawings. The technician then debugged and tested the fax machine. The more experienced technicians had their own wiring layouts and tricks to make the finished unit meet performance specifications. The design required a 1,800-Hz power amplifier to drive the synchronous motor. It was very difficult to prevent the motor's 1,800 Hz from being picked up by the low-level, high-impedance amplifiers for the fax signal. The most sensitive pickup point was the 1-inch-long wire connecting the photocell to the signal preamplifier. A metal cover on insulating posts was connected to ground at the point that gave best performance. It covered the wire, the socket pins, and a 2-MΩ load resistor. When examining a fax machine that had been delivered to the Signal Corps Laboratories, one of the government engineers found a wire that was connected only at one end. Removal of the "redundant" wire resulted in an undesirable pattern in the recorded copy, caused by a 1,800-Hz pickup from the motor amplifier. When the wire was put back, the fax machine worked properly again. The technician had perfected an antenna system of eliminating the pickup. He connected a short wire to the signal amplifier and moved the loose end to a position that balanced it out by introducing additional pickup of opposite phase and magnitude.

Production contract 1041-MPD-43 was awarded to TFC for 300 units of facsimile transceiver FX-1B, a 22-by-12-by-10-inch militarized transceiver (part of the RC-120B). The recently developed military specifications (MIL-SPEC) were a requirement. Rugged high-reliability components were specified, plus environmental testing at the Ft. Monmouth Signal Corps Squier Laboratories for proper operation during exposure to wide temperature ranges, vibration, and high humidity. The fax also had to operate after shock, more extreme temperature storage conditions, and fungus exposure. Interface was required with military radio and wire line channels, including field wire W-110-B (+26 dBM send level). The point-to-point wiring of the earlier design was forbidden. All electrical components had to be mounted on terminal boards or clamped to the chassis. Wires in preassembled laced cable harnesses would connect the components. That presented a design challenge in some areas where the original component connection of a fraction of an inch now needed an additional wire of 2 to 10 inches long to reach the component at a remote terminal board. "You'll need a bushel basket to hold all pieces after the tests," expressed one skeptic when the fax was delivered for testing. Contrary to the testers' expectations, the FX-1B stayed intact and became the first communications equipment to successfully pass the rugged tests.

Operation of the FX-1B was not as simple as that of modern fax equipment. When a fixed darkroom was not available for receiving photographs, it was operated through two hand holes in a small light-tight tent. Operators had to perform the steps needed to receive faxes without being able to see what they were doing inside the tent. The operator slipped a stainless steel daylight processing tank previously loaded with film through a hand hole. It took both hands to open the tank, wrap the film with the black paper separator around the drum, and then clamp it. If the film bulged, it had to be tightened near the clamp without touching the recording area of the film. The operator moved back a lever to disengage the half-nuts from the leadscrew, slid the drum to the right end, and asked the transmitting operator to send maximum signal level (by moving the transmitting drum for the lightest part of the picture).

The operator then rotated the magnetic coupling coil clamped on the telephone handset receiver for the highest reading on the decibel meter, extended to outside the tent on a cable. Next, the operator adjusted the fax receiver gain for proper recording level and pressed the start button, momentarily bringing the motor above synchronous speed. When the start button was released, the speed slowed and the motor usually locked into sync. If the drum stopped, the operation was repeated. Next, the operator asked for phasing pulses to be sent and held in the phase button. After hearing the drum stop before the first phasing pulse, the operator held the button until clicks were heard on succeeding pulses. After telling the transmitter operator that phasing was successful, the operator slid the half-nut holder sideways and engaged the lever to start the drum feeding. The operator had to be careful not to touch the drum itself to avoid slippage, which would distort the image. When no more fax signals were received, the operator set the selector switch to standby, removed the film, and inserted it into the light-tight tank. Daylight film processing was used with developer, water, and fixer. Fortunately, it was seldom necessary to receive photographs in the light-tight tent, which was not needed for receiving on direct recording paper or for sending faxes.

A second source contract for the RC-120 was intended for the J.V.L. Hogan Company. Hogan, a fax developer, said it would do the engineering but asked that the contract be awarded to Alden Products Company. Alden had previously produced fax machines for Hogan and was known for making cable harnesses. Soon after receiving the contract, Alden decided it could do the engineering work required, and Hogan was dismissed. Preproduction models were made and tested, but unfortunately the only fax engineer working for Alden died during the contract. Production was eventually terminated without producing machines that fully met requirements.

The Signal Corps Laboratories also bought tape fax machines from RCA and the *New York Times*. The machines sent and received messages that were hand printed or written on a narrow paper tape. To make a small fax unit that did not require synchronization or phasing, the received tape printed two images with the same information. One of the images could be read easily, but the other was usually split between the top and bottom of the tape as the received images drifted across the tape. These fax transceivers were designed for command communication with jeeps, tanks, and armored vehicles, where high ambient noise made voice communication difficult and continuous wave (CW) radio in a bouncing iron monster impossible. Operation over the FM radio sets reduced the effects of radio signal fading, interference, and noise.

In World War II, fax equipment used by combat troops was FX-1B, AN/TXC-1, and RC-58B. "During the early days of the Normandy invasion, maps, charts, and aerial photographs flashed from General Eisenhower's headquarters in England [by FX-1B] to General Bradley's command post in France over a cross-channel facsimile radio circuit. Pilots of the 9th Air Force swooped over Rommel's lines taking pictures of enemy fortifications and rail lines, and plotting on maps of the area their observations of movements on roads, tank parks, artillery positions, etc. Back at their English bases, film was developed, maps correlated, and within minutes they were on the air for the guidance of ground units battling the Hun" [11].

Some commercial (non-MIL-SPEC) fax equipment was used by the U.S. military services during World War II. Brigadier General Frank Stoner, then chief of the Army Communications Service, arranged for fax equipment. Modified Acme Model CNP fax was used in the antisubmarine program in 1942 for transmission of maps and charts between stations of the Navy and Army Air Force. They were also used for overseas radio circuits to Algeria, Brisbane, Port Moresby, Hollandia, London, Honolulu, Caserta, and Paris. HF CW radio circuits were used with carrier shift, later called frequency shift (FS), modulation.

The *New York Times* issue of May 31, 1943, ran a story about a "new technique for the international transmission and distribution of radio news photographs and facsimiles." The *Times* article said that the overseas branch of the Office of War Information (OWI) was achieving a greatly increased "flow of pictures of American victories and accomplishments to the neutral and warring countries" and "sending daily to an estimated 300,000 persons in Great Britain, Sweden, Russia, Spain, Switzerland, North Africa, Turkey, India,

China, and Australia. Propaganda experts have long recognized that news pictures are among the outstanding psychological weapons of modern war." Each side pushed to have photographic verification of its victories, mobilization, and industrial resources appear in world publications. Radio photos of special events such as the Casablanca conference of President Roosevelt and Prime Minister Churchill had extra coverage; the 43 Swiss newspapers alone had a circulation of 2,000,000.

Until the OWI began to experiment with blind shortwave radio broadcasts, the transmission of radio news photographs was made only after two-way communication was established with one receiving station, a very slow process. The broadcasts were very successful, with experienced operators all over the world simultaneously picking up as many as eight pictures in an hour. The portability of the FX-1B made it the equipment most preferred and used by the OWI. Its durability was verified in the China-Burma war theater, when it had to be delivered by jettisoning from an airplane because there was no alternative to reach the intended party. The case was severely damaged, but after replacement of a small casting, it worked perfectly.

Another application of the FX-1B was sending up-to-date weather maps from Washington, D.C., for the service that ferried warplanes to Europe via Presque Isle, Maine; Gander, Newfoundland; Goose Bay, Labrador; and Reykjavik, Iceland. Only a very small portion of a weather chart could be sent at a time because of the 7-by-8.5-inch picture size. International standards for weather map fax transmission were established when the problem motivated two men to devise a solution. After struggling for months with the limitation, Air Force Sergeant Chuck Halbrooks added "modified" to an order for FX-1Bs and slipped it in with other papers awaiting signature. He then visited TFC and told Cooley he would soon receive the order. Halbrooks and Cooley quickly altered the design. The drum diameter was increased from 2.75 to 6 inches, using aluminum tubing available without special order. That gave an 18.85-inch circumference for a usable copy width of 18 inches. The drum length was increased to 12 inches, and the sending rate was increased by 50%. The leadscrew of 96 lines/inch was retained. The new design worked very well and was standardized as TT-41/TXC-1 (p/o AN/TXC-1) by the military, with orders eventually placed for more than 1,000 units. Halbrooks paid for his rash action by being transferred to duty in the Far East, but the U.S. military had a much needed weather fax unit. Use of these facsimile machines for weather map transmission during World War II set the worldwide standards for weather facsimile. The International Telegraph and Telephone Consultative Committee (CCITT) had adopted somewhat different standards, but users abandoned them.

During World War II, the military wanted an even better frequency standard for its fax machines that sent weather maps. The larger image format required a higher stability with temperature extremes and immunity to atmospheric pressure and humidity variations. A stable oven might have extended the performance to lower temperatures, but size restraints and warm-up time made that impractical. Hermetic sealing was required to prevent frequency variations due to pressure and humidity changes. Other problems were salt water corrosion and fungus in the tropics. Test units were made with the tuning fork and coils inside a hermetically sealed cylinder. These designs required another compensation for the pressure variation inside the cylinder caused by temperature changes. Also, long-term frequency stability became a problem, probably caused by outgassing from the driving-pickup coil assembly slowly changing the operating pressure over time.

An improved design by Shonnard used a nonmagnetic stainless steel case with magnetically coupled drive and pickup coils mounted outside the case. The air was evacuated from the case before it was sealed. That eliminated the air loading on the tuning fork tines, providing better frequency stability. The entire fork assembly, including coils, was placed in a larger hermetically sealed steel can to protect the coils from humidity and corrosive conditions. That design took care of both problems. The processing of the tuning fork blanks went through painstaking development before long-term stability and other military requirements were met. Each production tuning fork was individually measured for temperature coefficient. The processing aging and testing required a month or more to make sure the tight specifications were met. The frequency stability and the aging frequency drift were far better than those with Invar™ steel only. To the best of our knowledge, the high precision of this standard with low drift rate, wide temperature range, and immunity to atmospheric pressure changes was not duplicated by others.

2.7.4 After World War II

Although production stopped on most military equipment at the end of World War II, the Army, the Navy, the Air Force, and the Weather Bureau needed many more Weatherfax units. The AP needed lightweight portable Wirephoto™ fax units similar to the FX-1B. In the next few years, new facsimile equipment such as Stenafax, Policefax, Pressfax, Messagefax, and other types of specialized fax units were designed and manufactured. Faced with an expanding need for the TFC capabilities, Sulzberger abandoned his plan for closing down the facsimile operation and kept it for another 14 years, until Litton Industries bought it and continued its operation. The name then became Westrex, and the fax operation was combined with a company recently purchased from AT&T. The fax operation was later combined with other Litton-owned companies, and the names changed to Westrex, Adler-Westrex, Litcom, RADCOM, Datalog, Maryland Division, and Amecom.

One of the Western Union developments resulted in a small-copy office fax unit called Deskfax. Garvice Rydings built the first Deskfax in his basement workshop using a fruit juice can for the drum. About 50,000 units were made to send telegrams between the customer office and the Western Union office, using voice-quality phone lines. The Deskfax was used in offices where the volume of telegram business was not sufficient to have a send-receive teleprinter with a trained operator. The simplified, low-cost transceiver was 12 by 12 by 7 inches. The customer wrapped a 4.5-by-6-inch page around a cylinder and rolled a drum-encircling "garter" spring down the drum to hold it in place. The page was a telegram if sending or a blank sheet of Teledeltos™ recording paper if receiving. The operation was automatic. Sending time was 2 min at 100 lines/inch. This very successful fax machine disappeared with the demise of the telegram. Western Union later developed fax for use on its new Broadband Exchange Service, a 4-wire, full-duplex phone and data service. A drum-type transceiver with solid state electronics operated at 300 rpm, sending an 8.5-by-11-inch page in 3 min. Neither a tuning fork nor other frequency standard was used for controlling the speed of the synchronous motor. Once a connection was made, the receiver sent to the transmitter a frequency of 2 kHz, AM modulated with the local 60-Hz power line frequency. After demodulation, the 60 Hz was amplified to drive the transmitter motor.

2.8 Fax Weather Map Broadcasting

RCA Communications initiated fax weather map transmissions in the United States, sending by HF radio to the *S.S. America* during its crossing to and from Europe in 1930 [12]. By 1933, it was broadcasting a weather map daily at 11 A.M. to ships in the North Atlantic Ocean. The 8.5-by-11-inch map was printed by a fax receiver on plain white paper with a carbon paper overlay. The paper fed automatically from long supply rolls of carbon and white paper. The map, prepared by the U.S. Weather Bureau in New York City, covered the area from the equator to 70 degrees north latitude, and from 110 degrees west longitude to 20 degrees east longitude (covering most of the United States and western Europe). These fax broadcasts provided information on storms at sea and were an important navigational aid for ships crossing the North Atlantic.

After World War II, networks were set up for broadcast of facsimile weather charts over telephone lines and by HF radio. A national weather facsimile broadcast network was set up with full-period leased telephone channels connecting Air Force bases, Naval air stations, and Weather Bureau forecast offices. TXC-1 fax machines at the Weather Bureau office in Silver Spring, Maryland, sent many types of weather charts 24 hours a day, 7 days a week. The operation later moved to Suitland, Maryland, which became the Weather central office (the Weather Bureau, the Air Force, and the Navy), where most of the fax weather maps originated. Some maps were faxed from the Weather Bureau at LaGuardia Airport and others from DWA, the Air Weather Service at Andrews Field. Eventually, there was not enough time available on the network to handle all the charts and images needed, and additional fax networks were set up by the Air Force and the Weather Bureau.

In early 1947, TFC started broadcasting weather charts by HF on experimental radio station W2XIK (later KE2XER). The 1,800-Hz carrier AM signal from a TXC-1 facsimile transmitter was changed into subcarrier frequency modulation (SCFM), a constant amplitude signal with 1,500-Hz white and 2,300-Hz black by a modified Wilcox-Gay facsimile converter CV-2/TX. Broadcasts were initially made using a war surplus Hallicrafters HT-4 100W HF radio transmitter with amplitude modulation using a doublet antenna on top of the Times building at 299 West 43rd Street, just off Times Square, in New York City. For reception, a standard HF radio receiver was used, with the 1,500/2,300-Hz signal being fed to another CV-2/TX. A limiter removed the amplitude variations caused by fading, and a discriminator converted the signal back to AM for the fax receiver. Ships at sea and other recipients throughout the world copied the fax transmissions [13].

The broadcasts were very successful, and by summer the operation was expanded. The original transmitter was replaced by a Wilcox 96C with three 1-kW channels. The frequencies 4,797.5, 12,862.5, and 17.310 kHz were used simultaneously to provide signals over a much wider area. At that time, there were almost no weather charts available for ships at sea. Permission was obtained from the Federal Communications Commission (FCC) to broadcast, on the experimental license, current weather charts as they were received on the Weather Bureau–Air Force–Navy (WBAN) network. By September 1, 1949, W2XIK was broadcasting 19 different weather charts live five days a week.

Experiments were also made using FS modulation in which the frequency of the RF carrier shifts 800 Hz. One frequency represents picture white, and the other represents picture black. Intermediate frequencies represent tonal values between white and black. A modified Radio Teletype™ Exciter Unit O-5/FR was combined with a modified CV-2/TX to produce the FS signal that substituted for the radio transmitter crystal. The technique had been devised and tested earlier by the Signal Corps. Later, the O-5/FR was replaced by a method devised by McConnell for directly shifting the frequency of the radio transmitter crystal. Simultaneous transmission of radio Teletype™ (RTTY) and SCFM fax on the same radio channel was successfully demonstrated in late 1947. About 40% modulation of the RF carrier was used by the SCFM fax signal to prevent interference between the two services. For reception of FS signals, another RF signal was introduced at the radio receiver, to beat with the transmitted carrier, producing an audio tone. If the beat frequency oscillator (BFO) was not stable enough, signal from a BC-221 frequency meter was introduced at the antenna. The frequency of the receiver output tone shifted the same amount as the RF carrier (800 Hz) in an audio band (1,500 to 2,300 Hz). Later, this receiving method was largely replaced by using a single-sideband radio receiver. Weather maps and weather satellite photos are still being broadcast today using similar standards. Reception of current weather maps and pictures is possible on personal computers using a single-sideband radio receiver and a fax adapter card. The pictures can be viewed on the screen or printed out.

The Weather Bureau wanted to automate the weather map reception and to save space in offices that required two or more fax receivers. That led to design of the relay rack-mounted drum-type facsimile recorder RG (RD-92/UX for military customers) that required only a 12-inch-high panel. An automatic start system later was added for a pair of weather map recorders, to relieve fax operators from monitoring a loudspeaker for start of the next fax chart and manually pressing a phase button before the phasing period ended. With this unit, the telephone line was automatically switched between fax recorders as a weather map finished, so the operator only had to change recording paper before the next map finished.

2.9 Remote Publishing

Methods of distributing some form of newspaper by fax were being explored by 1930, with emphasis on direct distribution to the home. By 1945, the search focused on distribution of newspapers and magazines from remote publishing plants linked by fax. For newspapers, a larger sales territory was the plum, with dreams of national or international distribution. For newsmagazines, later publication deadlines would be possible. This section describes some of the newspaper ventures.

2.9.1 Early Tests

In April 1945, the *New York Times* published a four-page facsimile edition for delegates to the United Nations Conference at San Francisco. At 2 A.M., the special edition was ready in New York. Pages were cut in half and sent with a standard AP Wirephoto™ unit to the fax receiver in San Francisco. Prints from the received negatives were matched to form full pages that were photoengraved and printed at the *Richmond Independent* publishing plant. The printing runs were 2,000 copies for each day of the conference. Although the resolution was only 100 lines/inch, one-fourth the picture elements (pels)/in^2 of Group 3 fax, the pages were very readable and the delegates were pleased to participate in a journalism first. In 1948, the *New York Times* published six editions daily of a four-page fax newspaper in a 8.2-by-11-inch format. Although it was broadcast by WQXR-FM using a multiplexed channel, it was not intended for home reception. The radio and fax receivers were at 14 department stores in New York and the Columbia University School of Journalism. Hogan fax equipment was used for the 1-month experiment [14].

2.9.2 Newspaper-Page-Size Fax

By 1956, interest in sending printing masters by facsimile for remote publishing of newspapers had grown to a point where the *Times* built special facsimile equipment to determine its feasibility. A 10-page edition of the *Times* was sent by fax from its offices in New York to San Francisco and printed there from Monday through Friday, August 20–24, when the Republican National Convention was taking place [15]. Each morning the *Times* was waiting at the hotel desks for the delegates when they got up for breakfast. Full-page negatives, received two at a time started arriving shortly after 1 A.M., San Francisco time, and were printed locally. The quality of the paper was such that few delegates realized they were looking at a facsimile reproduction.

Performance of some of the electrical components in the system had to be pushed well beyond their normal range. One of the largest problems was the recording lamp. A film density of 2.5 was needed while recording 120,000 elemental areas (dots) per second to meet the needed sending speed of 2 min/page. Tests were made with CRTs. They could easily meet the recording speed but were not bright enough for 8-μs/dot exposure time. An alternative, Sylvania recording lamp R1130-B, was more promising. Although it had a 35-ma maximum current rating, and 30,000 elemental areas (dots)/s recording rate, tests showed it performed well at much higher currents. Sylvania built experimental lamps with higher pressure for the gas filling and a much smaller crater diameter to increase the recording rate and brightness of the lamp and found that the needed rate could be met. To get adequate film density, the recording current was run at 90 ma. Life of the lamps was adequate, and no burnout slowed the operation. This lamp design later became the R1168. Two newspaper pages, sent in one transmission, were taped side by side on an 8-inch-diameter,

36-inch-long drum mounted on a shaft held in a lathe bed. A page was sent in 2 min at a rate of 175 in^2/min. Two transmitters and two receivers were used to allow one fax machine to be loaded while the other one was operating. Rotors of two synchronous motors made by TFC were mounted directly on the drum shaft. The head that normally held a lathe cutting tool carried a single aperture scanning head in the fax transmitter or a high-intensity crater lamp recording head in the fax receiver. Although a special wideband 480-kHz channel was planned initially, AT&T found it easier to furnish a television channel to carry the facsimile signals across the country. The 200-lines/inch fax equipment, specially designed for the experiment, was the first fax that would send a full newspaper page.

2.9.3 First Production Newspaper

A few years later, a much higher resolution Pressfax system was made for the *Wall Street Journal.* In Riverside, California, near Los Angeles, Dow Jones built a new printing plant whose printing operation depended on receiving full-size negatives of newspaper pages (there were no typesetting facilities). A television channel from San Francisco connected the transmitter there to the receiver in Riverside. Functioning as replacements for typesetting, the Pressfax receivers produced full-page negatives so sharp (1,000 lines/inch) that even their own technicians found it impossible to distinguish the papers printed in Riverside from those printed in San Francisco.

After the success of the *Wall Street Journal*, many Japanese newspapers ordered similar equipment. Years later, Japanese fax companies developed their own models using transistors instead of vacuum tubes. Production in the United States slowed and stopped. Because additional U.S. sales forecasts were low, arrangements were made to import from Japan rather than develop a new model. Today, many newspapers and magazines are printed in plants thousands of miles away from the place where the pages are composed. High-speed, high-resolution facsimile units can send pages up to 22 by 28 inches (two full-sized newspaper pages) over wideband satellite circuits or land lines. Both analog types and digital types of facsimile units are in use. Some units require a T-1 1.34-Mb or a T-2 6.34-Mb channel. Resolutions of 800 to 1,000 lines/inch are commonly used, but 1,800 lines/inch and higher are available. The facsimile recording is made on a film negative or directly on a printing plate. The *Wall Street Journal*, the *New York Times*, *USA Today*, and other newspapers send to many different printing plants via direct satellite broadcast. A satellite receiving station is located at each printing plant. *USA Today* covers the United States with 32 printing plants and has additional plants overseas. Color pages are sent by making three or four transmissions of the original color page using color separation filters.

2.9.4 Newspapers by Fax

The limited-edition newspaper broadcast experiments of 1948 led to the later sending of newspapers by Group 3 fax via stationary satellites such as Oceansat or Marisat for publishing abbreviated daily editions of many newspapers. Typically, eight-page standard letter-size editions were printed on cruise ships in the Caribbean and distributed to their passengers. The service later expanded to include hotels, resorts, and corporate clients at locations throughout the world. By the end of 1977, much of the satellite communication service for these newspapers was provided by Information Management Consultants (IMC). Papers using the service included *Die Welt* (Germany), *La Stampa* (Italy), and the *New York Times.* An example of one of these papers is *Timesfax,* a digest of top stories and editorial comment, plus sports, weather, business news, and the crossword puzzle. *Timesfax* alone reached 175,000 readers, including those on 65 cruise ships. Newspapers by fax also became available on a subscription basis directly to the customer's fax machine or on a call-in fax-back basis. Delivery is by personal computer or fax machine, with signals delivered by satellite or telephone line, including the Internet. Computer broadcast has replaced much of the earlier fax transmission. The paper is composed using Ventura Publisher or specially developed software. Transmission to the receiving computer takes 1.5 min for an eight-page newspaper. If the customer receives on a personal computer, the document is printed on a laser printer at 600 dpi instead of by the fax machine at 200 dpi.

2.10 High-Speed Fax

With the rapid postwar expansion of television, some visionaries thought that high-speed facsimile might utilize the new wideband channels and perform an important business purpose. Some of these systems proved valuable in regular fax service, while others, although breaking through existing speed limits, were unable to find a direct application in business. This section describes some of the early high-speed fax systems. Group 3 fax today can achieve on a standard PSTN voice channel speeds comparable with what some of the early high-speed fax machines could do on special wideband channels.

2.10.1 Ultrafax

On October 21, 1948, RCA demonstrated Ultrafax. It sent the entire 1,047-page book *Gone With the Wind* the 3-mile distance from radio station WNBW in Washington, D.C., to the Library of Congress over a wideband microwave relay channel in 2 min, 21 sec. RCA's President David Sarnoff heralded the dramatic event "as significant a milestone in communications as was the splitting of the atom in the world of energy." In advance of the demonstration, the book plus maps and many other kinds of documents had been copied to 16mm microfilm. A flying-spot CRT scanner generated a TV-type signal, and the information was sent over a 5-mHz TV channel. The receiving terminal recorded the image on 16mm film from a flying spot kinescope. In 40 sec, the film was ready for projection on a screen, having passed through a hot developing fluid and a fixing bath. Alternatively, it could be printed full size by a continuous process enlarger. Eastman Kodak had furnished a high-speed film processor and enlarger that it had developed for the military. Ultrafax, the first of a number of high-speed fax systems, was impressive but ahead of its time, and it never reached production.

2.10.2 Hogan Very High-Speed Facsimile System

In 1959, Hogan Laboratories produced a 100-line/inch, multistylus, high-speed fax recorder designed by George Stamps, John Smith, and Hugh Ressler. The 8-mil recording styli were spaced in a row at 100/inch across the paper, with a separate print transistor for each stylus. The recording paper feed rate was the recording rate for each stylus, and the 1,024-stylus design could print 1,024 times faster than earlier designs that had only one print amplifier. (This principle was later used in Group 3 fax machines with thermal print heads at much lower speeds.) Electrolytic recording paper was drawn from a supply roll over a 10-mil stainless steel tape. The tape moved slowly between two rollers as it supplied the small amount of metal used in the recording process. Insulation had been removed from the sides of the styli where they touched paper on the other side. Pressure applied to the wire assembly caused each wire to act as a spring, ensuring contact with the recording paper. Low-voltage recording paper allowed direct connection from the print transistor through the paper to the steel tape on the other side. The recording pulse time per stylus was about 4 μs, much too short to make a mark. The recording time was stretched out by a hold-and-mark circuit to almost the time it would take for a full scan across the page (400 μs in some cases). The only mechanical movement was the recording paper feeding through at about 2 ft/s. One machine design had a recording width of 10.24 inches, and a later design for the U.S. Navy Bureau of Ships weather map transmissions was 19 inches wide [16].

The facsimile scanner formed an image of the page being sent, on an opaque plate that had a very narrow slit across the image width. A second slit at almost a right angle on a disk just underneath let only a small spot (elemental area) of the image to pass through and be seen by a photomultiplier. As the disk rotated, the spot swept across the image to scan it. By having additional slits on the disk, another scan started just after the first one finished, and the image had moved one scan line width. Ten radial slits were used in some versions of this fax system, and one rotation of the disk made 10 sweeps across the image. Extreme accuracy of the slit position was not needed because each slit generated its own snyc pulse.

2.10.3 Xerox LDX

In late 1962, Xerox demonstrated LDX, a high-speed document facsimile transmission system that printed images on plain paper by the xerographic process. It sent at the rate of 5.5 to 66 sec per page. The equipment was delivered for 100, 135, or 190 lines/inch scanning. Operation was over either a group of 48 kHz (12 voice channels) or a supergroup of 240 kHz (60 voice channels) or suitable private microwave channels. Railroads found the LDX system particularly useful and economical to operate over their own microwave channels that parallel the railroad tracks. The system was based on the Xerox 914 office copier, which used the electrostatic printing process to produce high-quality permanent records on plain paper. In the scanner, a small intense spot of light rapidly swept across a CRT face, forming a line. A lens projected an image of the line onto the page being scanned, creating a series of pixels across the width of the page. A fiber optics array collected the light reflected from the page, delivering it to a photomultiplier tube (PMT). The brightness of the light on the PMT is determined by the density of the mark being illuminated on the page being scanned. Feed rollers advanced the page a step for each line scanned. The fax receiver printer had a CRT that operated the same way as the scanner, except that the light was projected onto a photosensitive selenium drum, and the received fax signal controlled the CRT spot brightness as black and white areas were scanned at the transmitter. The selenium drum rotated and produced an electrostatic image of the page being sent. The drum rotation brought the electrostatic image under a toner station, where black powder was applied to the charged areas and produced a visible image that transferred to white paper and fused by heat.

2.10.4 A.B. Dick Videograph

In 1962, the A.B. Dick Company made Videograph, a high-speed document fax system consisting of the 921 scanner and the 922 printer. Documents 8.5 inches wide fed through the equipment at 15 ft/min (less than 4 sec per 11-inch page) at a resolution of 96 lines/inch. In the scanner, a high-intensity lamp illuminated the page being sent. A rotating 20-sided prism deflected an image of the page across an aperture in front of a PMT at a scanning rate of 17,280 lines/min. The 922 printer had a CRT with a narrow 8.5-inch-wide faceplate and no phosphor coating. Instead, the electron beam swept across an 8.5-inch row of wires at 96 per inch for writing an electrostatic charge onto the white recording paper that touched them. As the paper moved over the wires, an electrostatic charge pattern was applied, forming a latent charge image of the page sent. A special coating on the paper held the charge until toner was applied and fused to the paper to make a permanent recording. Paper from a 1,500-foot roll was automatically cut at the end of each page. A 240-kHz supergroup channel was used between fax machines [17].

2.10.5 Matsushita Videofax

In the mid-1970s, Matsushita Graphic Communication Systems (MGCS) of Tokyo had a high-speed fax system that sent 7-by-10.2-inch pages over a 48-kHz group channel. The rate of sending was two to five pages per minute, depending on the resolution selected. Recording was electrophotographic, using a CRT to write on electrostatically charged white recording paper. Scanning was flatbed feed with a single photosensor and a fiber-optics line-to-circle converter. Liquid toner was used to develop the received image.

2.10.6 Satellite Business Systems Batch Document System

In mid-1979, Satellite Business Systems (SBS) demonstrated a high-speed document distribution system that used intelligent office copiers as fax machines with wideband satellite channels for interconnection of earth stations on customer premises. The primary use was between mail rooms of large and medium-sized organizations, to replace existing mail delivery systems. Documents were sent at 30 to 60 pages/min, with automatic printing, collating, and labeling of multiple recipient copies by the fax copier machine at each receiving mail station. High-quality copies at 300-dpi resolution were printed on plain white paper by the xerographic process. SBS was a partnership of Aetna, Comsat, and IBM.

2.11 Radio Facsimile

June 11, 1922, marked the onset of another phase in international fax transmission with the publication of a photograph of Pope Pius XI in the *New York World.* On that date, Dr. Korn sent the picture by radio fax from Rome to Bar Harbor, Maine. Occurring only 1 year after the start of commercial radio broadcasting, the success of the method showed how rapidly the technology was progressing. In 1924, RCA demonstrated transatlantic radiophoto transmission from New York to London and back again. Two years later, RCA offered radiophoto service commercially and expanded the service into key cities throughout the world. In the following years, the rapid growth and success of home radio broadcasting led some facsimile leaders to devote their energies to experimenting with home broadcasting. Commercial AM radio stations sent fax broadcasts from midnight to 6 A.M., when the stations were normally closed for the night. In the era before television and FM stations, public interest was high, with various experimental fax systems using hundreds and, in some cases, thousands of fax receivers in home locations.

In May 1927, WOR in Newark, New Jersey, was the first standard radio broadcast station to broadcast pictures by fax [18]. It used the Rayfoto system, built in 1926 by Austin Cooley. Fax signals from a home radio receiver were recorded on Azo #4 photographic print paper on a drum rotated by a spring-driven wind-up record player. The recording mechanism rested on the turntable center pin, where a phonograph record was normally placed. As the turntable rotated, a stylus needle in contact with the paper moved along the drum, tracing a spiral. Light from a spark at the stylus exposed the paper. There was much interest, and 27 radio stations experimented with it, but very few fax receivers were sold to the public. The high cost of the recording paper was one factor limiting its use [19].

"In 1929, Charles Young and Maurice Artzt at General Electric Company in the U.S. demonstrated a radio receiver using carbon paper as the medium for impact printing the received pictures onto plain paper" [20]. Carbon paper overlying white paper passed between a single-turn helix on a rotating drum and a magnetically operated printer blade. A pretuned radio receiver was set to the frequency for fax broadcasts, and a clock turned on the equipment at the time of broadcast. RCA Laboratories in Princeton, New Jersey, later took over this project, using improved equipment for AM radio broadcasting. In 1937, radio station KSTP in St. Paul, Minnesota, successfully delivered a special-edition newspaper by radio fax. A few other stations jumped into the home broadcast action, and hundreds of receivers were used for fax reception in the home.

Station W2XBF in New York City, owned and operated by William G.H. Finch, was the first to be granted an FCC experimental fax broadcasting license. Finch-designed fax transmitters and receivers were used in most of the AM fax broadcast trials that followed. The Finch fax receiver used Teledeltos™ dry recording paper, fed from a roll into a semicylindrical form at the recording line. A reciprocating arm with the same radius as the cylinder was driven by a cam as it swung back and forth with the recording stylus in contact with the paper. Crosley radio station WLW in Cincinnati sold about 10,000 of these kits to experimenters for $49.50 each.

War demands caused a decline in interest in AM fax home broadcast over the next few years. On June 9, 1948, the FCC established new standards for facsimile radio broadcast using FM radio channels instead of AM. Both 4- and 8-inch-wide fax formats were allowed. The fax signals were multiplexed at frequencies between 10 and 15 kHz in a manner that did not degrade the quality of music and voice programs. Home fax broadcast was now rejuvenated, and John V.L. Hogan's company Radio Inventions helped to establish a new fax broadcast service for newspapers. Hogan's fax recorded on a roll of wet electrolytic paper using a helix and blade mechanism and sent an 8.5-by-11-inch page in 3.5 minutes. Resolution was 105 lines/inch, and the signal was amplitude modulated on a 10-kHz carrier frequency. Synchronization was from the 60-Hz power line. Much of the equipment was manufactured by General Electric Company. The *New York Times* (WQXR-FM), the *Miami Herald* (WQAM-FM), the *Philadelphia Inquirer* (WFIL-FM), the *Atlanta Journal* (WSB), and the Columbia Broadcasting System broadcast experimental fax editions. They operated automatically during the sleep time of the households, with radio-dispatched newspapers ready by breakfast time. By late 1940, experimental fax newspaper broadcasts were successful enough to have around 40 commercial stations providing fax newspapers with at least 1,500 radio fax receivers in homes around the country [21].

Although many experimental systems were successful technically, none ever paid for their costs. Part of the problem was timing. Television held greater fascination for the public after World War II, and the enthusiastic visions of a fax receiver as common home equipment never materialized.

2.12 Connection to the PSTN

On July 23, 1938, AT&T issued Bell System Practices Section C55.711 for electrical connection of telephotograph apparatuses to telephone lines in either permanent or temporary installations, using coupling unit 104A or 105A. Both private lines (connected directly between user locations) and the PSTN utilized those couplers, which consisted of a 1:1 transformer for telephone line isolation. A neon lamp connected to a third winding would light at about 0 dBM on the telephone line to prevent sending signals at higher levels. C55.711 stated that "these arrangements are available to the Press ... For temporary installations the representative of the Press will have with him 104A Coupling Unit and usually, therefore, the Telephone Company will need only the facilities for connecting it to the telephone line." When covering news events not scheduled in advance, the fax operator might clip the 104A onto the phone line rather than waiting for an installation by the telephone company. Instead of the 104A, some fax operators used their own 1:1 line transformer, capable of passing 100-ma telephone line current. Private lines were leased for $3.50/mile/month for circuits of up to 100 miles. The rate dropped to $1.23 for circuits of 1,500 miles or longer. There was an extra charge of $25/month for a "Private Line Schedule 2 Telephoto Circuit," with limits on envelope delay distortion and noise.

2.12.1 Hush-A-Phone

On November 8, 1956, the regular switched telephone network opened up somewhat through an event that appeared to be unrelated to facsimile. The United States Court of Appeals of the District of Columbia overturned a previous FCC ruling that the Hush-A-Phone device impaired telephone service. The device snapped onto the talking end of a telephone handset and fit against the talker's face, screening out room noises and ensuring privacy of conversation. Fax use of the PSTN had been limited to the press and certain governmental agencies. The court decision was interpreted by some as eliminating that limitation for acoustic coupling of fax signals [22]. The PSTN provided the channel necessary to access fax units anywhere in the world by merely making a telephone call. Without such access, Group 3 fax would still be a small, insignificant service.

2.12.2 Dataphone

In early 1963, the FCC authorized sending facsimile over the PSTN using a coupling device with an analog modem. Bell System Data Set 602A connected between the fax machine and the PSTN, providing protection of the telephone system from incompatible signals and dangerous voltages unwittingly sent by the customer. The fax machine interface had 0–7V dc analog baseband fax signals. The 602A converted the baseband fax signal into 1,500- to 2,500-Hz audio frequency shift tones that were sent over the telephone line (PSTN) to the receiving 602A and converted back to 0–7V for the facsimile receiver.

Although the Bell System modem worked well, manufacturers resented being forced to change the fax machine design to bypass their own built-in modulator and to provide a new interface to send fax over the PSTN. Fax manufacturers had built the same type of modem with 1,500 to 2,300 Hz many years earlier for sending their fax signals over HF radio channels.

2.12.3 Carterfone

Radio amateurs, called ham radio operators, had long used phone patches to connect occasional telephone calls to foreign countries. This unauthorized use of the AT&T lines had been tolerated, but when a man named Carter who owned a small mobile radio communications operation took on the giant AT&T to extend regular telephone calls through his mobile radio equipment, no one thought he had a chance. Connecting any "foreign equipment" to the telephone network had been vigorously and successfully defended by AT&T, but they lost this fight, and a whole new era of telecommunication began. The so-called Carterfone decision in 1967 enabled fax machines and other terminals to use the PSTN. In November 1968, Bell System issued its Technical Reference PUB 41803, *Acoustic and Inductive Coupling for Data and Voice Transmission.* Although direct electrical connection to the telephone network was not allowed, acoustic or inductive coupling connected the signals through a regular telephone handset. The fax manufacturers then avoided the Bell System analog modem by using acoustic coupling, even though room noise and distortion caused by the carbon granules of the telephone transmitter caused some problems.

2.12.4 Bell Couplers

In July 1969, Bell System's Data Access Arrangement, later known as Data Coupler CBS, became available. Finally, the fax machine could electrically connect its signals to the PSTN, but only through a telephone company-provided device that isolated and protected the telephone network. Although the room noises and other distortions to the fax signal were eliminated, the customer still had to pay the telephone company a monthly fee for a device to access the telephone wires. Modulating and demodulating functions were performed by the customer's equipment, but the Bell System retained responsibility for network protection, including network control signaling. The CBS controlled the maximum signal power that could be sent on the PSTN. It also provided network control signaling telephone line isolation, a line-holding path for dc supervision, ring detection, customer's off-hook control, and a means for transmitting customer-originated dial pulses, surge and hazardous voltage protection, longitudinal imbalance protection, and remote test features. The fax machine could originate a call (equivalent to lifting a telephone handset from its cradle), dial, and automatically answer a call (equivalent to lifting a telephone handset from its cradle in response to ringing) or terminate a call (equivalent to hanging up). In the automatic answer mode, the customer's fax machine provided the logic necessary to answer a call.

2.12.5 FCC Part 68 Regulations

Nine years later, in June 1978, the FCC issued Part 68 regulations for direct electrical connection of fax machines and other devices to the PSTN (without paying extra for a device furnished by the telephone company). These regulations are about the same as the Bell System specifications for protection of the public telephone network. At first, only couplers in a self-contained box were permitted, and they had to be purchased from the same companies that manufactured CBS couplers for the telephone companies. Other manufacturers subsequently designed couplers and obtained FCC registration. Later, fax machine manufacturers designed their own equipment for meeting FCC regulations with the coupler inside the fax machine. At last, a fax machine could be connected directly to the telephone line by plugging into a standard RJ-11 telephone jack. For FCC approval, a sample of the fax equipment design is tested at an FCC-approved laboratory for compliance with FCC Part 68. Test results are submitted to the FCC and a registration number is assigned.

2.13 Age of Incompatible Office Fax

In 1940, the Institute of Radio Engineers (IRE) set up a Technical Committee on Facsimile to consider engineering problems and, through the IRE Standards Committee, to recommend technical facsimile standards for the United States. In 1942, a glossary of fax terms was approved and published, followed in 1943 by *Temporary Test Standards*. Standards for fax machines themselves were considered to be outside the IRE's scope of work. The IRE facsimile test chart was developed in 1955 and became a generally accepted testing standard for fax and other imaging applications. It was still in demand in 1990 as the Institute of Electrical and Electronics Engineers (IEEE) facsimile test chart but the negative has deteriorated too much for additional printing. The glossary was updated as *IRE Standards on Facsimile: Definitions of Terms, 1956*. The American Institute of Electrical Engineers (AIEE) sponsored technical papers on facsimile at its winter general meetings. Its annual conference paper, *Advancements in the Facsimile Art,* by Warren Bliss of RCA Laboratories, Princeton, New Jersey, became a journal of progress in the facsimile field.

After World War II, many companies designed and manufactured office facsimile equipment in the United States, England, France, Germany, and Japan. At first, the equipment included incompatible analog facsimile units that operated on different standards. Each manufacturer touted the advantages of its design and did not want compatibility because that would allow competition for its customers. Private telephone lines were used, because fax use of the regular switched telephone network was still limited to newspaper companies and certain government services. Most businesses either did not know facsimile existed or saw no use for it. This was a constantly evolving process with some fax designs, incompatible with everything else, being limited to 10 or fewer fax units.

In 1962, the Electronic Industries Association (EIA) established a technical committee, TR-29, on facsimile equipment and systems, with Ken McConnell as chair. It was evident to some that compatibility between fax machines of different manufacturers was needed. The U.S. manufacturers sent representatives to the meetings, at which engineers were reluctant to reveal much about their equipment specifications. They seemed to be restricted by their companies from agreeing to a fixed set of specifications with normal tolerances because their designs would have to be changed. It took until October 1966 before EIA Standard RS-328, *Message Facsimile Equipment for Operation on Switched Voice Facilities Using Data Communication Equipment,* was published. This was the first U.S. "standard" on office-type fax. The following is quoted from RS-328:

> Each manufacturer has designed his facsimile equipment used for message communications to fit the requirements of his customers and the natural characteristics of the types of mechanisms employed when operating on a private line basis. The same equipment was initially used when operation of facsimile equipment on a regular telephone call basis became available. Operation between some of the equipment was not possible due to differences in the facsimile equipment. This Standard specifies machine characteristics that will ensure interoperation.

Using RS-328, the received copy might be stretched from the original size or have parts near the edge of the page missing, but for the first time, diverse fax units could communicate. Each manufacturer stayed within the wide tolerances required to get the "standard" but did not convert to the "recommended for eventual standard." Obviously, there was an acute need for a facsimile standard to a single set of performance specifications instead of a low grade of interoperability. McConnell requested that all U.S. fax machine companies participate in TR-29 to develop meaningful standards. 1968 was the year of the CCITT Group 1 fax international standard, but most U.S. manufacturers were not even planning to comply with it. It was also 12 years after acoustic coupling to the PSTN was allowed and shortly after the Carterfone decision allowed direct connection to the PSTN (with a number of restrictions). Few of the office fax machines were then transceivers. Most mechanisms for scanning and printing were quite different, requiring separate drive motors. Physical movement of scanning and recording spots needed high mechanical precision to achieve acceptable results. Table 2.1 shows the compatibility problems that existed at the end of 1968.

Notice that the resolutions used for standard document transmission varied from 90 to 100 lines/inch. Scanning lines per minute varied from 120 to 360. Page sizes also varied. In most cases, compatibility of machines was not considered important by the manufacturers. Even within one manufacturing company, different models might be incompatible. Some models had power line synchronization with 60-Hz motors. The Alden system scanned an 8.5-inch page in the 11-inch dimension, providing the ability to transmit larger documents [23].

The EIA-recommended specifications under RS-328 got a boost when Magnavox met them all with its new fax transceiver and went on to produce thousands of fax machines. Scanning and recording were on the inside of a stationary half-cylinder. Two scan heads aligned 180 degrees apart pointed

Table 2.1
Incompatible Fax Machines

Company	Transceiver	Lines/inch	Lines/min	Min/page	Send	Receive
Alden11	Yes	96	120/240	6.8/3.4	10.2 by C	0.5 by C
Graphic Sciences	Yes	88	180	6	9 by 14	9 by 14
Litcom	Yes	91	180/300	6/4	8.5 by 14	8.5 by C
Magnavox	No	96	180	6	8.5 by 11	8.5 by C
Muirhead	Yes	90	200	6	8.6 by 14	8.6 by 14
Scanatron	Yes	96	180/360	6/3	8.5 by C	8.5 by C
Telautograph	Yes	100	180/300	6/4	8.5 by 11	8.5 by C
Xerox	No	96	180	6	8.5 by 11	8.5 by C

*C = Continuous feed (flat bed)

radially outward from a central shaft, with one focused on the page being scanned. As the heads rotated, one head traced a scan line across the page, which was held in semicylindrical shape. After a 180-degree rotation, the first head completed one scan line and the page moved one scan line width. As the heads continued to rotate, the second head scanned the next line, and the operation continued by switching back and forth between scan heads. In receiving mode, pressure-sensitive recording on white paper with a carbon paper overlay sheet was used for printing. Two magnetic recording heads mounted 180 degrees apart in the same plane as the scanning heads touched the surface of the recording paper set and marked the white paper while signals for black were received. Because the marketing of the Magnavox units was through Xerox, these standards were kept when Xerox started to manufacture its own units.

2.14 Group 1 Fax

European countries dominated the CCITT and the International Radio Consultative Committee (CCIR), and U.S. participation in international fax standardization was practically nonexistent. Many years earlier, the United States had official representation in the CCIR on fax standards through the State Department, but dropped it when there was minimal U.S. interest. When fax standards proposals came before the General Assembly, there was automatic approval unless there was a no vote. Thus, Group 1 was adopted without U.S. participation, making possible a more generalized business use of fax in Europe and other countries but not in the United States, where due to lack of compatibility, business fax users usually were limited to communicating within their own companies. Table 2.2 compares the CCITT and U.S. standards.

The U.S. engineers thought that 2,400- or 2,500-Hz black and 1,500-Hz white worked better, while the CCITT engineers thought 2,100-Hz black and 1,300-Hz white worked better over their phone lines. Both may have been right. Some of the European telephone lines limited the useful upper frequency more than those in the United States, and there were some other telephone line differences. Except for the black and white frequencies and some variation in U.S. page format, most other specifications were the same as Group 1. The U.S. salespeople mistakenly called the U.S. fax "Group 1." Companies such as Graphic Sciences and 3M kept their own different standards even after the CCITT Group 1 facsimile standard was adopted. The EIA TR-29 standards work was still not a team effort to develop a standard that all companies would

Table 2.2
Comparison of Group 1, U.S., and Group 2 Fax

	Group 1	U.S. Fax	Group 2
Lines/min	180	180	360
Modulation	FM	FM	VSB AM/PM*
Carrier frequency	—	—	2,100 +10 Hz
White signal	1,300 Hz	1,500 Hz	Maximum carrier
Black signal	2,100 Hz	2,400 Hz	26 dB minimum lower

*Vestigal sideband amplitude/phase modulation

follow, insofar as the fax engineers from each company defended their own company's fax machine designs. Fax machines were unable to communicate very well with those from another company, and none were compatible with European designs. The U.S. fax machines still had differences that distorted the received copy shape or created operational problems between machines of different manufacturers. Later, compatibility between U.S. fax machines improved, and some 6-min fax machines, by throwing a switch, would work with the CCITT Group 1 standard. Many units offered an optional mode that moved the original page through in 4 min instead of 6 min. That option produced lower resolution in the paper feed direction for even lower copy quality and was thus a poor attempt to increase the transmission speed. Although the Group 1 and the U.S. units generally worked better than the earlier office fax units, they still were unreliable 6- or 4-min analog units, which gave fuzzy copy and required much manual attention.

At the time, a few systems existed for increasing the transmission speed over existing telephone facilities. Electronic Image Systems (an Addressograph-Multigraph subsidiary) had an image-coding scheme in their demonstration model of Telikon II. For each line scanned, the unit selected the most efficient of several stored digital codes for minimum transmission time. The short code was used for an all-white line. This scheme was reported to have up to a 12:1 transmission time advantage over conventional techniques, with deliveries expected in 1970. Scanatron had developed send and receive "bandwidth compression" modems called Pacfax for operation of fax over Schedule 2 conditioned telephone lines at double speed. Those modems operated with their 391-1 graphic transmitter and recorder, but were also available for use with other fax systems.

2.15 Group 2 Fax

Both Xerox and Graphic Sciences saw the need for a faster system for users who had accepted the quality of Group 1. Trying to gain the competitive edge, both developed and marketed new fax units capable of 3 min/page instead of 6, or 2 min/page instead of 4. Xerox had a frequency modulation (FM) scheme, and Graphic Sciences had an amplitude modulation (AM) scheme. The AM system had the capability to send gray scale, while the FM system did not.

At the TR-29 meetings, the engineers enjoyed a camaraderie and seemed more loyal to development of the facsimile art than following the competitive line that their companies wanted. They had seen how unsatisfactory the U.S. approach was when Group 1 was developed and knew it was time for a new worldwide standard instead of a competitive race. They decided to cooperate to develop a new standard based on the competing systems that would be best technically and that could be submitted for CCITT adoption. Up to this time, the fax standards on which TR-29 had been working were for use in the United States. The EIA headquarters gave approval for the committee to extend its operations, and contacts were made with the U.S. State Department. Arrangements were made for TR-29 to work through an existing State Department CCITT group that would officially represent the United States for fax standardization. The U.S. companies were already making fax units that met the 3-min speed requirement of the Group 2 standard just starting. Both U.S. systems were proposed to the CCITT, which agreed that comparison tests should be made to determine the better modulation system. The British Post Office did the evaluation by using black-white copy, testing many different lines, and switching between the two systems on each call. The AM system gave somewhat better results overall and was selected for Group 2. The 3-min fax equipment already in service was not compatible with the new CCITT standard.

The new Group 2 standard, based on 3-min machine designs that had a customer performance record, was a definite improvement. Adoption of the standard by the CCITT and worldwide implementation opened the door to universal fax machines. A good standard resulted from the excellent teamwork among the TR-29, the British Facsimile International Consultative Committee (BFICC), and engineers from other countries and provided the impetus for pursuing a standard for digital fax. Because the few digital fax machines operating on private standards were complex and expensive, it was expected that Group 2 would be the dominant fax standard for many years.

The Japanese established efficient facsimile manufacturing facilities that, combined with low labor costs, allowed them to take over the design and manufacture of most facsimile equipment. CCITT standards opened up worldwide markets for the Japanese equipment. The much lower price of Group 2 fax units allowed them to compete favorably against the expensive private-standard digital facsimile units, which became available in 1974. Even after the Group 3 standards existed, the Group 2 price advantage remained for a few years; digital fax, however, was showing its worth in higher speed and better quality fax copies for those customers who could afford expensive machines that worked only with other machines from the same manufacturer. The same engineering committees that successfully developed Group 2 fax machine standards set to work on Group 3 standards.

2.16 Development of Group 3 Standards

The earliest digital fax units used the adaptive run-length coding algorithm patented by Donald Weber. A digital code word represents the number of successive white picture elements along the scanning line before the next black pixel. The next code word represents the number of black pixels following the white run. Because the code words usually are much shorter than the number of pixels, it takes fewer bits to send the page. This coding removes redundancy from the page being sent and thus shortens the transmission time. Single-line coding under this patent was used in Dacom's Rapifax 100. A three-line technique under the same patent was later used for coding newspaper pages for fax transmission. The digital modem for the Rapifax 100 was the first modern large-scale integrated circuit (LSI) high-speed modem to be used successfully on regular telephone lines (PSTN). The modem reduced transmission time by coding 4 bits at a time, requiring only 1,200 symbols/s (changes of the signal state on the telephone line) to send 4,800 bps. Symbols were sent as an analog tone signal (PM/AM) that passed through the regular dialup telephone system in a manner similar to voice. The receive modem synchronized on the symbol rate to properly decode the received signal.

Many of the key items in the Group 3 standard were developed by the TR-29 fax committee as the following information from the minutes of their meetings shows [24].

- January 1975: The CCITT planned for the Group 3 specification to be generated by French and English postal services and manufacturers. Primary concern for TR-29 was the generation of G3 sub-minute digital fax machine standards.
- June 1975: A completely new scheme was proposed for G1, G2, and G3, involving binary signaling for handshaking between sending

and receiving fax machines as opposed to the tonal technique currently under CCITT consideration.

- August 1975: Planned binary signaling with the V.21 (or V.23) modem for facsimile handshake.
- September 1975: It was agreed that the receiver should tell its option capabilities to the transmitter following the 2,100-Hz answer tone as proposed by Graphic Sciences.
- October 1975: A Special Rapporteur's Group on T.4 recommended rates up to 9.6 Kbps for Group 3. Most of the Group 3 standard was developed between TR-29 and the British Post Office, with cooperation from England, France, Germany, Japan, and other countries.
- September 1976: Some proposals by other groups "might be interpreted as requiring an external modem. EIA members plan to provide the $V.27_{ter}$ modulation system internal to the machines. To do otherwise would be difficult economically and technically in light of T.4 requirements."

During the generation of the Group 3 standard, many long day and evening sessions took place at the CCITT meetings in Geneva. Each country had its own concerns, and many proposals were made to add protective details to the standard. The chair of CCITT Study Group XIV was conscious of the wrangling that might occur. He used a slogan that was successful in preventing undue complexity: "If in doubt, leave it out." The Germans would not allow any modem to be connected to their lines unless it was furnished by them. The French wanted to limit equipment connected to their lines to fax machines made in France. One of the most important tasks in developing the Group 3 fax standards was selection of a coding scheme to reduce the number of uncoded bits per page from about 2 million to 400,000 or fewer, allowing faster transmission of fax pages. That process is called redundancy reduction or source encoding.

Concern was expressed in the CCITT about selecting a code that would not require royalty payments. The Huffman code was considered because it was a very efficient run-length code and had no active patent. Huffman coding requires a separate code word for each white or black run length encountered. The code words are assigned with the shortest length words for the run lengths that occur most often. The number of pels per scan line was 1,728 using the charge-coupled device (CCD) scanner chips then available. That meant 3,456 code words because the run-length statistics are different for black runs and white runs. Bob Krallinger, then chair of TR-29, worked out a modified

Huffman (MH) system that greatly reduced the number of code words by coding each run as two words. All run lengths up to 1,728 were reduced to only 92 code words using the MH code tables. This system was much more easily implemented. To select good code tables, the probability of each run length needed to be determined. Eight test images were selected by the CCITT as representative of documents that might be sent for business use of fax. These were scanned 10 to 20 times with histograms made of each run length. The multiple scanning reduced the variation in run lengths on a given page caused by page placement and skew. Even a small skew reduces some of the white run lengths between printed lines.

After the MH code was accepted, it became evident that most of the fax machine manufacturers planned to furnish their own private two-dimensional code to achieve faster transmission rates. Private codes were allowed on a nonstandard provision of T.30, but those codes worked only between machines from the same manufacturer. Each company claimed a superior code, but when comparison tests were run, they varied only a few percentage points in transmission time per page. A relative address (READ) code was advocated for a standard option by the Japanese, and it was offered royalty free if accepted as a standard. Fortunately, international cooperation prevailed again. The code was studied by the British Post Office, and Harry Robinson came up with the modified read (MR) code, which was adopted by the CCITT.

One of the sticking points in the Group 3 standard was incorporation of an escape code, a nonstandard facilities (NSF) call, that allowed fax machines from the same manufacturer to operate with any private enhancement as long as the feature did not interfere with communication with a standard Group 3 machine. The TR-29 fax committee argued that NSF was essential to prevent freezing of the standard and making it prematurely obsolete. After many arguments, NSF was finally put into the specification, allowing undocumented private codes. It has allowed manufacturers to invest development money for improvements in the Group 3 standard. Some of the best NSF ideas were later incorporated into the Group 3 standard as a recognized option, allowing use between fax machines of different manufacture.

In 1980, the first Group 3 fax machines cost three or four times as much as Group 2 fax machines, but over the next few years the price differential between Groups 2 and 3 diminished rapidly. The Group 3 standard proved far superior and finally manufacture of the Group 2 machines slowed down and stopped.

Table 2.3 shows the technical progress made in facsimile and some other communication systems since the first fax unit was conceived. The list would be much too long if all the many other developments and contributors to the facsimile art were included.

Table 2.3
Facsimile Technical Progress, 1843–1998

1843	First facsimile patent: Bain, pendulum
1844	Morse telegraph patent
1850	Rotating drum fax: Bakewell (England)
1864	First drum and helix recorder (used ink): Meyer (France)
1865	First commercial fax: Caselli (France)
1876	First telephones: Bell and Gray
1902	Optical scan: Dr. Korn (Germany)
1906	Newsphoto service: Dr. Korn (Munich to Berlin)
1907	Vacuum tube (Audion): Deforest
1915	Telephone repeaters: long distance
1917	Teletype: AT&T
1920	Digital fax on transatlantic telegraph cable: Bartlane
1924	Telephotos in United States: AT&T, RCA, and Western Union
1926	Transatlantic radio newsphoto service: RCA
1928	Photo recording without optics (Rayfoto): Cooley
1929	Drum and helix carbon paper recorder: GE, Young and Artzt
1934	Photofax, AP news picture service
1936	Drum and helix electrolytic paper recorder: Dr. Hogan
1936	Teledeltos (dry direct recording paper): Western Union
1936	*New York Times* manufactures fax, Times Facsimile Corporation: Cooley
1943	Signal Corps Military Fax FX-1B
1943	Timefax (dry direct recording paper)
1945	TV (regular broadcast in United States)
1948	Western Union Deskfax (40,000 units)
1948	Ultrafax (very high speed fax on TV channel): RCA
1956	Hush-A-Phone: acoustic coupling on PSTN lines in United States
1958	Stewart-Warner Datafax
1959	Telautograph Quickfax
1959	Very high speed fax: Hogan
1962	Videograph high speed fax: A. B. Dick
1962	LDX high-speed fax: Xerox
1962	EIA committee TR-29 starts standard for office fax
1965	EIA issues fax standard RS-328

Table 2.3 (continued)

1965	Magnavox Telecopier; first U.S. large-production office fax
1967	"Carterfone" decision: direct connection to PSTN
1967	Xerox manufactures Telecopier
1967	Graphic Sciences DEX 1
1968	CCITT starts Group 1 recommendation
1968	Scanatron fax with bandwidth compression
1969	Digital fax starts on telephone lines: DACOM (Weber run-length code)
1970	Xerox 400
1972	3M enters fax market; VRC 600 (made by MGCS/Matsushita)
1973	3M 603; Visual Sciences (MGCS KD-211)
1974	Rapifax 100 digital fax: Dacom
1974	Qwip 1000: Exxon
1974	Xerox 410
1974	CCITT Group 1 fax adopted
1975	Xerox 200
1975	100,000 fax units in United States
1976	Qwip 1200
1976	CCITT Group 2 fax adopted
1976	CCITT starts Group 3 fax standards
1977	3M 9600: "Near Group 3 fax" digital fax
1978	Panafax UF-20 and UF-320: "Near G3"
1978	DEX 5100: "Near G3"
1979	Southern Pacific: DMS 2000 ("near G3") (made by Hitachi)
1979	Telautograph Omnifax: "Near G3" (made by Oki; thermal recording)
1979	NEC 6200: "Near G3"
1980	CCITT Group 3 fax adopted
1984	CCITT Group 4 fax adopted
1988	Error-correction mode for Group 3 fax adopted
1988	Group 4 Classes 2 and 3 rescinded
1991	V.17 14.4-Kbps modem for Group 3 fax adopted
1991	MMR coding for Group 3 fax adopted
1992	FCC Telecommunications Consumer Protection Act (12/20/92): fax requirements
1993	64-Kbps Group 3 fax adopted for digital channels (ISDN)
1994	VoiceView™ (switching voice, data, fax in one phone call)

Table 2.3 (continued)

1994	T.42 JPEG CIELAB color modes for Group 3 fax adopted
1994	T.611 programming communication interface (PCI) for Groups 3 and 4 fax adopted
1995	T.31 digital interface Class 1 and T.32 Class 2 adopted
1995	T.85 JBIG fax adopted
1996	V.34 33.6 Kbps modem for full-duplex Group 3 fax adopted
1996	T.434 and TIA/EIA-614 Binary File Transfer Standard for Group 3 fax adopted
1996	TIA 465-A and 466A conformity test procedures for Group 3 fax adopted
1997	T.36 "Commercial secure" modes for Group 3 fax adopted
1997	T.43 JBIG color modes for Group 3 fax adopted
1997	Simultaneous voice and fax
1997	Signal verification for Group 3 fax adopted
1998	TIA 668 PC-VGA HF radio fax adopted
1998	Internet fax via e-mail for Group 3 adopted
1998	High resolution for Group 3 fax adopted (300 and 400 pels/inch)
1998	Shortened handshake protocol for Group 3 fax adopted
1998	T.38 real-time Internet fax

References

[1] http://www.antique-radio.org/bios/henry/html

[2] http://www.cnam.fr/museum/ref/r11a07.html

[3] Evans, E. R., "History of Facsimile," *Times Facsimile Service Bulletin*, Vol. 5, No. 4, pp. 91–92.

[4] Ranger, R. H., "Transmission and Reception of Photoradiograms," *Radio Facsimile*, RCA Institutes Technical Press, October 1938, p 3. (Originally presented IRE, New York City, June 3, 1925.)

[5] Jones, C. R., *Facsimile*, New York: Murray Hill Books, 1949, p. 187.

[6] Berger, M., *The Story of the New York Times*, New York: Simon and Shuster, 1951, p. 410.

[7] Ibid., p. 412.

[8] Barck, O. T., and N. M. Blake, *Since 1900*, New York: Macmillan, 1952, p. 421.

[9] Ibid., p. 422.

[10] Cooley, A. G., "Datalog in the Beginning," *Litton Datalog News*, Vol. 4, No. 4, Oct. 1980, p. 7.

[11] Davies, R. C., and P. Lesser, "Facsimile Equipment," *Electronic Industries Magazine*, Feb. 1945, pp. 96–170.

[12] Byrnes, I. F., and C. J. Young, "Radio Weather Map Service to Ships," *RCA Institutes Technical Press*, pp. 294–304.

[13] "Weather Map Broadcasts," *Times Facsimile Service Bulletin*, Vol. 1, No. 10, Jan. 1947.

[14] Berger, p. 506.

[15] "Facsimile Edition of the *New York Times*," *Times Facsimile Service Bulletin*, Vol. 1., No. 10, Jan. 1947.

[16] Stamps, G. M., and H. C. Ressler, "A Very High Speed Facsimile Recorder," *IRE Trans. Professional Group on Communications Systems*, Vol. CS-7, No. 4, Dec. 1959. (Originally presented at the 4th Natl. Aero-Com Symp., Utica, NY, October 20, 1958.)

[17] Bliss, W., "Advancements in the Facsimile Art During 1962," AIEE Winter General Meeting, New York, Jan. 27, 1963.

[18] Evans, E. R., "History of Facsimile," *Times Facsimile Service Bulletin*, Vol. 6, No. 10, p. 69.

[19] Hills, L., and T. Sullivan, *Facsimile*, New York: McGraw-Hill, p. 7.

[20] Costigan, D. M., *Electronic Delivery of Documents and Graphics*, New York: Van Nostrand, 1978, p. 10.

[21] Ibid., p. 12.

[22] Stamps, G. M., *A Case History of Telephone-Coupled Facsimile*, Report for FCC, Contract 0647-7(30E25.14), Jan. 1977.

[23] Tewlow, J., *Research Institute Bulletin No. 90*, American Newspaper Publishers Association, p. 978.

[24] EIA, *Technical Committee on Facsimile TR-29 minutes*, Jan. 1975–Sept. 1976 meetings.

3

Group 3 Facsimile

3.1 Architecture

To give the reader an understanding of the architecture of a Group 3 facsimile machine, Figure 3.1 illustrates the system as basic blocks for the transmitting and receiving functions. Figure 3.2 is a block diagram of the transmitter of a typical Group 3 facsimile machine.

A typical facsimile transmitter has a charge-coupled device (CCD) scanner that works something like a camcorder, but the image of the page being sent is focused on a CCD chip that has 1,728 photosensors in a single line. The CCD reads brightness of spots in a very narrow line, 0.01 inch (0.254 mm) high, across the width of the page being sent. That generates a pulse for each photosensor, 1,728 pulses per line. The pulse amplitude represents image brightness at that point. After each line is completed, the image steps to the next line. The analog-to-digital (A/D) converter block changes the signal from analog to digital. What were pixels become 1-bit picture elements (pels). A two-line memory stores each pel for two adjacent scanning lines. Modified Huffman (MH), modified READ (MR), or modified modified READ (MMR) coding then compresses the bit-pattern information into a small fraction of the number of bits required before coding. This process is called source encoding or redundancy reduction.

A buffer memory stores the output of the MH/MR/MMR coder for use by the modem. The coder block generates code words containing the picture information in a compressed format that needs only one-fifth to one-twentieth as many bits. In the modem block, the compressed digital signal is further coded and converted into an analog signal that can be sent over the telephone

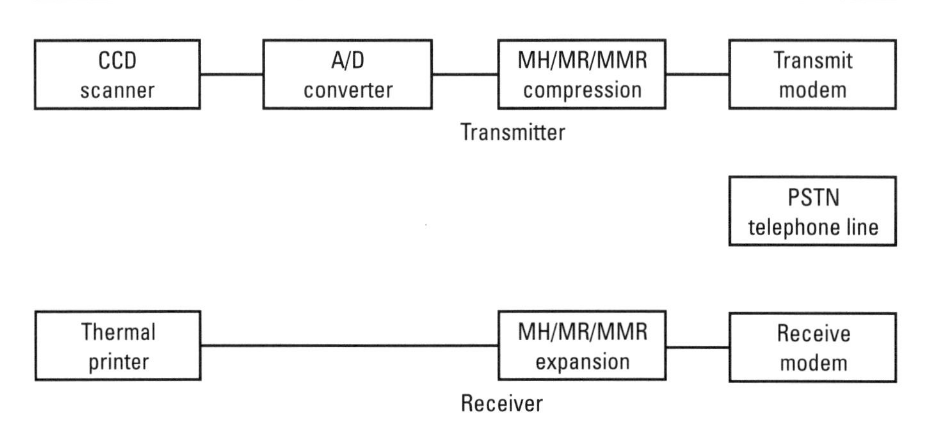

Figure 3.1 Block diagram of the Group 3 facsimile architecture.

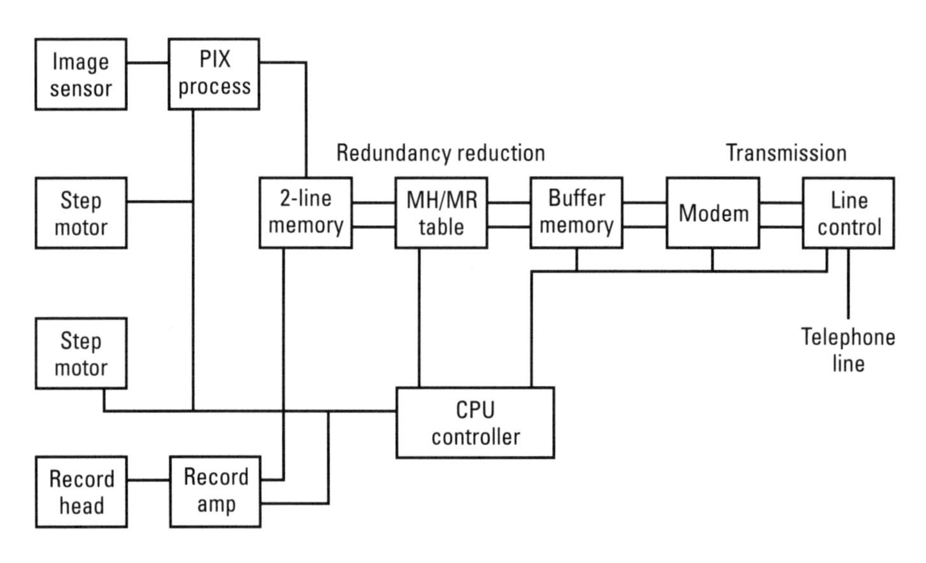

Figure 3.2 Block diagram of a Group 3 transmitter.

line. When operating at 14.4 Kbps, the modem takes 6 bits at a time and represents them as one of 128 different pulse amplitude modulation (PAM) states for transmission as an analog signal. The telephone line has only 2,400 baud (changes per second) to convey 14.4 Kbps. The modem coding thus achieves a further 6:1 compression of the signal sent. An FCC, Canadian Standards Association (CSA), or other Post Telephone and Telegraph (PTT)-approved built-in line connection unit (LCU) provides a standard miniature jack RJ-11 or other standard connector for direct connection to regular telephone lines

(PSTN). Another built-in jack allows use of a telephone for dialing or voice communication.

The receiver modem decodes the received analog facsimile signal, regenerating the digital signal sent by the facsimile transmitter. The MH/MR/MMR block then expands the facsimile data to black-white pel information for printing. The thermal printer has wires spaced at 203/inch touching the temperature-sensitive recording paper. Heat is generated in a small high-resistance spot on each wire when high current for black marking is passed through it. To mark a black spot, the wire heats from nonmarking temperature (white) to marking temperature (black) and back to white in a few milliseconds. The thermal printer converts the bit stream into a copy of the original page. Almost all Group 3 facsimile machines either send or receive at one time (half-duplex). Some circuitry and mechanical components serve either sending or receiving functions. The same LCU, modem, buffer memory, coder, and line memories are used for the receiving function in half-duplex mode.

Two slightly different sizes are normal for a page: the American letter size, 8.5 by 11 inches (216 by 279 mm), and the International Organization for Standardization (ISO) size A4, 210 by 297 mm (8.27 by 11.69 inches). Most Group 3 facsimile machines are furnished for the normal page size. The same basic facsimile machine design is used for both paper sizes. For most Group 3 facsimile machines, the paper length is of secondary consequence since the paper used for recording is in roll form. For laser-printing facsimile machines using cut sheets, the length can be a problem, because printing of one European A4 or U.S. legal paper facsimile may take two 11-inch pages. Facsimile machines that can scan the wider B4 paper (255 mm, or 10 inches) and A3 paper (303 mm, or 11.9 inches) are also available. Not all these facsimile machines record on a matching wider paper, but the recording page size can be reduced in both directions to print on the normal 8.5-inch paper width. When a wider scan facsimile machine sends normal 8.5-inch-wide pages, it sends only the normal width scan of 1,728 pels.

3.2 Digital Image Compression

3.2.1 Resolution and Pel Density

The resolution of the scanners and printers used in facsimile apparatus (and the associated transmitted pel density) has a direct effect on the resulting output image quality. The highest pel density specified for the original Group 3 terminal was approximately 204 by 196 pels per 25.4 mm. The actual specification is 1,728 pels per 215 mm by 7.7 lines/mm, referred to as metric-based pel density. The Group 4 recommendations support the following standard and optional pel densities: 200 by 200, 240 by 240, 300 by 300, and 400 by 400 pels per 25.4 mm (referred to as inch-based pel density). Note that the pel densities specified for Group 3 are "unsquare," that is, not equal horizontally and vertically. Group 4 pel densities are square. The difference causes a compatibility problem. If the Group 3 pel densities were to be extended in multiples of their current pel densities, for example, to 408 by 392, then a distortion of approximately 2% horizontally and vertically occurs during communication with a 400-by-400 "square" machine. The International Telecommunications Union–Telecommunications Standardization Sector (ITU-T) has decided to accept the distortion and encourage a gradual migration to the square pel densities. Accordingly, Group 3 has been enhanced to include higher pel densities, including both multiples of the original Group 3 pel densities and square pel densities. Specifically, optional Group 3 pel densities of 408 by 196, 408 by 392, 200 by 200, 300 by 300, and 400 by 400 pels per 25.4 mm have been added.

The metric-based pel densities and their pels are given in Table 3.1, as are specific values for the number of pels per line for all the Group 3 pel densities for ISO A4, ISO B4, and ISO A3.

Table 3.2 lists the optional inch-based pel densities and their pels. Specific values for the number of pels per line are given for all the Group 3 pel densities for ISO A4, ISO B4, and ISO A3.

3.2.2 One-Dimensional Coding Scheme: MH Code

A digital image to be transmitted by facsimile is formed by scanning a page from left to right and top to bottom, producing a bit map of pels. A scan line is made up of runs of black and white pels. Instead of sending bits corresponding

Table 3.1
Metric-Based Pel Densities

Pel Density (Approximate) (pels/25.4 mm)		**Tolerance**	**Number of Pels Along a Scan Line**		
Horizontal	Vertical		ISO A4	ISO B4	ISO A3
204	98	±1%	1,728/215 mm	2,048/255 mm	2,432/303 mm
204	196	±1%	1,728/215 mm	2,048/255 mm	2,432/303 mm
408	392	±1%	3,456/215 mm	4,096/255 mm	4,864/303 mm

Table 3.2
Inch-Based Pel Densities

Pel Density (Approximate) (pels/25.4 mm)		**Tolerance**	**Number of Pels Along a Scan Line**		
Horizontal	Vertical		ISO A4	ISO B4	ISO A3
200	100	±1%	1,728/219.45 mm	2,048/260.10 mm	2,432/308.86 mm
200	200	±1%	1,728/219.45 mm	2,048/260.10 mm	2,432/308.86 mm
300	300	±1%	2,592/219.45 mm	3,072/260.10 mm	3,648/308.86 mm
400	400	±1%	3,456/219.45 mm	4,096/260.10 mm	4,864/308.86 mm

Note that an alternative standard pel density of 200 pels/25.4 mm horizontally by 100 lines/25.4 mm vertically may be implemented provided that one or more of 200 by 200 pels/25.4 mm, 300 by 300 pels/25.4 mm, and 400 by 400 pels/25.4 mm are included.

to black and white pels, coding efficiency can be gained by sending codes corresponding to the lengths of the black and white runs. A Huffman procedure uses variable-length codes to represent the run lengths; the shortest codes are assigned to those run lengths that occur most frequently [1]. Run-length frequencies are tabulated from a number of "typical" documents and are then used to construct the code tables. True Huffman coding would require twice 1,729 code words to cover a scan line of 1,728 pels. To shorten the table, the Huffman technique was modified for Group 3 facsimile to include two sets of code words, one for lengths of 0 to 63 (Table 3.3) and one for multiples of 64 (Table 3.4). Run lengths in the range of 0 to 63 pels use terminating codes. Run lengths of 64 pels or greater are coded first by the appropriate makeup code word specifying the multiple of 64 less than or equal to the run length, followed by a terminating code representing the difference. For example, a 1,728-pel white line would be encoded with a makeup code of length 9, which represents a run of length 1,728, plus a terminating code of length 8, which represents a run of length zero, resulting in a total code length of 17 bits (without the synchronizing code). When the code tables for Group 3 were constructed, images containing halftones were deliberately excluded, so as not to skew the code tables and degrade the compression performance for character-based documents.

For those machines that choose to accommodate larger paper widths or higher pel densities, the makeup codes in Table 3.5 can be used.

MH code is mandatory for all Group 3 machines, providing a basis for interoperability. It is relatively simple to implement and produces acceptable results on noisy telephone lines. To ensure that the receiver maintains color

Table 3.3
Terminating Codes

White-Run Length	Code Word	Black-Run Length	Code Word
0	00110101	0	0000110111
1	000111	1	010
2	0111	2	11
3	1000	3	10
4	1011	4	011
5	1100	5	0011
6	1110	6	0010
7	1111	7	00011
8	10011	8	000101
9	10100	9	000100
10	00111	10	0000100
11	01000	11	0000101
12	001000	12	0000111
13	000011	13	00000100
14	110100	14	00000111
15	110101	15	000011000
16	101010	16	0000010111
17	101011	17	0000011000
18	0100111	18	0000001000
19	0001100	19	00001100111
20	0001000	20	00001101000
21	0010111	21	00001101100
22	0000011	22	00000110111
23	0000100	23	00000101000
24	0101000	24	00000010111
25	0101011	25	00000011000
26	0010011	26	000011001010
27	0100100	27	000011001011
28	0011000	28	000011001100
29	00000010	29	000011001101
30	00000011	30	000001101000

Table 3.3 (continued)

White-Run Length	Code Word	Black-Run Length	Code Word
31	00011010	31	000001101001
32	00011011	32	000001101010
33	00010010	33	000001101011
34	00010011	34	000011010010
35	00010100	35	000011010011
36	00010101	36	000011010100
37	00010110	37	000011010101
38	00010111	38	000011010110
39	00101000	39	000011010111
40	00101001	40	000001101100
41	00101010	41	000001101101
42	00101011	42	000011011010
43	00101100	43	000011011011
44	00101101	44	000001010100
45	00000100	45	000001010101
46	00000101	46	000001010110
47	00001010	47	000001010111
48	00001011	48	000001100100
49	01010010	49	000001100101
50	01010011	50	000001010010
51	01010100	51	000001010011
52	01010101	52	000000100100
53	00100100	53	000000110111
54	00100101	54	000000111000
55	01011000	55	000000100111
56	01011001	56	000000101000
57	01011010	57	000001011000
58	01011011	58	000001011001
59	01001010	59	000000101011
60	01001011	60	000000101100
61	00110010	61	000001011010
62	00110011	62	000001100110
63	00110100	63	000001100111

Table 3.4
Makeup Codes

White-Run Length	Code Word	Black-Run Length	Code Word
64	11011	64	0000001111
128	10010	128	000011001000
192	010111	192	000011001001
256	0110111	256	000001011011
320	00110110	320	000000110011
384	00110111	384	000000110100
448	01100100	448	000000110101
512	01100101	512	0000001101100
576	01101000	576	0000001101101
640	01100111	640	0000001001010
704	011001100	704	0000001001011
768	011001101	768	0000001001100
832	011010010	832	0000001001101
896	011010011	896	0000001110010
960	011010100	960	0000001110011
1024	011010101	1024	0000001110100
1088	011010110	1088	0000001110101
1152	011010111	1152	0000001110110
1216	011011000	1216	0000001110111
1280	011011001	1280	0000001010010
1344	011011010	1344	0000001010011
1408	011011011	1408	0000001010100
1472	010011000	1472	0000001010101
1536	010011001	1536	0000001011010
1600	010011010	1600	0000001011011
1664	011000	1664	0000001100100
1728	010011011	1728	0000001100101
EOL	000000000001	EOL	000000000001

Table 3.5
Makeup Codes for Larger Paper Widths or Higher Pel Densities

Run Length (Black and White)	Makeup Codes
1792	00000001000
1856	00000001100
1920	00000001101
1984	000000010010
2048	000000010011
2112	000000010100
2176	000000010101
2240	000000010110
2304	000000010111
2368	000000011100
2432	000000011101
2496	000000011110
2560	000000011111

Note that run lengths in the range of lengths longer than or equal to 2,624 pels are coded first by the makeup code of 2560. If the remaining part of the run (after the first makeup code of 2560 is 2,560 pels or greater), additional makeup code(s) of 2560 are issued until the remaining part of the run becomes less than 2,560 pels. Then the remaining part of the run is encoded by terminating code or by makeup code plus terminating code according to the range listed in Tables 3.3, 3.4, and 3.5.

synchronization, all coded lines begin with a code word for a white-run length. If the actual scan line begins with a black run, a white-run length of zero is sent. Black- or white-run lengths, up to a maximum length of one scan line, are defined by the code words in Tables 3.3, 3.4, and 3.5. Note that there is a different list of code words for black- and white-run lengths.

Each coded line begins with an end-of-line (EOL) code. It is a unique code word that can never be found within a valid coded scan line; therefore, resynchronization after an error burst is possible.

A pause can be placed in the message flow by transmitting FILL. FILL is a variable-length string of zeros that can be placed only after a coded line and just before an EOL but never within a coded line. FILL must be added to ensure that the transmission time of the total coded scan line is not less than the minimum transmission time established in the premessage control procedure. The end of a document transmission is indicated by sending six consecutive EOLs.

3.2.3 Two-Dimensional Coding Scheme: MR Code

The MR coding method makes use of the vertical correlation between black (or white) runs from one scan line to the next (called vertical mode). In vertical mode, transitions from white to black (or black to white) are coded relative to the line above. If the transition is directly under (zero offset), the code is only 1 bit. Only fixed offsets of zero and ±1, ±2, and ±3 are allowed. If vertical mode is not possible (e.g., when a nonwhite line follows an all-white line), then horizontal mode is used. Horizontal mode is simply an extension of MH code; that is, two consecutive runs are coded by MH code and preceded by a code indicating horizontal mode. To avoid the vertical propagation of transmission errors to the end of the page, a one-dimensional (MH code) line is sent every *K*th line. The factor *K* is typically set to 2 or 4, depending on whether the vertical scanning density is 100 or 200 lines per inch. The *K*-factor is resettable, which means that a one-dimensional line may be sent more frequently (than 2 or 4) when considered necessary by the transmitter. The synchronization code (EOL) consists of 11 zeros followed by a 1, followed by a tag bit to indicate whether the following line is coded one dimensionally or two dimensionally.

The two-dimensional coding scheme is an optional extension of the one-dimensional coding scheme specified in Section 3.2.2. It is defined in terms of changing pels (Figure 3.3) as follows:

- A changing element is defined as an element whose color (i.e., black or white) is different from that of the previous element along the same scan line.
- a_0: The reference or starting changing element on the coding line. At the start of the coding line, a_0 is set on an imaginary white changing element situated just before the first element on the line. During the coding of the coding line, the position of a_0 is defined by the previous coding mode.
- a_1: The next changing element to the right of a_0 on the coding line.
- a_2: The next changing element to the right of a_1 on the coding line.

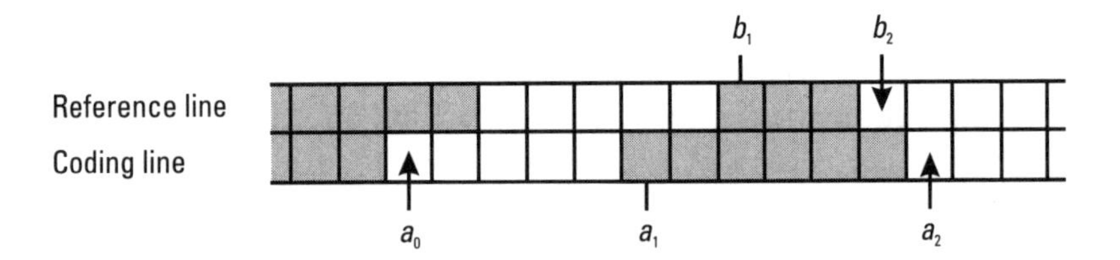

Figure 3.3 Changing pels.

- b_1: The first changing element on the reference line to the right of a_0 and of opposite color to a_0.
- b_2: The next changing element to the right of b_1 on the reference line.

Code words for the two-dimensional coding scheme are given in Table 3.6.

3.2.4 Extended Two-Dimensional Coding Scheme: MMR Code

The basic facsimile coding scheme specified for Group 4 facsimile (MMR) can be used as an option in Group 3 facsimile. This coding scheme must be used with the error correction mode (ECM) option. This coding scheme, described in ITU-T Recommendation T.6 [2], is similar to that of Group 3 (Recommendation T.4). The same modified READ code is used, but only two-dimensional lines are transmitted, with no EOL codes for synchronization. An error-free communication link makes that possible. A white line is assumed before the first actual line in the image. No fill is used; adequate buffering is assumed to provide a memory-to-memory transfer.

The objective of each image-compression scheme is to reduce the number of bits transmitted and thus the transmission time (and cost). The more aggressive schemes provide the most compression, at a cost of increased implementation complexity. In general, T.6 coding outperforms T.4 two-dimensional coding, which outperforms T.4 one-dimensional coding. The

Table 3.6
Two-Dimensional Codes

Mode	Elements to be Coded		Notation	Code Word
Pass	b_1, b_2		P	0001
Horizontal	a_0a_1, a_1a_2		H	H, 001 + M(a_0a_1) + M(a_1a_2)*
Vertical	a_1 just under b_1	$a_1b_1 = 0$	V(0)	1
	a_1 to the right of b_1	$a_1b_1 = 1$	$V_R(1)$	011
		$a_1b_1 = 2$	$V_R(2)$	000011
		$a_1b_1 = 3$	$V_R(3)$	0000011
	a_1 to the left of b_1	$a_1b_1 = 1$	$V_L(1)$	010
		$a_1b_1 = 2$	$V_L(2)$	000010
		$a_1b_1 = 3$	$V_L(3)$	0000010

*Code M( ) represents the code words in Tables 3.3, 3.4, and 3.5 in Section 3.2.2.

differences tend to be smaller on "busy" images and greater on images containing more white space. On dithered images and halftones, the compression is very poor (or even negative), and one-dimensional coding outperforms the other methods in some cases.

3.3 Protocol

Protocol is a set of rules that governs communication in international diplomacy, in physical communication such as mail, and in electronic communication such as telephone, data, and facsimile. To illustrate, envision the U.S. Postal Service protocol for sending a letter. After preparing the letter, the sender must put the pages in an envelope that is within certain size limits. The address has a line for the recipient's name, another line for the street address, and a line for the city, state, and zip code. Further requirements are proper postage and depositing the letter in prescribed locations. The postal system checks that the mailing protocol has been followed, sorts for the proper destination area, transports the letter, and then delivers it. The steps in the postal system need not be understood by the sender, but it must be understood by those in the system, and the proper protocol must be followed for the letter to be delivered. A mismatch in the postal system protocol can result in no delivery.

In a similar manner, Group 3 facsimile protocol is the set of rules that governs communication between a facsimile transmitter and receiver. The document to be sent, with certain size limits, is placed in the hopper of a facsimile machine. Pressing a button calls the facsimile number of the receiving facsimile machine and starts the electronic protocol, often called handshaking. The two facsimile machines have an electronic "chat" back and forth to make sure the connection is good enough and that the transmitter selects only those options the receiver can handle properly. Rarely is there a mismatch, but if one were to occur, the call would be terminated without any of the message being sent. Again, the sender need not know the details of the electronic portion of the facsimile protocol, but the protocol must be rigidly adhered to for proper delivery of the facsimile message. The format used for selecting options and for formatting the facsimile data signal sent from the transmitter must meet an exacting set of rules.

3.3.1 Signaling for Calling Party Sends

ITU-T Recommendation T.30 specifies the procedures for document facsimile transmission using the PSTN. Making a facsimile call and sending a document is divided into five time phases.

1. *Phase A: Call setup.* The calling facsimile dials the telephone number of the receiving facsimile machine. The ring signal and the calling tone (CNG) are received at the called facsimile machine. The CNG tone beeps indicate the call is from a facsimile machine instead of a voice call. The called facsimile machine answers the ring signal by going off-hook and wakes up the rest of the facsimile machine. The facsimile machine can be designed to have the ac power off until the ring signal arrives. Some facsimile machines answer an incoming call immediately, so the ring may not be heard. After a 1-sec delay, the called facsimile sends a 3-sec 2,100-Hz tone, back to the calling facsimile machine.
2. *Phase B: Premessage procedure.* The called facsimile machine sends its digital identification signal (DIS) at 300 bps identifying its capabilities, including optional features. On hearing this distinctive signal, the caller presses the send button to connect the facsimile machine to the telephone line. (An automatic calling facsimile machine does this unattended.) The calling facsimile automatically sends a digital command signal (DCS), locking the called unit into the capabilities selected. The calling facsimile sends a high-speed training signal for the data modem. The called facsimile sends a confirmation to receive (CFR) signal to confirm that the receiving modem is trained (adjusted for low-error operation) and that the facsimile machine is ready to receive.
3. *Phase C: Message transmission.* The calling facsimile sends a training signal and then picture signals for the entire page being sent.
4. *Phase D: Postmessage procedure.* The calling facsimile sends a return to control (RTC) command, switching the facsimile modem back to 300 bps, then an end of procedure (EOP) signal. The called facsimile sends message confirmation (MCF), indicating the page was received successfully.
5. *Phase E: Call release.* The calling facsimile sends a disconnect (DCN) signal, and both facsimile machines disconnect from the telephone line.

After answering the ring, the called facsimile machine waits 1 sec, sends a 3-sec 2,100-Hz CED tone, followed by 75-ms silence, and then a short burst of 300-bps DIS code that tells the calling facsimile machine what standard and optional features are available.

If the call is made manually, the caller presses the send button when the CED tone starts to connect the calling facsimile machine to the telephone line. The calling facsimile machine then decodes the next burst of DIS it receives, identifying the capabilities of the called facsimile machine (Figure 3.4 and Figure 3.5). DIS and other handshake signals are repeated at 3-sec intervals if no response is detected.

Originally, if the call in Figure 3.4 was made from an automatic dialer, the calling unit would send a CNG tone after dialing. CNG is a 0.5-sec 1,100-Hz beep, sent every 3 sec. In the current standard, facsimile machines send CNG on all calls. If the call is answered as a voice call, the CNG tone

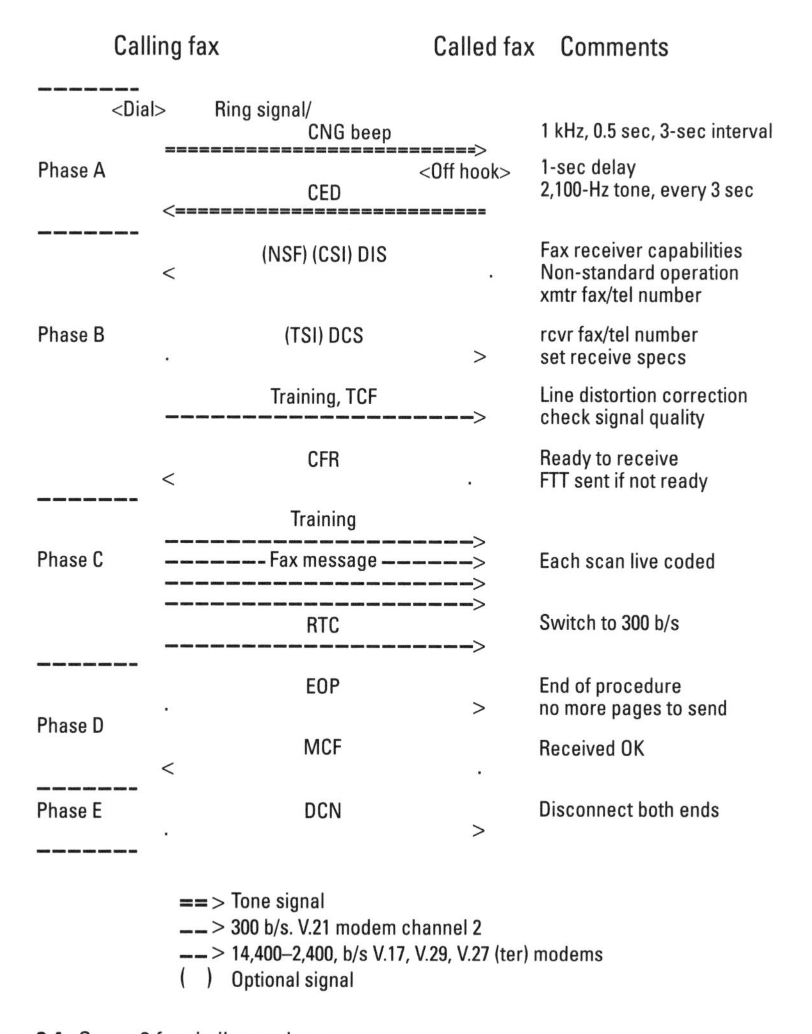

Figure 3.4 Group 3 facsimile sends one page.

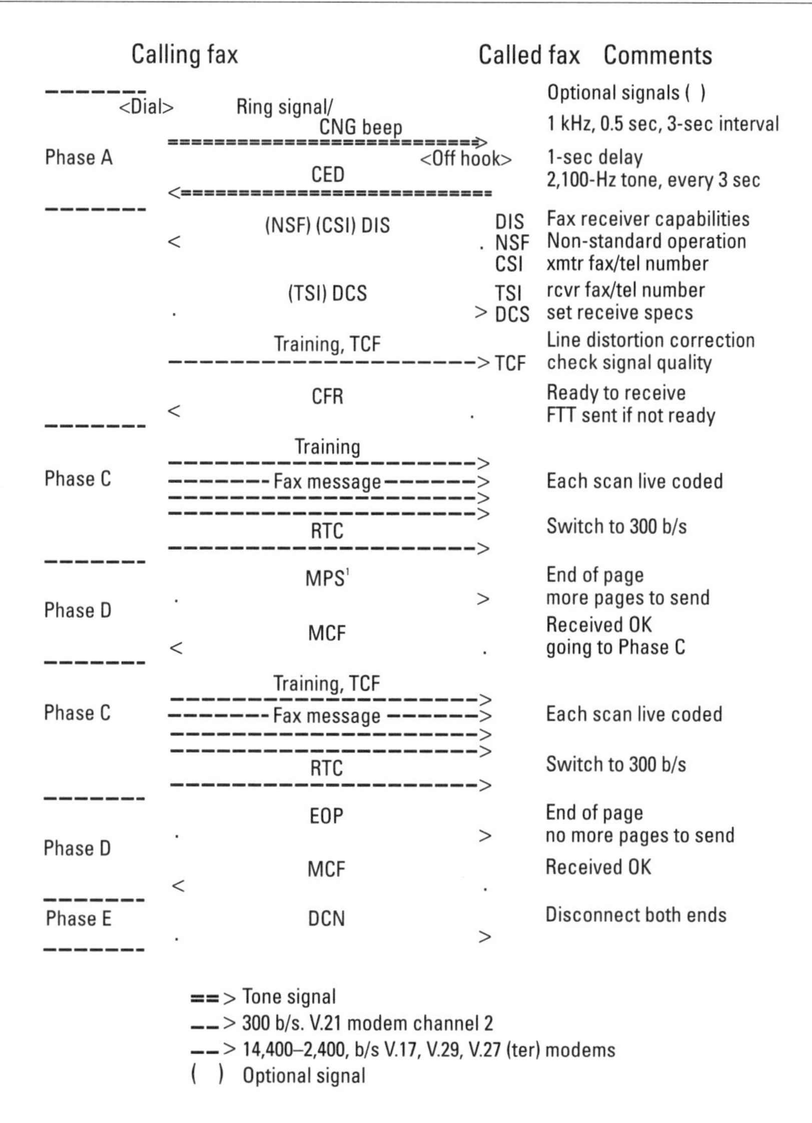

Figure 3.5 Group 3 facsimile sends two pages.

indicates that a machine instead of a person is calling. CNG is also used by some facsimile/voice automatic switching units to connect an incoming call to the facsimile machine instead of to a telephone. See Table 3.7 for handshake abbreviations.

Optional features, such as subscriber identification, polling, and nonstandard facilities (NSF) private options, have been added.

Table 3.7
Handshake Abbreviations

Abbreviation	Function	Signal Format
CED	Called station identification	2,100 Hz
CFR	Confirmation to receive	X010 0001 (1,850 or 1,650 Hz for 3 sec)
CRP	Command repeat	X101 1000
CIG	Calling subscriber identification	1000 0010
CNG	Calling tone	1,100 Hz on 0.5, off 3 sec
CSI	Calling subscriber identification	0000 0010
CTC	Continue to correct	X100 1000
CTR	Response to continue to correct	X010 0011
DCN	Disconnect	X101 1111
DCS	Digital command signal	X100 0001
DIS	Digital identification signal	0000 0001
DTC	Digital transmit command	1000 0001
EOL	End-of-line	000000000001 (11 zeroes)
EOM	End of message	X111 0001 (1,100 Hz)
EOP	End of procedure	X111 0100
EOR	End of retransmission	X111 0011
ERR	Response to end of retransmission	X011 1000
FCD	Facsimile coded data	0110 0000
FCF	Facsimile control field	—
FIF	Facsimile information field	—
FTT	Failure to trail	X010 0010
GC	Group command	G1 1,300 Hz for 1.5–10.0 sec G2 2,100 Hz for 1.5–10.0 sec
GI	Group identification	G1/G2 1,650/1,850 Hz on 0.5, off 3 sec
HDLC	High level data link control	—
LCS	Line conditioning signals	1,100 Hz
MCF	Message confirmation	X011 0001 (1,650 or 1,850 Hz)
MPS	Multipage signal	X111 0010
NSC	Nonstandard facilities command	1000 0100
NSF	Nonstandard facilities	0000 0100

Table 3.7 (continued)

Abbreviation	Function	Signal Format
NSS	Nonstandard setup	X100 0100
PIN	Procedure interrupt negative	X011 0100
PIP	Procedure interrupt positive	X011 0101
PIS	Procedure interrupt signal	462 Hz for 3 sec
PPS	Partial page signal	X111 1101
PPR	Partial page request	X011 1101
PRI-EOM	Procedure interrupt-EOM	X111 1001
PRI-EOP	Procedure interrupt-EOP	X111 1100
PRI-MPS	Procedure interrupt-MPS	X111 1010
RCP	Return to control for partial page	0110 0001
RNR	Receive not ready	X011 0111
RR	Receive ready	X111 0110
RTN	Retrain negative	X011 0010
RTP	Retrain positive	X011 0011
RTC	Return to control	six EOLs
TCF	Training check	Zeros for 1.5 sec
TSI	Transmitting subscriber identification	X100 0010
[...]	Alternative tone signaling for Group 1 or 2	

3.3.2 Nonstandard Facilities Call

Valuable nonstandard proprietary features are permitted in the protocol for Group 3 facsimile machines. These features are invoked during handshake in a manner that does not interfere with the basic Group 3 service.

NSF and CSI are added ahead of DIS. NSF means the facsimile receiver has the ability to use proprietary features not covered by the Series T recommendations. After receiving NSF, the calling facsimile machine returns nonstandard setup (NSS) to lock in the called facsimile machine to the nonstandard operation specified. Some manufacturers had a private error-correction system or 14.4-Kbps modem speed before these features were added as standardized options. Group 3 facsimile machines with NSS capabilities also must have standard capabilities to operate with other standard facsimile machines.

CSI sends a coded signal with the telephone number of the called facsimile machine. Some facsimile machines display that number for the facsimile operator to check that they are sending to the correct unit, and the facsimile machine records the number in its electronic log. The sending facsimile machine uses transmit subscriber identification (TSI) to send its telephone number to the facsimile receiver.

3.3.3 Polling Called Party

Polling allows a central facsimile machine to be programmed to call the sending facsimile machines sequentially and command each one to send documents from memory or its automatic document feeder. Polling is often used for an orderly sending of once-a-day information from many facsimile machines to a single facsimile machine at a central point.

Without polling, it would be difficult to prevent line-busy conditions at the central point. Many Group 3 facsimile machines have delayed calling programs with an automatic dialer that allows polling at night when there are no operators around (Figure 3.6).

After the central calling facsimile machine receives the called facsimile DIS response, it sends a nonstandard command (NSC) and the other facsimile machine takes over the session until it has sent all of the pages destined for the calling facsimile machine. Calling subscriber identification (CIG) is sent instead of TSI, with the telephone number coded in the same format. The transmitter can check the subscriber identification codes to make sure the documents are being sent to the right caller.

3.3.4 Handshake Signal Formats

At the start of the initial handshake, the receiver uses its 300-bps V.21 modem to send information back to the transmitter. Because the modem does not need a training signal to operate with a low error rate, the handshaking procedure is simplified. High-level data link control (HDLC) frames are used for the binary coded handshaking. An optional handshake at 2.4 Kbps is possible but is seldom used. The following sequence is sent:

1. Preamble signal of flag fields (0111 1110) for 1 sec precedes DIS, making sure the telephone line echo suppressers are operating in the right direction. Frame synchronization is also derived from this signal.

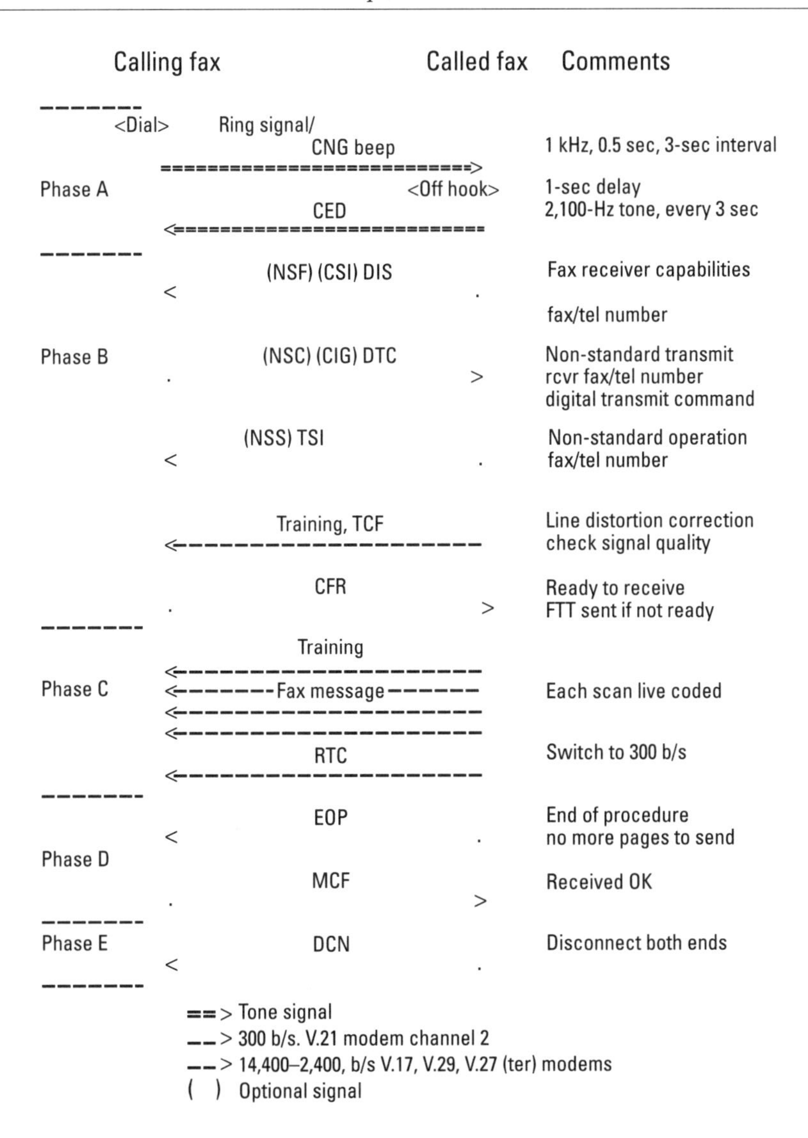

Figure 3.6 Polling a called facsimile.

2. The first byte, the HDLC address field, identifies specific stations in a multipoint network. For calls on the PSTN, the code 1111 1111 is always used.
3. The HDLC control field is 1100 1000 for the last frame sent prior to an expected response from the distant station or 1100 0000 for

other frames. The frame error checking sequence is a 2-byte signal calculated from the content of the data being sent. The receiving unit calculates the same digital number when there are no errors in the handshake signal data.

3.3.5 DIS/DTC Signal

DIS is a 300-bps handshake signal sent by the called facsimile machine when it answers a call. It notifies the calling facsimile machine of its capabilities in standard, optional, and nonstandard specifications. The same codes are used for DIS from the facsimile receiver and the command DTC from the facsimile transmitter. The following options (and possibly more) are available:

- Modem types available;
- Number of pels (dots) per scan line;
- Number of pels per recording line;
- Number of scanning lines per inch (or millimeter) in the paper feed direction;
- Coding scheme for compression of the information to be sent;
- Maximum page length;
- Minimum time needed to record a scanned line at the facsimile receiver;
- ECM or not;
- Nonstandard mode or not;
- V.17 14.4-Kbps or V.34 33.6-Kbps modem.

CSI, CIG, and TSI are each 1 byte, followed by 1 byte for a plus sign (+) and 1 byte for each digit of the international telephone number of the facsimile machine. The telephone number should be programmed into the facsimile machine after it is installed on the telephone line and reprogrammed if moved to another telephone number. On receipt of the number, the other facsimile machine may display it, record it in its log, print it on received copy, or use it to screen calls. Another method sometimes used for subscriber identification is to send the desired information as a facsimile signal added to the page sent.

On decoding DIS, the calling facsimile machine automatically sends a burst of DCS, selecting the compatible features in the facsimile receiver to be used for the call. Appended to DCS is the training signal for the receiving modem and a 1.5-sec TCF signal (all zeros) to check whether the modem is receiving error-free bits. If the signals are received correctly, a CFR signal is sent from the facsimile receiver, indicating it is properly set and waiting for the facsimile signals. The calling station then again sends a training signal followed by the facsimile signals. At the end of the facsimile signals for one page, an RTC command is sent to switch back to the 300-bps modem, followed by EOP. The facsimile receiver sends MCF to indicate the page has been received correctly. The facsimile transmitter notes a success in its log and sends DCN. Both facsimile machines then disconnect from the telephone line.

"It is acknowledged that existing equipments may not conform in all aspects to this recommendation. Other methods may be possible as long as they do not interfere with recommended operation" [2]. One example of this is the CED tone. Some facsimile machines have 1,100 Hz for the first portion of the signal and 2,100 Hz for the second part.

Table 3.8 lists the bit assignments, including new features beyond the 40-bit 1988 Blue Book version, handled by T.30 protocol. Older Group 3 facsimile machines do not have the extra features, but they are compatible for the original options. Flexibility was provided by incorporating unassigned bits and by allowing extension of the field to add new features. Some of these bits were assigned years after the standard was adopted (e.g., ECM and V.17 modem).

3.3.6 NSF

The Group 3 facsimile standard permits a facsimile transmitter to request that the facsimile receiver switch to a nonstandard mode of operation if the receiver is equipped with the appropriate proprietary capability. The nonstandard mode of operation permits facsimile manufacturers to provide improved technology and service beyond those provided by the basic standard.

NSF (0000 0100) is used in the DIS signal to invoke nonstandard features between two facsimile machines made by the same manufacturer. ITU-T Recommendation T.35 gives the details. NSF must be followed immediately by 1 byte of country code, provider code, and at least 1 more byte coded in any way desired to describe the nonstandard features to the other facsimile machine. Formerly these codes were not furnished to others. Table 3.9 lists U.S. provider codes.

The features specified by NSF are not recognized by facsimile machines made by other manufacturers. If handshake is completed in the NSF mode, the format for facsimile transmission could be changed for a proprietary mode of operation, for example, nonstandard compression algorithms, encrypted signals, or higher speed modems, or for other purposes.

Table 3.8
Bit Assignments Handled by T.30 Protocol

Bit No.	DIS/DTC	Note	DCS	Note
1	Reserved	1	Reserved	1
2	Reserved	1	Reserved	1
3	Reserved	1	Reserved	1
4	Reserved	1	Reserved	1
5	Reserved	1	Reserved	1
6	V.8 capabilities	23	Invalid	24
7	"0" = 256 octets preferred "1" = 64 octets preferred	23, 42	Invalid	24
8	Reserved	1	Reserved	1
9	Ready to transmit a facsimile document (polling)	18	Set to "0"	
10	Receiver fax operation	19	Receiver fax operation	20
11, 12, 13, 14	Data signaling rate	3	Data signaling rate	33
0, 0, 0, 0	V.27 *ter* fall-back mode		2,400 bps, Rec. V.27 *ter*	
0, 1, 0, 0	Rec. V.27 *ter*		4,800 bps, Rec. V.27 *ter*	
1, 0, 0, 0	Rec. V.29		9,600 bps, Rec. V.29	
1, 1, 0, 0	Recs. V.27 *ter* and V.29		7,200 bps, Rec. V.29	
0, 0, 1, 0	Not used		Invalid	31
0, 1, 1, 0	Reserved		Invalid	31
1, 0, 1, 0	Not used		Reserved	
1, 1, 1, 0	Invalid	32	Reserved	
0, 0, 0, 1	Not used		14,400 bps, Rec. V.17	
0, 1, 0, 1	Reserved		12,000 bps, Rec. V.17	
1, 0, 0, 1	Not used		9,600 bps, Rec. V.17	
1, 1, 0, 1	Recs. V.27 *ter*, V.29, and V.17		7,200 bps, Rec. V.17	
0, 0, 1, 1	Not used		Reserved	
0, 1, 1, 1	Reserved		Reserved	
1, 0, 1, 1	Not used		Reserved	
1, 1, 1, 1	Reserved		Reserved	
15	R8 × 7.7 lines/mm and/or 200 × 200 pels/25.4 mm	10, 11, 13, 25	R8 × 7.7 lines/mm or 200 × 200 pels/25.4 mm	10, 11, 13

Table 3.8 (continued)

Bit No.	DIS/DTC	Note	DCS	Note
16	Two-dimensional coding capability		Two-dimensional coding	
17, 18	Recording width capabilities	27	Recording width	27
(0,0)	Scan line length 215 mm ± 1%		Scan line length 215 mm ± 1%	
(0,1)	Scan line length 215 mm ± 1% and scan line length 255 mm ± 1% and scan line length 303 mm ± 1%		Scan line length 303 mm ± 1%	
(1,0)	Scan line length 215 mm ± 1% and scan line length 255 mm ± 1%		Scan line length 255 mm ± 1%	
(1,1)	Invalid	6	Invalid	
19, 20	Recording length capability	2	Recording length	2
(0,0)	A4 (297 mm)		A4 (297 mm)	
(0,1)	Unlimited		Unlimited	
(1,0)	A4 (297 mm) and B4 (364 mm)		B4 (364 mm)	
(1,1)	Invalid		Invalid	
21, 22, 23	Minimum scan line time capability at the receiver	4, 8, 23	Minimum scan line time	8, 24
(0,0,0)	20 ms at 3.85 l/mm: $T_{7.7} = T_{3.85}$		20 ms	
(0,0,1)	40 ms at 3.85 l/mm: $T_{7.7} = T_{3.85}$		40 ms	
(0,1,0)	10 ms at 3.85 l/mm: $T_{7.7} = T_{3.85}$		10 ms	
(1,0,0)	5 ms at 3.85 l/mm: $T_{7.7} = T_{3.85}$		5 ms	
(0,1,1)	10 ms at 3.85 l/mm: $T_{7.7} = 1/2\ T_{3.85}$			
(1,1,0)	20 ms at 3.85 l/mm: $T_{7.7} = 1/2\ T_{3.85}$			
(1,0,1)	40 ms at 3.85 l/mm: $T_{7.7} = 1/2\ T_{3.85}$			
(1,1,1)	0 ms at 3.85 l/mm: $T_{7.7} = T_{3.85}$		0 ms	
24	Extend field	5	Extend field	5
25	Reserved	1, 41	Reserved	1, 41
26	Uncompressed mode		Uncompressed mode	
27	Error correction mode	9, 17, 23, 25	Error correction mode	9, 17, 24, 34
28	Set to "0"		Frame size 0 = 256 octets Frame size 1 = 64 octets	7, 24

Table 3.8 (continued)

Bit No.	DIS/DTC	Note	DCS	Note
29	Reserved	1	Reserved	1
30	Reserved	1	Reserved	1
31	T.6 coding capability	9, 17	T.6 coding enabled	9, 17
32	Extend field	5	Extend field	5
33	Field not valid capability		Field not valid capability	
34	Multiple selective polling capability		Set to "0"	
35	Polled SubAddress	26, 44, 45	Set to "0"	
36	T.43 coding	17, 25, 34, 35, 37, 39, 40	T.43 coding	17, 25, 34, 35, 37, 39, 40
37	Plane interleave	25, 46	Plane interleave	25, 46
38	Reserved	1	Reserved	1
39	Reserved	1	Reserved	1
40	Extend field	5	Extend field	5
41	R8 × 15.4 lines/mm	10	R8 × 15.4 lines/mm	10, 34
42	300 × 300 pels/25.4 mm	34	300 × 300 pels/25.4 mm	34
43	R16 × 15.4 lines/mm and/or 400 × 400 pels/25.4 mm	10, 12, 13	R16 × 15.4 lines/mm and/or 400 × 400 pels/25.4 mm	10, 12, 13, 34
44	Inch based resolution preferred	13, 14	Resolution type selection "0": metric based resolution "1": inch based resolution	13, 14
45	Metric based resolution preferred	13, 14	Don't care	
46	Minimum scan line time capability for higher resolutions "0": $T_{15.4} = T_{7.7}$ "1": $T_{15.4} = 1/2\ T_{7.7}$	15	Don't care	
47	Selective polling	26, 44	Set to "0"	
48	Extend field	5	Extend field	5
49	Subaddressing capability		Subaddressing transmission	26
50	Password	26	Sender Identification transmission	26

Table 3.8 (continued)

Bit No.	DIS/DTC	Note	DCS	Note
51	Ready to transmit a data file (polling)	17, 21	Set to "0"	
52	Reserved	1	Reserved	1
53	Binary File Transfer (BFT)	16, 17, 21	Binary File Transfer (BFT)	16, 17
54	Document Transfer Mode (DTM)	17, 21	Document Transfer Mode (DTM)	17
55	Electronic Data Interchange (EDI)	17	Electronic Data Interchange (EDI)	17
56	Extend field	5	Extend field	5
57	Basic Transfer Mode (BTM)	17, 21	Basic Transfer Mode (BTM)	17
58	Reserved	1	Reserved	1
59	Ready to transmit a character or mixed mode document (polling)	17, 22	Set to "0"	
60	Character mode	17, 22	Character mode	17, 22
61	Reserved	1	Reserved	1
62	Mixed mode (Annex D/T.4)	17, 22	Mixed mode (Annex D/T.4)	17, 22
63	Reserved	1	Reserved	1
64	Extend field	5	Extend field	5
65	Processable mode 26 (Rec. T.505)	17, 22	Processable mode 26 (Rec. T.505)	17, 22
66	Digital network capability	43	Digital network capability	43
67	Duplex and half duplex capabilities		Duplex and half duplex capabilities	
(0)	Half duplex operation only		Half duplex operation only	
(1)	Duplex and half duplex operation		Duplex operation	
68	JPEG coding	25, 34, 35, 39, 40	JPEG coding	25, 34, 35, 39, 40
69	Full color mode	25, 35	Full color mode	25, 35
70	Set to "0"	36	Preferred Huffman tables	25, 36
71	12 bits/pel component	25, 37	12 bits/pel component	25, 37
72	Extend field	5	Extend field	5
73	No subsampling (1:1:1)	25, 38	No subsampling (1:1:1)	25, 38
74	Custom illuminant	25, 39	Custom illuminant	25, 39
75	Custom gamut range	25, 40	Custom gamut range	25, 40

Table 3.8 (continued)

Bit No.	DIS/DTC	Note	DCS	Note
76	North American Letter (215.9 × 279.4 mm) capability	28	North American Letter (215.9 × 279.4 mm)	
77	North American Legal (215.9 × 355.6 mm) capability	28	North American Legal (215.9 × 355.6 mm)	
78	Single-progression sequential coding (Rec. T.85) basic capability	17, 29, 30	Single-progression sequential coding (Rec. T.85) basic	17, 29
79	Single-progression sequential coding (Rec. T.85) optional L0 capability	17, 29, 30	Single-progression sequential coding (Rec. T.85) optional L0	17, 29
80	Extend field	5	Extend field	5
81	HKM key management capability		HKM key management selected	
82	RSA key management capability		RSA key management selected	47
83	Override mode capability		Override mode selected	
84	HFX40 cipher capability		HFX40 cipher selected	
85	Alternative cipher number 2 capability		Alternative cipher number 2 selected	
86	Alternative cipher number 3 capability		Alternative cipher number 3 selected	
87	HFX40-I hashing capability		HFX40-I hashing selected	
88	Extend field	5	Extend field	5
89	Alternative hashing system number 2 capability		Alternative hashing system number 2 selected	
90	Alternative hashing system number 3 capability		Alternative hashing system number 3 selected	
91	Reserved for future security features	1	Reserved for future security features	1
92	Reserved	1	Reserved	1
93	Reserved	1	Reserved	1
94	Reserved	1	Reserved	1
95	Reserved	1	Reserved	1
96	Extend field	5	Extend field	5

Table notes:

NOTE 1 – Bits that are indicated as "Reserved" shall be set to "0".

NOTE 2 – Standard facsimile terminals conforming to Recommendation T.4 must have the following capability: Paper length = 297 mm.

Table 3.8 (continued)

NOTE 3 – Where the DIS or DTC frame defines V.27 *ter* capabilities, the terminal may be assumed to be operable at either 4800 or 2400 bps.
Where the DIS or DTC frame defines V.29 capabilities, the terminal may be assumed to be operable at either 9,600 or 7,200 bps per Recommendation V.29; where it defines Recommendation V.17, the terminal may be assumed to be operable at 14,400 bps, 12,000 bps, 9,600 bps or 7,200 bps per Recommendation V.17.NOTE 4 – $T_{7.7}$ and $T_{3.85}$ refer to the scan line times to be utilized when the vertical resolution is 7.7 lines/mm (or 200 lines/25.4 mm or 300 lines/25.4 mm) or 3.85 lines/mm, respectively (see bit 15 above). $T_{7.7}$ =1/2 $T_{3.85}$ indicates that when the vertical resolution is 7.7 lines/mm or 200 lines/25.4 mm or 300 lines/25.4 mm, the scan line time can be decreased by half. NOTE 5 – The standard FIF field for the DIS, DTC and DCS signals is 24 bits long. If the "extend field" bit(s) is a "1", the FIF field shall be extended by an additional eight bits.

NOTE 6 – Existing terminals may send the invalid (1.1) condition for bits 17 and 18 of their DIS signal. If such a signal is received, it should be interpreted as (0,1).

NOTE 7 – The values of bit 28 in the DCS command is valid only when the indication of the Recommendation T.4 error correction mode is invoked by bit 27.

NOTE 8 – The optional T.4 error correction mode of operation requires 0 ms of the minimum scan line time capability. Bits 21-23 in DIS/DTC signals indicate the minimum scan line time of a receiver regardless of the availability of the error correction mode.
In case of error correction mode, the sender sends DCS signal with bits 21-23 set to "1, 1, 1" indicating 0 ms capability.
In case of normal transmission, the sender sends DCS signal with bits 21-23 set to the appropriateness according to the capabilities of the two terminals.

NOTE 9 – T.6 coding scheme capability specified by bit 31 is valid only when bit 27 (error correction mode) is set as a "1".

NOTE 10 – Resolutions of R8 and R16 are defined as follows:
R8 = 1,728 pels/(215 mm ± 1%) for ISO A4, North American Letter and Legal.
R8 = 2,048 pels/(255 mm ± 1%) for ISO B4.
R8 = 2,432 pels/(303 mm ± 1%) for ISO A3.
R16 = 3,456 pels/(215 mm ± 1%) for ISO A4, North American Letter and Legal.
R16 = 4,096 pels/(255 mm ± 1%) for ISO B4.
R16 = 4,864 pels/(303 mm ± 1%) for ISO A3.

NOTE 11 – Bit 15, when set to "1", is interpreted according to bits 44 and 45 as follows:

bit 44	bit 45	Interpretation
0	0	(invalid)
1	0	200 × 200 pels/25.4 mm
0	1	R8 × 7.7 lines/mm
1	1	R8 × 7.7 lines/mm and 200 × 200 pels/25.4 mm

"1" in bit 15 without bits 41, 42, 43, 44, 45 and 46 indicates R8 × 7.7 lines/mm.

NOTE 12 – Bit 43, when set to "1", is interpreted according to bits 44 and 45 as follows:

bit 44	bit 45	Interpretation
0	0	(invalid)
1	0	400 × 400 pels/25.4 mm
0	1	R16 × 15.4 lines/mm
1	1	R16 × 15.4 lines/mm and 400 × 400 pels/25.4 mm

NOTE 13 – Bits 44 and 45 are used only in conjunction with bits 15 and 43. Bit 44 in DCS, when used, shall correctly indicate the resolution of the transmitted document, which means that bit 44 in DCS may not always match the indication of bits 44 and 45 in DIS/DTC. Cross selection will cause the distortion and reduction of reproducible area.
If a receiver indicates in DIS that it prefers to receive metric based information but the transmitter has only the equivalent inch-based information (or vice versa), then communication shall still take place.

NOTE 14 – Bits 44 and 45 do not require the provision of any additional features on the terminal to indicate to the sending or receiving user whether the information was transmitted or received on a metric-metric, inch-inch, metric-inch, inch-metric basis.

NOTE 15 – $T_{15.4}$ refers to the scan line times to be utilized when the vertical resolution is 15.4 lines/mm or 400 lines/mm.
$T_{15.4}$ = 1/2 $T_{7.7}$ indicates that when $T_{7.7}$ is 10, 20 or 40 ms the scan line time can be decreased by half in higher resolution mode. When $T_{7.7}$ is 5 ms [i.e., (bit 21, bit 22, bit 23) = (1, 0, 0), (0, 1, 1)] or 0 ms [i.e. (1, 1, 1)], bit 46 in DIS/DTC should be set to "0" ($T_{15.4} = T_{7.7}$).

NOTE 16 – The binary file transfer protocol is described in Recommendation T.434.

NOTE 17 – When either bit of 31, 36, 51, 53, 54, 55, 57, 59, 60, 62, 65, 78 and 79 is set to "1", bit 27 shall also be set to "1".

NOTE 18 – Bit 9 indicates that there is a facsimile document ready to be polled from the answering terminal. It is not an indication of a capability.

Table 3.8 (continued)

NOTE 19 – Bit 10 indicates that the answering terminal has receiving capabilities.

NOTE 20 – Bit 10 in DCS is a command to the receiving terminal to set itself in the receive mode.

NOTE 21 – Bit 51 indicates that there is a data file ready to be polled from the answering terminal. It is not an indication of a capability. This bit is used in conjunction with bits 53, 54 and 57.

NOTE 22 – Bit 59 indicates that there is a character coded or mixed mode document ready to be polled from the answering terminal. It is not an indication of a capability. This bit is used in conjunction with bits 60, 62 and 65.

NOTE 23 – When the optional procedure defined in Annex C/T.30 is used, in DIS/DTC bits 6 and 7 shall be set to "0" and bits 21 to 23 and 27 shall be set to "1".

NOTE 24 – When the optional procedure defined in Annex C/T.30 is used, in DCS bits 6, 7 and 28 shall be set to "0" and bits 21 to 23 and 27 shall be set to "1".

NOTE 25 – The optional continuous-tone color mode and gray-scale mode (JPEG mode) protocols and the optional lossless encoded color and gray-scale mode (T.43 mode) are described in Annexes E/T.30 and I/T.30 respectively. If bit 68 in the DIS/DTC frame is set to "1", this indicates JPEG mode capability. If bits 36 and 68 are set to "1", this indicates that the T.43 capability is also available. Bit 36 in the DIS/DTC frame shall only be set to "1" when bit 68 is also set to "1". Additionally, then bits 15 and 27 in the DIS/DTC frame shall also be set to "1", if bit 68 or bits 36 and 68 are set to "1". Bit 15 indicates 200 × 200 pels/25.4 mm resolution capability, which is basic for color facsimile. Bit 27 indicates error correction mode capability, which is mandatory for color facsimile. Bits 69 to 71 and 73 to 75 are relevant only if bit 68 is set to "1". Bit 73 is relevant only for JPEG mode. Bits 69, 71, 74 and 75 are relevant for JPEG mode and/or T.43 mode. Bit 37 is relevant only when bit 36 is set to "1" – see also Notes 39 and 40.

NOTE 26 – To provide an error recovery mechanism, when PWD/SEP/SUB/SID/PSA frames are sent with DCS or DTC, bits 49 and 50 in DCS or bits 47 and 50 and 35 in DTC shall be set to "1". For bit 47, setting "1" for DTC means selective polling transmission and for DIS means selective polling capability. For bit 50, setting "1" for DTC means password transmission and for DIS means password or Sender ID capability. For bit 35, setting "1" for DTC means Polled SubAddress transmission and for DIS means Polled SubAddress capability. Terminals conforming to the 1993 versions of this Recommendation may set the above bits to "0" even though PWD/SEP/SUB frames are transmitted.

NOTE 27 – The corresponding scan line lengths for inch-based resolutions can be found in 2.2/T.4.

NOTE 28 – While using bits 76 and 77 in DIS/DTC, the terminal is required to be able to receive ISO A4 documents in every combination of bits 76 and 77. A4, B4 and A3 transmitters may ignore the settings of bits 76 and 77.

NOTE 29 – The coding scheme indicated by the bits 78 and 79 is defined in Recommendation T.85.

NOTE 30 – When bit 79 in DIS is set to "1", bit 78 shall also be set to "1".

NOTE 31 – Some terminals which conform to the 1994 and earlier versions of this Recommendation may have used this bit to indicate use of the V.33 modulation system.

NOTE 32 – Some terminals which conform to the 1994 and earlier versions of this Recommendation may have used this bit sequence to indicate V.27 *ter*, V.29 and V.33 capabilities. In order to maintain compatibility with such terminals, a terminal which has the capability to receive using the modulation system defined in Recommendation V.17 must also be capable of receiving using the modulation system defined in Recommendation V.33. Further, a terminal which has the capability to receive using the modulation system defined in Recommendation V.33 must also be capable of receiving using the modulation system defined in Recommendation V.29.

NOTE 33 – When the modulation system defined in Recommendation V.34 is used, bits 11-14 in DCS are invalid and should be set to "0".

NOTE 34 – Setting bit 68 to "0" indicates that the called terminal's JPEG mode and T.43 mode are not available and it cannot decode JPEG or T.43 encoded data. In a DCS frame, setting bit 68 to "1" indicates that the calling terminal's JPEG mode is used and JPEG encoded image data are sent. Setting bit 68 to "0" and bit 36 to "1" indicates that the calling terminal's T.43 mode is used and T.43 encoded image data is sent. If bit 68 or 36 in the DCS is set to "1" then bits 41 or 42 or 43, and 27 in the DCS frame shall also be set to "1". Bits 42 and 43 indicate 300 × 300 and 400 × 400 pels/25.4 mm resolution respectively. Setting bit 68 and 36 to "0" indicates neither the JPEG mode nor the T.43 mode is used, image is not encoded using JPEG nor Recommendation T.43.

NOTE 35 – In DIS/DTC frame, setting bit 69 to "1" indicates that the called terminal has full color capability. It can accept full color image data in CIELAB space. If bit 36 is also set to "1", it can also accept color image data defined in Recommendation T.43. Setting bit 69 to "0" and bit 68 or bits 68 and 36 to "1" indicates that the called terminal has gray-scale mode only, it accepts only the lightness component (the L* component) in the CIELAB representation for JPEG mode and for T.43 mode respectively. In a DCS frame, setting bits 68 and 69 to "1" indicates that the calling terminal sends image in full color representation in the CIELAB space in JPEG mode. In a DCS frame, setting bits 36 and 69 to "1" indicates that the calling terminal sends color image in T.43 mode. Setting bit 68 or 36 to "1" and bit 69 to "0" indicates that the calling terminal sends only the lightness component (the L* component) in the CIELAB representation for JPEG or T.43 mode respectively. Note that color image will be transmitted only when bits 68 and 69 or 36 and 69 are both set to "1".

NOTE 36 – Bit 70 is called "Indication of default Huffman tables." A means is provided to indicate to the called terminal that the Huffman tables are the default tables. Default tables are specified only for the default image intensity resolution (8 bits/pel/component). The default Huffman tables are to be determined (for example, Tables K.3/T.81-K.6/T.81). In a DIS/DTC frame, bit 70 is not used and is set to "0". In a DCS frame, setting bit 70 to "0" indicates that the calling terminal does not identify the Huffman tables that it uses to encode the image data as the default tables. Setting bit 70 to "1" indicates that the calling terminal identifies the Huffman tables that it uses to encode the image data as the default tables.

Table 3.8 (continued)

NOTE 37 – In a DIS/DTC frame, setting bit 71 to "0" indicates that the called terminal can only accept image data which has been digitized to 8 bits/pel/component for JPEG mode. This is also true for T.43 mode if bit 36 is also set to "1". Setting bit 71 to "1" indicates that the called terminal can also accept image data that are digitized to 12 bits/pel/component for JPEG mode. This is also true for T.43 mode if bit 36 is also set to "1". In a DCS frame, setting bit 71 to "0" indicates that the calling terminal's image data are digitized to 8 bits/pel/component for JPEG mode. This is also true for T.43 mode if bit 36 is also set to "1". Setting bit 71 to "1" indicates that the calling terminal transmits image data which has been digitized to 12 bits/pel/component for JPEG mode. This is also true for T.43 mode if bit 36 is also set to "1".

NOTE 38 – In a DIS/DTC frame, setting bit 73 to "0" indicates that the called terminal expects a 4:1:1 subsampling ratio of the chrominance components in the image data, the a* and b* components in the CIELAB color space representation are subsampled four times to one against the L* (Lightness) component. The details are described in Annex E/T.4. Setting bit 73 to "1" indicates that the called terminal, as an option, accepts no subsampling in the chrominance components in the image data. In a DCS frame, setting bit 73 to "0" indicates that the called terminal uses a 4:1:1 subsampling ratio of the a* and b* components in the image data. Setting bit 73 to "1" indicates that the called terminal does no subsampling.

NOTE 39 – In a DIS/DTC frame, setting bit 74 to "0" indicates that the called terminal expects that the CIE Standard Illuminant D50 is used in the color image data as specified in Recommendation T.42. Setting bit 74 to "1" indicates that the called terminal can also accept other illuminant types besides the D50 illuminant. Setting bit 68 to "1" indicates that the terminal has the JPEG coding capability as described in Annex E/T.4. Setting bit 36 to "1" indicates that the terminal has the color coding capability as described in Recommendation T.43. In a DCS frame, setting bit 74 to "0" and bit 68 or bit 36 to "1", indicates the calling terminal uses the D50 illuminant in the color image data representation a specified in Recommendation T.42. Setting bit 74 to "1" indicates that another type of illuminant is used. When bits 68 and 74 are set to "1" the specification is embedded into the JPEG syntax as described in Annex E/T.4. When bits 36 and 74 are set to "1" the specification is embedded into the T.43 syntax as described in Recommendation T.43.

NOTE 40 – In a DIS/DTC frame, setting bit 75 to "0" indicates that the called terminal expects that the color image data are represented using the default gamut range as specified in Recommendation T.42. Setting bit 75 to "1" indicates that the called terminal can also accept other gamut ranges. Setting bit 68 to "1" indicates that the terminal has the JPEG coding capability, as described in Annex E/T.4. Setting bit 36 to "1" indicates that the terminal has the color coding capability, as described in Recommendation T.43 . In a DCS frame, setting bit 75 to "0" and bit 68 or bit 36 to "1", indicates that the calling terminal uses the default gamut range as specified in Recommendation T.42. Setting bit 75 to "1" indicates that the calling terminal uses a different gamut range. When bits 68 and 75 are set to "1", the specification is embedded into the JPEG syntax as described in Annex E/T.4. When bits 36 and 75 are set to "1", the specification is embedded into the T.43 syntax as described in Recommendation T.43.

NOTE 41 – Some terminals which conform to the pre-1996 versions of this Recommendation may set this bit to "1". Such terminals will give an answering sequence as shown in Figure III.2/T.30.

NOTE 42 – It is understood that for backwards compatibility, a transmitting terminal may ignore the request for the 64 octet frame and therefore the receiving terminal must be prepared to handle 256 octet frames by some means.

NOTE 43 – See C.7.2/T.30.

NOTE 44 – Clarification on the use of selective polling based on the settings of bit 47 and bit 35 is given in 5.3.6.1.2 5)/T.30.

NOTE 45 – Clarification on the use of subaddress for polling based on the setting of bit 35 is given in 5.3.6.1.2 6)/T.30.

Table 3.9
U.S. Provider Codes

Company	City	State	NSF Code	Date Assigned
Netexpress Communications, Inc.	Vienna	VA	0084	04/20/1988
Wang Laboratories, Inc.	Lowell	MA	0082	09/06/1988
Castelle	Santa Clara	CA	0066	04/30/1989
Brooktrout Technology Inc.	Wellesley Hills	MA	0062	06/01/1989
Xerox Corporation	Lewisville	TX	0098	06/02/1989
Murata Business Systems, Inc.	Dallas	TX	00A2	06/02/1989
Gammalink Graphics Communications	Palo Alto	CA	0064	06/02/1989
Hybrid Fax, Inc.	Redwood City	CA	0068	06/02/1989

Table 3.9 (continued)

Company	City	State	NSF Code	Date Assigned
Fujitsu Imaging Systems of America	Danbury	CT	009B	06/02/1989
Adobe Systems Inc.	Mountain View	CA	0044	07/07/1989
Bogosian Engineering	Fremont	CA	0042	07/07/1989
Fremont Communications Company	Fremont	CA	0046	08/30/1989
Hayes Microcomputer Products	Atlanta	GA	0048	10/25/1989
Data Race	San Antonio	TX	004C	07/13/1990
TRW Electronic Products, Inc.	San Luis Obispo	CA	004E	07/13/1990
Audiofax	Marietta	GA	0052	10/03/1990
Computer Automation	Richardson	TX	0054	11/19/1990
Serca Communication	Corvallis	OR	0056	12/19/1990
Fujitsu Imaging Systems of America	Danbury	CT	009A	03/22/1991
Octocom Systems Inc.	Wilmington	MA	0058	04/24/1991
Power Solutions, Inc.	Ann Arbor	MI	005C	06/19/1991
Digital Sound Corp.	Carpinteria	CA	005A	06/19/1991
Pacific Data Products	San Diego	CA	005E	07/10/1991
Compaq Computer Corporation	Plano	TX	0074	01/31/1992
Speaking Devices Corp.	Santa Clara	CA	0072	01/31/1992
Cylink Corporation	Sunnyvale	CA	0078	02/17/1992
Pitney Bowes	Stratford	CT	007A	02/17/1992
Digiboard, Inc.	Eden Prairie	MN	007C	02/17/1992
General Kinetics Incorporated (Cryptek)	Herndon	VA	0076	02/17/1992
Codex Corporation	Mansfield	MA	007E	07/22/1992
Omnifax	Los Angeles	CA	006A	07/22/1992
Hewlett Packard	San Diego	CA	006C	11/03/1992
Microsoft	Redmond	WA	006E	02/17/1993
Rockwell International	Newport Beach	CA	0020	06/16/1993
Commetrex Corporation	Norcross	GA	0060	06/16/1993
Comsat Technology Services	Clarksburg	MD	0022	07/08/1993
Octel Communications Corp.	Milpitas	CA	0024	10/25/1993
Rolm	Santa Clara	CA	0026	12/14/1993
Sofnet	Marietta	GA	0028	12/23/1993

Table 3.9 (continued)

Company	City	State	NSF Code	Date Assigned
STF Technologies, Inc.	Concordia	MO	002A	12/24/1993
HKB, Inc.	San Jose	CA	002C	01/24/1994
Delrina	Toronto	ONT	002E	02/07/1994
Dialogic Corporation	Parsippeny	NJ	0030	02/07/1994
Applied Synergy	Fort Worth	TX	0032	04/26/1995
Syncro Development Corp.	Langhorne	PA	0034	04/26/1995
Picturetel Corporation	Peabody	MA	0001	05/02/1995
TIA TR-29 Committee	—	—	0029	05/11/1995
Genoa Technology Inc.	Moorpark	CA	0036	06/16/1995
Texas Instruments	Austin	TX	0038	09/27/1995
C/o IBM Westlake Programming Lab	Roanoke	TX	003A	09/29/1995
Viasat	Carlsbad	CA	003C	11/15/1995
Ericsson Messaging Systems	Woodbury	NY	003E	09/19/1996
Cable-sat Systems, Inc.	San Jose	CA	0086	10/03/1996
MFPA	El Cajon	CA	0088	12/23/1996
Telogy Networks	Germantown	MD	008A	12/23/1996
Telecom Multimedia Systems, Inc.	Irvine	CA	008E	02/13/1997
AT&T Bell Laboratories	Middletown	NJ	008C	03/04/1997
Nuera Communications, Inc.	San Diego	CA	0092	03/04/1997
Lucent Technologies	Holmdel	NJ	004A	03/04/1997
K56flex	Newport Beach	CA	0094	05/08/1997
Mibridge, Inc.	Eatontown	NJ	0096	08/01/1997
Natural Microsystems	Framingham	MA	009C	08/01/1997
Copytele	Huntington Station	NY	009E	10/20/1997
Lanier Worldwide, Inc.	Atlanta	GA	00A4	03/06/1998
Qualcomm	Washington	DC	00A6	05/29/1998

3.3.7 ECM

The switched telephone network is prone to error during the transmission of digital data. When such a transmission error occurs, a streak typically is created on the output page, which can be disturbing to the observer. Before the optional ECM was approved by the ITU, error-concealment techniques were used to minimize the visual effect of transmission errors on the output copy. Error concealment is possible because an EOL code is transmitted between scan lines. If the receiver does not detect the EOL code at the expected location in the data stream, an error likely has occurred on the line following the last correctly received EOL code. Following are some examples of possible error conditions.

- EOL occurs before 1,728 pels have been written.
- More than 1,728 pels have been written before EOL is received.
- No word in the applicable code tables matches the received bit pattern.
- The current line decoding references a run that does not exist in thc previous line.

Error-concealment techniques that were used by many machines and that might still be used by some low-end machines include the following:

- The first erroneous line is printed white, and all subsequent lines are printed white until a one-dimensional MH line is correctly received.
- The first erroneous line (x) is replaced by the previous correctly received line ($x-1$), and all subsequent lines are replaced by $x-1$ until a one-dimensional MH line is correctly received.
- This processing technique is a combination of the previous two. The first erroneous line (x) is replaced by the previous correctly received line ($x-1$), and all subsequent lines are printed white until a one-dimensional MH line is correctly received.
- The first erroneous line is decoded and printed in the normal MH or MR manner up to the point in the line where the error is detected. From that point, the remainder of the first erroneous line is replaced by the corresponding pels in the "previous line." The resultant "correct" line is then used as a new reference "previous" line, and the process is repeated until an MH line is correctly decoded. This error-processing technique should be particularly advantageous in those instances when a transmission error occurs near the end (right side of the page) of a scan line period. When that occurs, it should be possible to decode correctly most of the scan line that was "hit" as well as most of the subsequent scan lines before a correct MH line is received.

The ITU-T originally defined two error-control options. One option merely limited the effect of errors, without correcting them. The second option employs true error-correction technology. Subsequently the error-limiting option was rescinded.

The optional ECM applies to one-dimensional and two-dimensional coding and provides true error correction. The primary objective of ECM is to perform well against burst errors. Additional objectives included backward compatibility with existing facsimile machines and minimizing the transmission overhead in channels with low error rates. Service providers in countries with high-quality telephone networks did not want to suffer the extra overhead that error correction typically entails.

The error correction scheme is known as page selective repeat automatic repeat request (ARQ). The compressed image data are embedded in HDLC frames of length 256 octets or 64 octets and transmitted in blocks of 256 frames. The communication link operates in a half-duplex mode, that is, the transmission of image data and the acknowledgment of the data are not sent at the same time. The technique can be thought of as an extension to the Group 3 protocol. The protocol information is also embedded in HDLC frames but does not use selective repeat for error control. Every Group 3 facsimile machine must have the mechanism to transmit and receive the basic HDLC frame structure, including flags, address, control, and frame check sequence. Thus, the use of an extended HDLC scheme helped to minimize changes to existing facsimile designs.

The transmitting terminal divides the compressed image data into 256-octet or 64-octet frames and sends a 256-octet block of frames to the receiving terminal. (The receiving terminal must be able to receive both frame sizes.) Each transmitted frame has a unique frame number. The receiver requests retransmission of bad frames by frame number. The transmitter retransmits the requested frames. After four requests for retransmission for the same block, the transmitter may stop or continue, with optional modem speed fallback.

The page selective repeat ARQ is a good compromise that balances complexity and throughput [3]. A continuous selective repeat ARQ provides slightly higher throughput but requires a modem back channel. Forward error correction (FEC) schemes typically have higher overhead on good connections, can be more complex, and may break down in the presence of burst errors. In addition to providing the capability of higher throughput on noisy lines, the error correction mode option supports an error-free environment that has enabled many new features. Examples of these are T.6 encoding, binary file transfer, and operation over the ISDN.

3.3.8 Character Mode

Character mode is an optional feature of Group 3 that permits transmission of character-coded documents by means of the Group 3 protocol. The objective is to achieve very high compression of the message to be transmitted while allowing the receiver the flexibility to render the page. Because character-coded documents must be reliably transferred, ECM is mandatory for this option.

The character repertoire that represents and describes the graphic characters allowed for character mode is that of Standard ISO 8859-1 [4] together with the box-drawing character repertoire that is a subset of registered ITU-T set ISO 72. The coding of the graphic characters is *not* that of the code table given in ISO 8859-1; it follows the coding rules of Recommendation T.51 [5]. To be coded, some graphic characters represented in ISO 8859-1 need 2 bytes of the 8-bit code table. For example, diacritical characters require 2 bytes, the diacritical mark followed by the basic character.

Character-coded pages have a fixed format of 55 lines with 77 characters in each line. The maximum length of a page is 55 lines. Shorter pages are permitted. The displaying of the coded characters is assumed to be from left to right. The position of the first character line on the facsimile page is the 105th pel on the 131st scanning line (at 3.85 lines/mm). The size of each character box is 20 pels wide by 16 lines/mm high.

3.3.9 Mixed Mode

The mixed mode (MM) option allows pages containing both character-coded and facsimile-coded information to be transferred. As with character mode, ECM is mandatory for this option.

With MM, the page is divided into slices horizontally across the page, and each slice contains either facsimile- or character-coded information but not both. The first slice may be either facsimile or character coded. Subsequent slices are alternatively character or facsimile coded.

Facsimile code slices must be transmitted in integral multiples of 16 scanning lines. Each character-coded line is equivalent to 16 scanning lines (at standard resolution). The width of each coded character is equivalent to 20 pels (at standard resolution). A maximum of 77 characters per line is allowed. The total length of each page cannot exceed 1,024 scanning lines (at standard resolution). That means the maximum length of a coded character slice is 64 character lines.

3.4 Sending Gray-Scale Images With Dither Coding

When the Group 3 facsimile recommendations were approved in 1980, only bilevel data could be transmitted. Dithering techniques could be used to provide a pseudogray-scale capability, but the data transmitted were still bilevel data. Over the years, a number of options have been added to accommodate color and gray-scale images with varying degrees of functionality. This section and the following ones describe these techniques.

Most of the images transmitted through facsimile systems contain only bilevel (black or white) information such as text and line drawings. Because the transmission system and the printing system are binary, the output quality for documents of this type is excellent. When a gray-scale image is scanned and transmitted as if it were a bilevel image, however, the output image usually is severely distorted. The purpose of dither coding is to transmit gray-scale images through facsimile systems employing binary transmission and binary printing techniques with improved output quality.

3.4.1 The Basic Dither Process

The threshold circuit in a bilevel facsimile system is the key element in dither coding. For this discussion, assume that conventional bilevel systems employ a fixed threshold at the midpoint between peak black and peak white. Figure 3.7 illustrates such a coding process. Note that an input gray level near the threshold is drastically altered in the output image. In dither coding, the threshold is varied, or dithered, in amplitude from pel to pel, as shown in Figure 3.8. If the threshold is dithered uniformly over the gray-scale range, the average value of the output image over a number of neighboring pixels will approximate the input gray value. The eye will tend to perform the averaging function, and the observer will perceive the input gray-scale.

3.4.2 Clumped Dither

The clumped dither technique is an electronic approximation of the photomechanical halftone screening process. This technique employs a matrix of fixed thresholds that are arranged in such a way that a "dot" grows outward from the center as successively darker gray levels are encountered in low-contrast regions of the image, as shown in Figure 3.9 [2]. The numbers in the matrix in Figure 3.9 represent threshold values out of 256 possible brightness levels. The 8-bit gray level of each input pel is compared to one of the matrix thresholds;

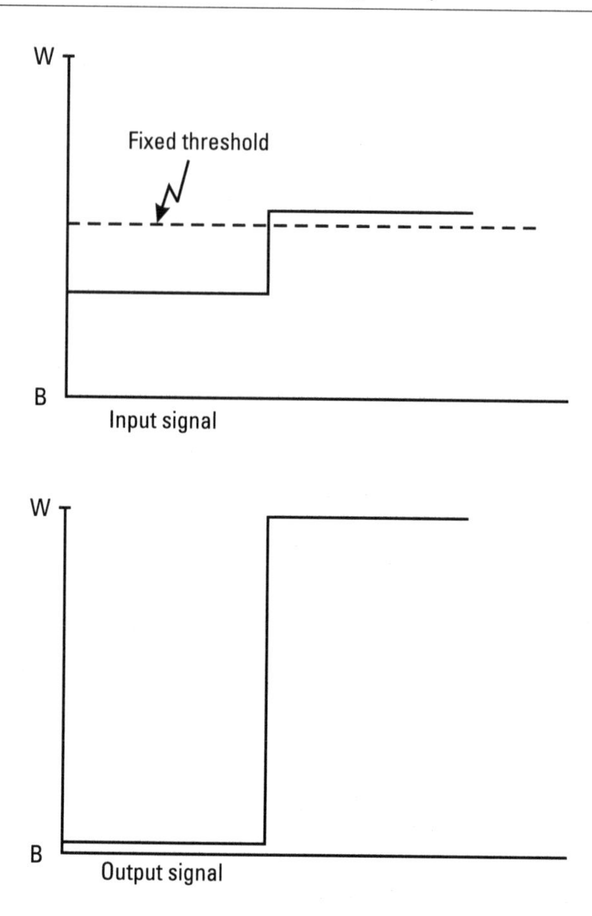

Figure 3.7 Fixed threshold sampling.

which threshold is used depends on the position of the pel in the image. Figure 3.10 shows how the irregular matrices interlock.

3.4.3 Ordered Dither

Ordered dithering employs an *n*-by-*n* matrix of fixed thresholds that is repeated throughout the image, in the same way as clumped dithering. The distribution of the thresholds in the matrix is designed to provide acceptable rendition and edge sharpness while producing a minimum of visible patterns or artifacts. An example of an ordered dither matrix is shown in Figure 3.11. This is an 8-by-8 matrix in which the thresholds are arranged symmetrically. The threshold values are related to a maximum of 32 shades of gray from black to white. The even-numbered thresholds (2 through 30) are arranged spirally in the upper

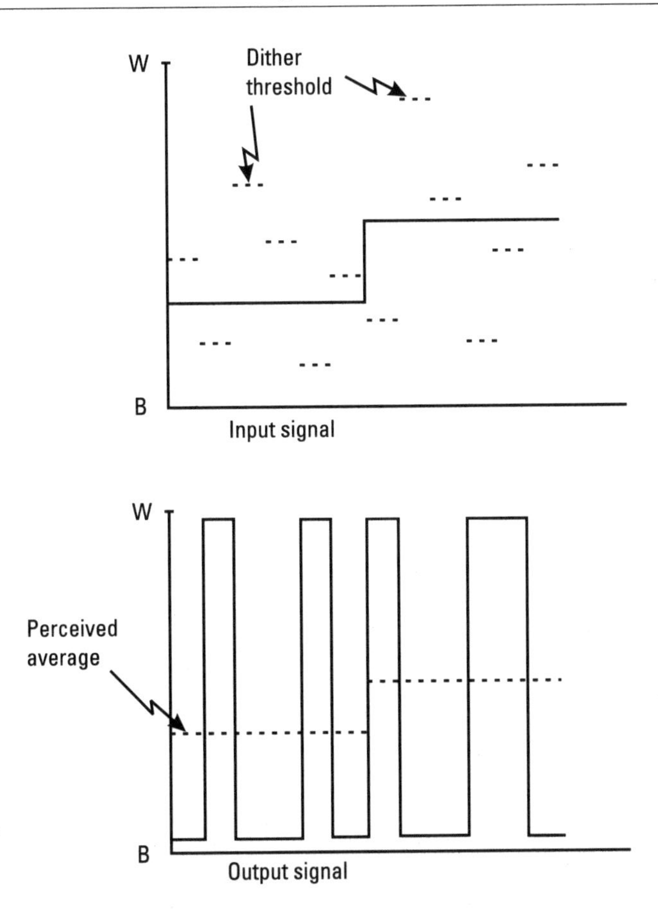

Figure 3.8 Dithered sampling.

right and lower left corners, and the odd-numbered thresholds (1 through 31) are arranged spirally in the upper left and lower right corners. This arrangement provides 32 levels with a pattern of four dots growing in each of the four corners of the matrix as successively darker gray levels are encountered in low-contrast regions of the image.

All the dithering techniques used with the basic MR and MMR Group 3 compression share the same shortcoming: the length of the black and white runs of pels is usually very short, and therefore the compression is poor. Scanned images containing halftones suffer for the same reason. The original Group 3 compression algorithms just were not designed for halftones but for text and line art. One of the newer Group 3 options, based on the Joint Bi-level Image Group (JBIG) compression algorithm, does handle the compression of halftones quite well.

		167	200	
230	210	94	72	
153	111	36	52	193
	216	181	126	222
	242	232		

Figure 3.9 Fixed threshold matrix.

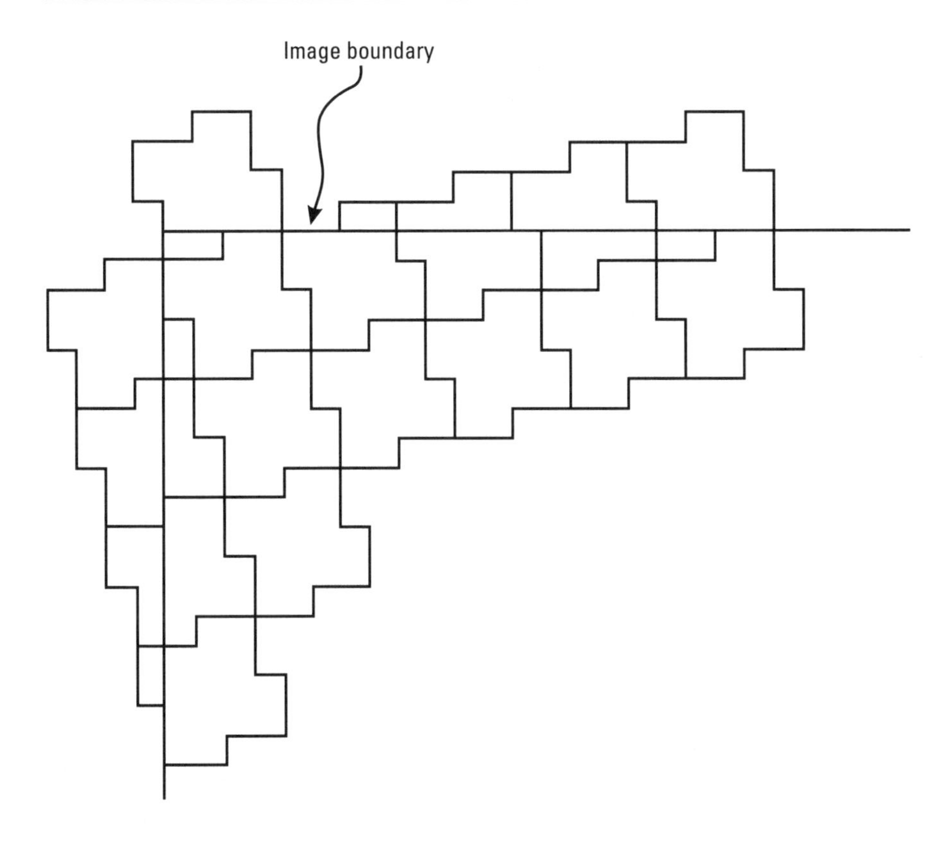

Figure 3.10 Clumped dither matrix lattice.

25	19	11	27	24	18	10	2
9	1	3	21	8	1	2	20
7	7	5	13	16	6	4	12
31	15	23	29	30	14	22	28
24	18	10	26	25	19	11	27
8	1	2	20	9	1	3	21
16	6	4	12	17	7	5	13
30	14	22	18	31	15	23	29

Figure 3.11 Ordered dither matrix.

3.5 Coding Continuous-Tone Color and Gray-Scale Images

A method to transmit continuous-tone color was added to Group 3 facsimile in 1994, with the approval of Recommendation T.42. To represent continuous-tone color data accurately and uniquely, a device-independent interchange color space is required. The color image space must be able to encode the range of hard copy image data when viewed under specified conditions. In addition to the basic color space, the reference white point, illuminant type, and gamut range also must be specified. An example of the use of this model is as follows: A sender scans an original color image using a specific device-dependent color space, which may depend on the illuminant or filters of a particular scanner system. The sender converts the device-dependent color data to the interchange color representation, compresses the data, and transmits them. The receiver receives the encoded data, decompresses them, and converts them to the color space of its printer, which is device dependent. The method chosen to compress the continuous-tone color images was the popular Joint Photographic Experts Group (JPEG) standard.

3.5.1 JPEG Overview (ITU-T T.81/ISO IS 10918-1)

A broad set of service requirements was established for the JPEG algorithm to meet, including both progressive and sequential buildup, soft copy and hard copy, and a wide range of image compressions. The range included very lossy highly compressed images to lossless (with lower compression). Probably as a result of the diverse requirements, the JPEG compression algorithm is not a single algorithm, but a collection of techniques often referred to as a toolkit. The intent is that applications, such as facsimile, will have a customized subset of the JPEG components. A detailed description of JPEG with a comprehensive bibliography is given in [6].

JPEG specifies two classes of coding processes and lossy and lossless processes. The lossy processes are all based on the discrete cosine transform (DCT), and the lossless are based on a predictive technique. There are four modes of operation under which the various processes are defined: the sequential DCT-based mode, the progressive DCT-based mode, the sequential lossless mode, and the hierarchical mode.

In the sequential DCT-based mode, 8-by-8 blocks of pixels are transformed, and the resulting coefficients are quantized and then entropy coded (losslessly) by Huffman or arithmetic coding. The pixel blocks typically are formed by scanning the image (or image component) from left to right and then block row by block row from top to bottom. The allowed sample precisions are 8 and 12 bits per component sample. All decoders that include any DCT-based mode of operation must provide a default decoding capability, referred to as the baseline sequential DCT process. This is a restricted form of the sequential DCT-based mode, using Huffman coding and 8 bits per sample precision for the source image. The application of JPEG to facsimile is based on the baseline sequential DCT process.

The formal definitions of the DCT and its inverse are well documented [7]. The forward DCT transforms a square block of image pixel values into a similar block of spatial frequency "coefficients." The inverse DCT transforms the coefficients back into the block of image pixels. To achieve data compression, the coefficients are quantized. Each coefficient is linearly quantized according to a step size assigned to that coefficient, the assigned value being just small enough so the distortion resulting from quantizing that coefficient is barely noticeable to a human observer. The resulting quantum step numbers are then ranked into an encoding order with the object of placing those quantum numbers most likely to have values of zero last, thus reducing the data to be encoded. A simple zigzag order that arranges the quantum numbers in order of increasing spatial frequency is used.

The positions and values of the nonzero quantum numbers are then transmitted losslessly by Huffman coding. The receiver decodes the quantum numbers, multiplies each quantum number by the step size associated with that coefficient to obtain quantized versions of the original coefficients, and then performs the inverse DCT to obtain an approximation of the original image. The compression-versus-distortion tradeoff can be controlled by a single quantization scale factor that scales all the step sizes assigned to the coefficients by a single multiplicative constant. The larger the scale factor is, the greater the compression but also the greater the distortion.

3.5.2 Application of JPEG to Facsimile

The Group 3 option for the transmission of continuous-tone color and gray-scale images is based on the ISO/ITU-T JPEG standard [8]. This option is defined in Recommendation T.42 [9], Annex E of T.30 and Annex E of T.4. The image pixel values are represented in the CIE 1976 (L* a* b*) color space, often referred to as CIELAB. This color space, defined by the Commission Internationale de l'Éclairage (CIE), has approximately equal visually perceptible differences between equispaced points throughout the space. The three components are L*, or luminance, and a* and b* in chrominance. The luminance-chromaticity spaces offer gray-scale compatibility and higher DCT-based compression than other spaces. The human eye is much more sensitive to luminance than chrominance; thus, it is easier to optimize the quantization matrix where luminance and chrominance are separate components. Subsampling the chrominance components provides further compression.

The pixel image data consist of block interleaved L*, a*, and b* data. Blocks are entropy-encoded DCT-transformed 8-by-8 arrays of image data from a single image component. When a gray-scale image is transmitted, only the L* component is represented in the data structure. The number of image components is either one (for a gray-scale image) or three (for a color image). The data are block-interleaved when a color image is transmitted. The blocks are organized in minimum coding units (MCUs) such that an MCU contains a minimum integral number of all image components. For the default (4:1:1) subsampling case, an MCU consists of four blocks of L* data, one block of a* data, and one block of b* data. The data are ordered L*, L*, L*, L*, a*, b* in the MCU. The four L* blocks proceed in the same scan order as the page: left to right and top to bottom. Therefore, the L* blocks are transmitted first upper left, then upper right, then lower left, then lower right. The default (4:1:1) subsampling is specified as a four-coefficient (tap) filter with coefficients (1/4, 1/4, 1/4, 1/4). Thus, a* and b* are computed from nonsubsampled data by averaging the four values of chrominance at the luminance locations. The location of the subsampled chrominance pixels is shown in Figure 3.12. The option is to have no subsampling.

The basic resolution is 200 pels/25.4 mm. Allowed values include 200, 300, and 400 pels/25.4 mm, with square (or equivalent) pels. At 200 by 200 pels/25.4 mm, a color photograph (A4 paper size) compressed to 1-bit/pixel and transmitted at 64 Kbps would require about 1 minute to send. The selection of 8-bit or 12-bit data precision also can affect the data compression. Subsampling and data precision, as well as the ability to send color, are negotiated before image transmission begins.

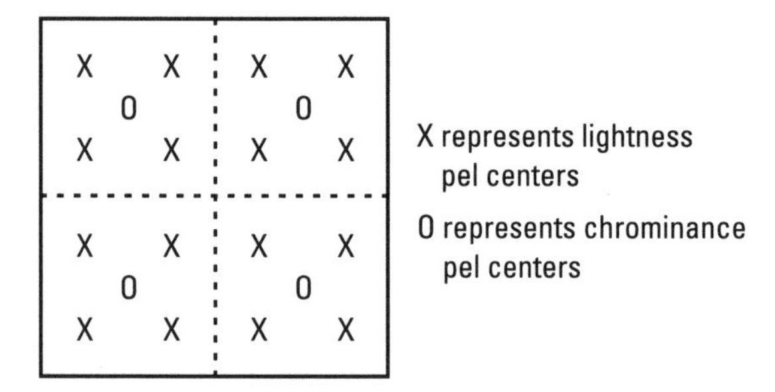

Figure 3.12 Position of lightness and chrominance samples (4:1:1 subsampling) in the MCUs. Each of the smaller squares represents an MCU.

3.6 Lossless Coding of Color and Gray-Scale Images

When Recommendation T.42 (described in Section 3.5) was approved, it provided the capability to transmit color using JPEG. Although that was a good first step toward a color facsimile capability, it was not an efficient solution for text or line art. JPEG can compress text and line art, but to get good compression efficiency, image quality suffers. Recommendation T.42 works best on photographic content. Recommendation T.43, on the other hand, was intended to accommodate text and line art with some limited color capability. Recommendation T.43 defines a lossless color data representation method using Recommendation T.82 [10], which was prepared by JBIG. In that recommendation, three types of images are treated. The first type is a 1 bit per color cyan, magenta, yellow, black (CMYK) or red, green, blue (RGB) image. The second type is a palettized color image in which palette tables are specified with the CIELAB color space defined in Recommendation T.42. The last type is a continuous-tone color and gray-scale image also specified with the CIELAB color space. Images can be created in a variety of ways, including conventional scanning, computer generation, or image processing techniques such as one of the dither methods. Recommendation T.43 was approved in July 1997. The rest of this section presents an overview of JBIG compression, followed by a description of the three image types treated by Recommendation T.43.

3.6.1 JBIG Overview (ITU-T T.82/ISO11544)

The progressive bilevel coding technique consists of repeatedly reducing the resolution of a bilevel image, R_0, creating images R_1, R_2, ..., R_n, image R_i having one-half the number of pels per line and one-half the number of lines of image R_{i-1}. The lowest-resolution image, R_n, called the base layer, is transmitted losslessly (free of distortion) by binary arithmetic coding. Next, image R_{n-1} is transmitted losslessly, using pels in R_n and previously transmitted (causal) pels in R_{n-1} as predictors in an attempt to predict the next R_{n-1} pel to be transmitted. If prediction is possible (both transmitter and receiver are equipped with rules to tell whether that is the case), the predicted pel value is not transmitted. The progressive buildup is repeated until image R_0 has been losslessly transmitted (or the process stopped at the receiver's request). A sequential mode of transmission also exists. It consists of performing the entire progressive transmission on successive horizontal stripes of the original image. The algorithm performs image reduction, typical prediction, deterministic prediction, and binary arithmetic encoding and decoding.

3.6.1.1 Image Reduction

Each low-resolution pel is determined by the values of several high-resolution pels and low-resolution pels that already have been determined. The objective of the reduction algorithm is to preserve as much detail as possible in the low-resolution image under the constraint that the latter be half as wide and high as the high-resolution image. Resolution reduction could be achieved with subsampling, which is simple but yields poor results, especially on thin lines and dithered images. JBIG recommends a reduction algorithm that has given excellent results, but any algorithm can be used.

3.6.1.2 Prediction

When a difference layer is being encoded or decoded, much of the compression is achieved by predicting new pel values from the values of pels in a predictor template. The predictor template contains pels from the reference layer and pels already predicted or encoded from the difference layer. When the predictor state is such that the prediction is known to be correct (the receiver must know this also), the predicted pel value need not be encoded or decoded. The JBIG algorithm employs two kinds of prediction: typical prediction and deterministic prediction.

Typical Prediction

Typical prediction (TP) refers to prediction in which the predicted value is almost always, but not necessarily always, correct. Because in bilevel imagery each pel carries only 1 bit of information, it would be wasteful for the transmitter to inform the receiver of whether the prediction is correct for each pel predicted. Instead, the transmitter looks ahead for and reports TP errors (exceptions).

Deterministic Prediction

Deterministic prediction (DP) refers to prediction in which the predicted value is *always* correct. DP is tightly bound to the image-reduction rules. Whether a pel is or is not deterministically predictable is determined by looking up a rule in a table indexed by the state of the predictor pels. Deterministically predictable pels are flagged and not encoded by the arithmetic coder. Provision is made for an encoder to download DP tables to a decoder if it is using a private resolution reduction algorithm.

3.6.1.3 Binary Arithmetic Coding

The data compression achieved by a binary arithmetic coder is best when the probabilities of the two symbols are near 1 and 0, and worst when they are near

1/2. In any practical application, the probability of a 1 or a 0 at any given time is frequently dependent on the conditions under which the symbol is being encoded or decoded. Therefore, best compression is achieved by keeping separate probability estimates for those conditions under which the encoded symbol probabilities are the most strongly skewed. These conditions are called contexts.

Consider, for example, a bilevel image containing line drawings and text. As the image is scanned, if the previous pel was white, then there is a high probability that the current one also will be white. Therefore, if one uses the previous pel value as a predictor, there are two contexts, one for each color of the previous pel. The probabilities for each usually are much nearer 1 and 0 than is the single probability with the previous pel value ignored. In the JBIG system, there is a separate context for every possible combination of pel values in a context "template."

3.6.1.4 Adaptive Context Templates

The purpose of adaptive context templates is to take advantage of horizontal periodicity, which often occurs in halftone and dithered images. Data compression is best if at least one of the pels in a context template is a good predictor of the pel being encoded. An adaptive context template (AT), contains a "floating" pel; all other pels in the template are fixed in position relative to the encoded pel. There are also other pels designated as candidate floating pels, not currently a part of the context template. If one of the candidates becomes a much better predictor than the current floating pel, the candidate and the current floating pels swap status, so that the better predictor becomes a part of the context template. This test is made infrequently, and the swap is made only if the candidate is a much better predictor. These restrictions are imposed because, when a swap is made, compression is temporarily degraded until the binary arithmetic coder has time to adapt to the swap.

3.6.1.5 Performance

Studies have shown that TP, DP, ATs, and arithmetic coding all contribute to compression in varying amounts, depending on image content and resolution. TP contributes the most compression in line drawings and text images, including handwriting, but is of negligible importance in halftone images. DP and AT both contribute significantly in halftone images but negligibly in drawings and text. Note that these compression results are based on the application of all parts (TP, DP, AT) of the algorithm. That does not mean similar overall compression could not be achieved by omitting part (e.g., TP) and making up the compression with another part (e.g., the arithmetic coder). However, to achieve the highest compression in a progressive system designed to be independent of image type, all components should be applied. For database storage, browsing, and retrieval, with various resolutions from icons to full-scale images, the full JBIG algorithm is appropriate. If the full functionality were added to terminals that were intended to provide soft copy interactive capability, then users of such terminals would realize the full benefits of the progressive algorithm.

For bilevel point-to-point facsimile transmission, a single-layer subset of the full JBIG algorithm might be more applicable. The JBIG algorithm is computationally complex. The complexity can be reduced by the use of a proper subset to forego the progressive functionality and eliminate resolution reduction, TP, and DP. The ITU has taken this approach to provide an optional bilevel compression algorithm for Group 3 and Group 4 facsimile terminals with the creation of Recommendation T.85. That recommendation is an application profile for bilevel facsimile that specifies that the image be coded in a single layer made up of stripes of 128 lines.

In terms of compression, JBIG typically outperforms Recommendation T.6, especially on dithered images. For text and line drawings, JBIG improves compression by 20–30% over Recommendation T.6. For dithered images, the improvement is much greater, approximately 20 to 1.

3.6.2 Recommendation T.43

3.6.2.1 One-Bit-Per-Color Mode

The 1-bit-per-color mode was intended to represent images with primary colors using the CMY(K) or RGB color space, depending on whether the images are scanned or computer generated. Typically, a sender might scan a color image, producing three- or four-color components, and create a 1-bit-per-color image using processing techniques such as dither. Each bit plane indicates the existence of one of the primary colors. The image is encoded with JBIG and transmitted. The sender then sends the encoded data with an indication of this mode. The receiver represents the image on a CRT (soft copy) or a printer (hard copy) using the receiver's own primary colors. The colors of the document may not be represented accurately at the receiver in the 1-bit-per-color mode because neither RGB nor CMY(K) is a device-independent color space. A typical application of this mode is to transmit business correspondence containing colored logos.

3.6.2.2 Palettized Color Mode

The palettized color mode expands the possible number of colors that may be used to characterize an image; in addition, it provides the capability for accurate color reproduction. Both features are achieved by using color palette table data as specified by the device-independent interchange color space (CIELAB) defined in Recommendation T.42. The price for this added capability is coding (and transmission) efficiency, that is, the compression is less, so the facsimile transmission takes longer. An example of the palettized color mode follows.

A sender creates an original color image using a color palette in a device-dependent color space. This color space may depend on the primaries, white point, and gamma of the sender's CRT. The sender converts the device-dependent color palette to the palette interchange color space as defined in Recommendation T.42. The sender then transmits the interchange color palette together with the image data encoded with JBIG. The receiver uses the interchange color palette to decode the image data and convert the interchange color palette to the receiver-specific color palette, which is determined by the receiver's device-dependent hard copy or soft copy color space. The image can then be rendered.

3.6.2.3 Continuous-Tone Color Mode

The continuous-tone color mode provides the highest color accuracy of the three modes in Recommendation T.43 and the lowest compression efficiency. In this mode, an original continuous-tone color or gray-scale image is generated in the color space defined in Recommendation T.42 (CIELAB). This mode provides lossless encoding using JBIG bit-plane compression. Gray code conversion on the bit planes is used to improve the compression efficiency.

3.7 Coding Images With Mixed Raster Content

The mixed raster content (MRC) option was initiated to correct a weakness of the existing color facsimile options that use JPEG and JBIG. Before MRC, there was no standard way of efficiently coding a page that contained both text and color photographic content. The MRC option, as implemented by ITU-T Recommendation T.44, is a way of describing raster-oriented (scanned or synthetic) documents with both bilevel and multilevel data within a page. A rasterized page may contain one or more image types, such as multilevel continuous-tone or palettized colors (contone) usually associated with naturally occurring images, bilevel detail associated with text and line art, and multilevel colors associated with the text and line art. The goal of MRC is to make the exchange of raster-oriented mixed-content color documents possible with higher speed, higher image quality, and modest computing resources. This efficiency is realized through segmentation of the image into multiple layers (planes), as determined by image type, and applying image-specific encoding and spatial and color resolution processing. The MRC method defines no new image compression methods, but does require that all previously defined compression methods used in Group 3 facsimile be supported.

The best approach to achieve high compression ratios and retain quality is to compress the different segments of the raster data according to their individual attributes. Text and line art data (bilevel data) are compressed with an approach that emphasizes maintaining the detail and structure of the input. Pictures and color gradients (multilevel data) are compressed using an approach that emphasizes maintaining the smoothness and accuracy of the colors. These different data types (bilevel and multilevel) often are conceptualized as being on separate layers within the page. This separation of the data by content (spatial detail versus color) suggests the use of different resolutions for the different data, with high spatial resolution used for text and line art and high color resolution for images and gradients.

3.7.1 Three-Layer Model

This concept of data separation by importance of content led to the development of the three-layer model on which the MRC recommendation is built. The three-layer model identifies three basic data types that can be contained within a page: multilevel data associated with "contone" color (continuous-tone and/or palettized color) image for which mid-to-low spatial and high color resolution is typically appropriate for good reproduction; bilevel data associated with high detail of text or line art for which high spatial and low color resolution is typically appropriate; and multilevel data associated with multilevel colors of the text or line art data for which mid-to-low spatial and high color resolution is typically appropriate. Each page within the MRC model is processed independently. The data types within each page are represented in distinct layers (also referred to as planes) to be compressed and transmitted independently. Multilevel contone data are represented in the lower layer, bilevel in the middle layer, and multilevel data of text or line art colors in the upper layer. The lower layer is referred to as the background layer, and the upper layer is referred to as the foreground layer. Figure 3.13 illustrates a three-layer model. The process of image reconstruction is controlled by the middle bilevel layer that acts as a mask to select whether pixels from the background contone layer or the foreground text or line art color layer will be reproduced. Due to its selection function this layer is referred to as the mask layer. When the value of a mask layer pixel is 1, the corresponding pixel from the foreground is selected and reproduced. When the value of the mask layer pixel is 0, the corresponding pixel from the background is selected and reproduced. Figure 3.14 illustrates the selection process.

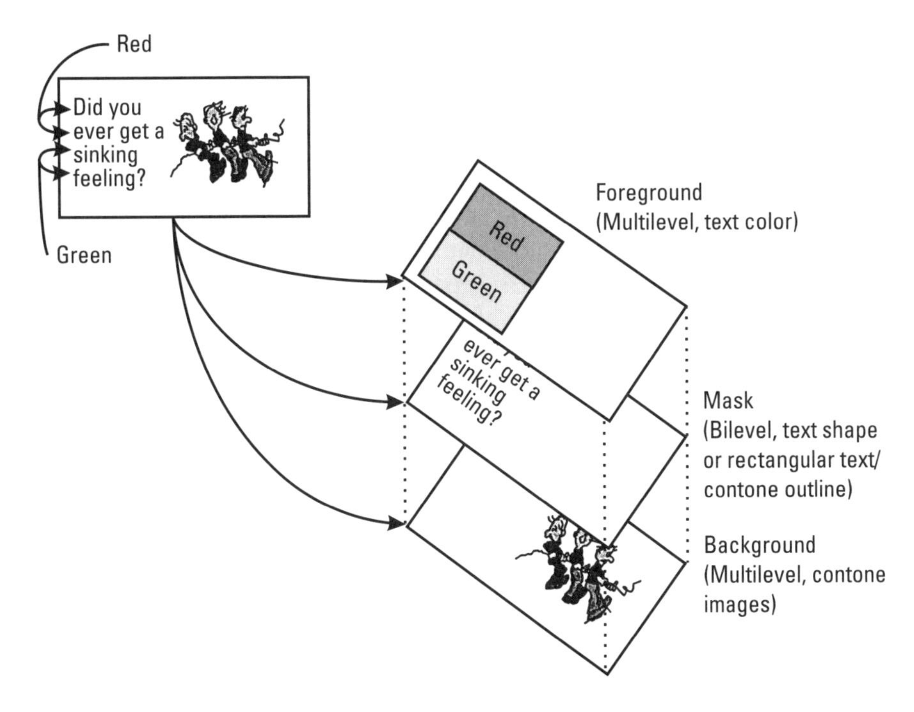

Figure 3.13 Three-layer model.

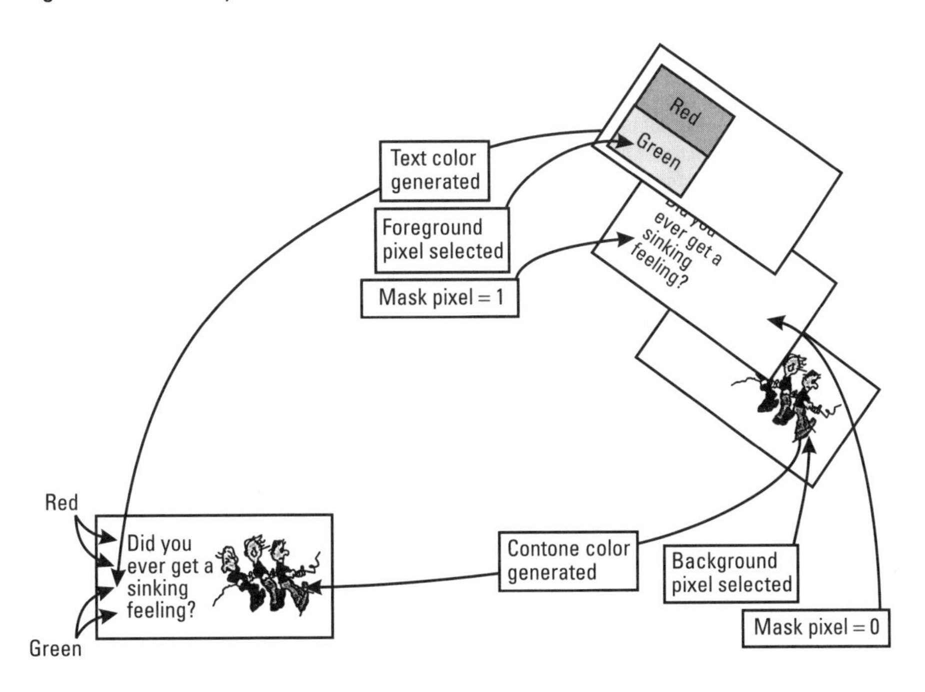

Figure 3.14 Selection process.

3.7.2 Page Subdivision

To overcome the limited device memory in many facsimile implementations, there are provisions to subdivide the page into horizontal strips that span the entire width of the page and isolate individual regions (Figure 3.15). Strips are composed of one or more layers as determined by the image type within the strip. The mask layer must span the entire width and height of the strip. The background and foreground layers need not span the width and height of the strip. Reduction in the amount of white space coded in the background or foreground layers can be realized by taking advantage of the image width and height data included in the layer data stream, such as JPEG, and a horizontal and vertical offset provision. A default foreground color of black is defined such that at mask pixel locations (value = 1) where a corresponding foreground pixel is not present, the foreground default is applied. A default background color of white (default can be change to any color) is defined such that at mask pixel locations (value = 0) where a corresponding contone image is not present, the default background color is applied. Each of the defaults can be changed to any color (Figure 3.16).

The maximum strip size can be negotiated between the default size of 256 lines and the full length of the page. The three-layer model has three types of horizontal strips that are implemented according to the type of data they contain.

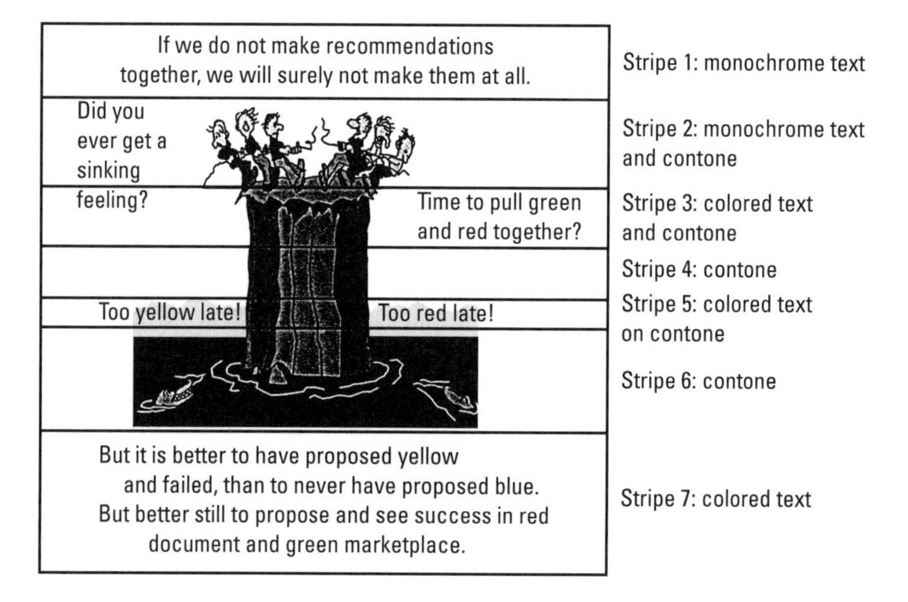

Figure 3.15 Subdivision into horizontal strips.

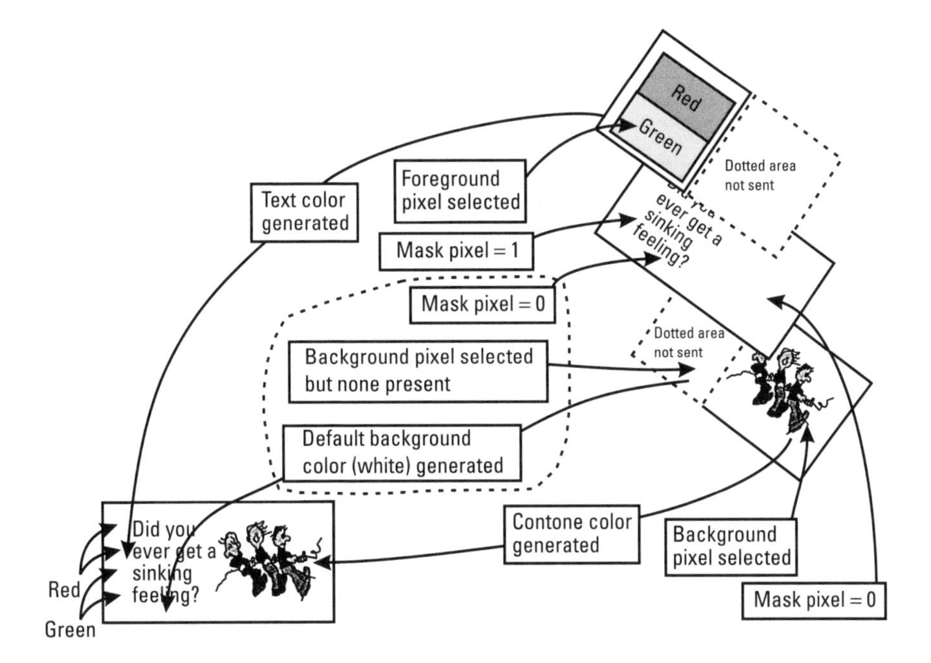

Figure 3.16 Default operation.

- Three-layer strip (3LS), so referenced because it contains all three of the foreground, mask, and background layers, as shown in Figure 3.13. The 3LS is appropriate for an image that contains multicolored text/line art and contone, or monochrome text/line art on colored background and contone.
- Two-layer strip (2LS), so referenced because it contains coded data for two of the three layers, with the third set to a fixed value. The two layers can be the mask and background layers or the mask and foreground layers. All combinations of multiple layers include the mask layer. The 2LS is appropriate for an image that contains monochrome text/line art and contone, or colored text/line art and no contone.
- One-layer strip (1LS), so referenced because it contains coded data for only one of the three layers, with the other two set to fixed values. The one layer can be mask, background, or foreground. The 1LS is appropriate for an image that contains monochrome text/line art, contone, or possibly richly colored graphics.

The three-layer model requires application of a multilevel coding scheme to the background and foreground layers. Any ITU-T multilevel coding (such as JPEG or JBIG, as defined in Recommendation T.81 and T.43, respectively) can be used for the background or foreground. A bilevel coding scheme is required for the mask layer. Any ITU bilevel coding (such as JBIG or MMR, as defined in Recommendation T.85 and T.6, respectively) can be used. See Figure 3.17.

3.7.3 Marker Segments

Each MRC page begins with a start-of-page marker segment, which is followed by page data and terminated with an end-of-page marker. The page data consist of strips. During transmission, strips are sent sequentially from the top of the page, strip 1 through *N*. Within a strip, the mask layer is transmitted first, followed by the background and then the foreground, as appropriate. The specific coding used in the mask layer is also provided by parameters in the start-of-page marker segment. The spatial resolution of the mask layer, to be used throughout the page, is also identified. Layers with varied spatial resolutions can be combined within a strip, but the resolution of the foreground and background layers must be integral factors of the mask resolution layer. Resolutions must be square (same resolution in the horizontal and vertical directions). To meet that requirement, an optional resolution of 100 by 100 pels/inch was added to the Group 3 recommendations to handle the low-resolution color

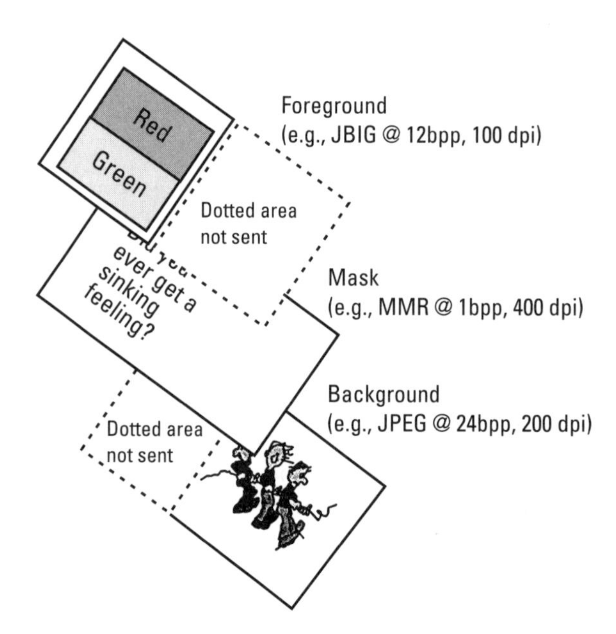

Figure 3.17 Application of coding schemes.

plane. The specific resolution being used in the foreground and background layers is identified within the multilevel coder marker segment at the start of each layer within a strip. A start-of-strip marker segment contains parameters that indicate type of strip (1LS, 2LS, or 3LS), foreground and background default colors, offset of the foreground and background, strip height, and mask layer length.

3.7.4 Negotiation

The use of ECM to provide error-free transmission is required for MRC. The negotiation to transmit and receive pages using the MRC procedure is carried out in the usual way with the DIS/DTC and DCS frames of the T.30 protocol. Note that the MRC procedure is available only when the base color encoding mode as defined in Recommendation T.42, T.4, and T.30 is also available. The current use of MRC in Group 3 facsimile makes provision to apply one encoding scheme, one spatial resolution, and one color resolution within each of the three layers. Future MRC modes may permit multiple discrete image elements and encoding schemes within each layer.

3.7.5 Planned Enhancements

Work on Recommendation T.44 is continuing in Study Group 8 of the ITU-T in two areas. First, the number of layers will be extended from three to any number. Although three layers accommodate scanned images reasonably well, computer-generated graphics can take advantage of more layers. Also, it will be possible to compose pages made up of overlapping images. Another area of work concerns the new JBIG2 compression algorithm (discussed in Chapter 12), which is currently under development. To achieve the highest compression efficiency afforded by JBIG2, the interpage dependencies in the algorithm must be exploited. That means Recommendation T.44 must be extended to satisfy this requirement.

3.8 Secure Facsimile

This section describes commercial secure facsimile as implemented in Group 3 facsimile. (Government and military secure facsimile is addressed in Chapter 11.) Two fundamental security issues are associated with the typical point-to-point communication between a facsimile sender and a facsimile receiver. The first security issue is the possibility that the facsimile data can be intercepted or changed without either party knowing that the information has been compromised. The second issue is authentication, that is, ensuring that the sender is the true originator of the document and the receiver is the intended recipient of the document.

Historically, the solution to securing the data being transmitted is to use a cryptosystem to encrypt, or encipher, the message to be sent in an extremely complicated way, rendering it unintelligible. The cryptosystem consists of encryption and decryption functions together with a cryptographic key. The encryption function, using the key, scrambles the original data (also known as plaintext) into what appears to be nonsense, or ciphertext. The decryption function uses the same key on the ciphertext to restore the original data. The key is not an integral part of the encryption and decryption functions, so the same functions can be used with different keys. The key must be kept secret to prevent an unauthorized person from decrypting intercepted ciphertext. The encrypt and decrypt functions can be made public, because without the appropriate key ciphertext cannot be decrypted.

When evaluating commercial secure facsimile systems, it is important to distinguish between machine-to-machine security and person-to-person security. Several people often share a single facsimile machine, especially in the business world. Therefore, a secure facsimile message sent to a facsimile machine could be read by many people in addition to the intended recipient. Person-to-person security requires the storage of the encrypted document in the memory of the receiving facsimile machine until the intended recipient retrieves the facsimile. Person-to-person secure facsimile also suggests the use of smart cards or the equivalent so the recipient does not have to enter long key strings into the facsimile machine. Each qualified recipient would have a smart card containing a private key or shared secret key.

Currently, there are two major types of cryptosystems, single-key (secret-key) systems and public-key systems, each with advantages and disadvantages. The Group 3 recommendations now contain both single-key and public-key cryptosystems as options.

3.8.1 Secret-Key Cryptosystems

Prior to the late 1970s, all generally known cryptosystems were single-key systems. A single-key cryptosystem is one in which the encryption and decryption keys are the same (or readily derived from each other). These cryptosystems are also referred to as secret-key or symmetric systems. Single-key cryptosystems provide authenticity because only the holders of the common single or secret key are able to create ciphertext that decrypts into meaningful plaintext. The main problem associated with single-/secret-key cryptosystems is the secure exchange of the secret key between the sender and the receiver. That means a key must be transmitted via secure channels so both parties can know it before encrypted messages can be sent over insecure channels. That is not always convenient. If the key cannot be exchanged securely, any subsequent transmissions using the key can be intercepted and decrypted. The generation, transmission, and storage of keys together are called key management and are common to all cryptosystems.

3.8.2 Public-Key Cryptosystems

Public-key (or asymmetric) cryptosystems were invented in 1976 by Whitfield Diffie and Martin Hellman to solve the key management problem. In a public-key system, every person gets a pair of keys, one public, one private. Everyone publishes the public key and keeps the private key secret. Each key unlocks the code that the other key makes. Knowing the public key does not help to deduce the corresponding secret key. The need for the sender and the receiver to share a secret key is eliminated because all communications are encrypted by the sender using the public key of the receiver. Because private key distribution is not necessary, interception of exchanged keys is not a problem. Anyone can send a confidential message using public information only. This message can be decrypted only with the secret key held by the intended recipient. The RSA algorithm is an implementation of a public-key cryptosystem (see Section 3.8.3).

Although message authenticity is automatically provided by a single-key cryptosystem (only the holders of the secret key can encrypt or decrypt the message), the situation in a public-key cryptosystem is somewhat more complicated. In a public-key cryptosystem, authentication is achieved through the use of digital signatures. A digital signature in digital communications plays the same role as a handwritten signature for printed documents. A digital signature is a block of data added to a digital message that binds the message to a particular individual or entity. A digital signature can be generated by the public-key cryptosystem itself (by using the private keys of the sender/receiver) or by a digital signature system. A digital signature system is similar to a public-key cryptosystem in that each user has a public and a private key. The sender signs a message by sending a block of data (typically a "hash" of the message itself) encrypted with the sender's private key, generating the digital signature. The receiver verifies the message by decrypting the signature block with the sender's public key and comparing the contents with the receiver's own hash value of the message. That proves that the sender was the originator of the message and that the message has not been altered by anyone else, because only the sender has the secret key that made the signature. Forgery of a signed message is almost impossible, and the sender cannot later deny his signature.

A public key cryptosystem does have an important vulnerability. It is important to protect public keys from tampering, to make sure that a public key really belongs to whom it appears to belong. One way to accomplish that is to trust a new public key from someone else only if you got it directly from its owner or if it has been signed by someone you trust. The "someone you trust" could be a centralized key server or registration authority with an available known good copy of the registration authority's public key.

3.8.3 Secure Facsimile Standardization Efforts

Work on a secure facsimile extension to the ITU-T Group 3 facsimile recommendations began in 1994. This work has resulted in new and amended recommendations for Group 3 facsimile. Those recommendations, developed in Study Group 8 of the ITU-T, accommodated two proposals. One proposal, the RSA algorithm, initially was from France and was based on a public key management system devised by Ron Rivest, Adi Shamir, and Leonard Adleman. (It is called RSA after the initials of its inventors.) The other proposal, from the United Kingdom, is based on the Hawthorne Key Management (HKM) system, the Hawthorne Facsimile Cipher (HFX40), and the HFX40-I message integrity system. Hereafter, this proposal is referred to as the HKM/HFX cryptosystem. Both secure facsimile proposals were incorporated into the facsimile protocol. Group 3 secure facsimile, including Annex G and Annex H to T.30 and new Recommendation T.36, was formally approved by the ITU-T in July 1997. The work in security standards for Group 4 facsimile was left for future study.

3.8.3.1 Procedures Using the HKM/HFX Cryptosystem

The proposal advanced by the United Kingdom consists of the HKM system, the HFX40 carrier cipher (encryption algorithm), and the HFX40-I message integrity system (hashing algorithm). The HKM key management algorithm includes a registration procedure and a secure transmission of a secret key that enables subsequent transmissions to be provided securely. The procedures are defined in ITU-T Recommendation T.36.

In the registration mode, the two terminals exchange information that enables them to uniquely identify each other. Identification is based on the agreement between the users of a secret one-time key that must be exchanged securely (not defined by the recommendations). Each terminal stores a 16-digit number that is uniquely associated with the terminal with which it has carried out registration. The number is used with a challenge procedure to provide mutual authentication and document confirmation. The procedure is also used to transmit the session key to be used for document encryption and hashing.

An override mode is also provided that bypasses the exchange of security signals between two terminals. Just as in the classic secret-key cryptosystem approach, the override mode depends on the secure exchange of a secret key outside the system. This key is used by the transmitting terminal to encrypt the document and by the receiving terminal to decrypt the document.

To implement the HKM/HFX cryptosystem, 10 new signals were added to the T.30 Recommendation, including transmitter not ready (TNR), transmitter ready (TR), transmitter key (TKY), and receiver key (RKY). Altogether, there are five transmitter keys and three receiver keys. The facsimile information fields (FIFs) of these signals are used to exchange a total of 22 security parameters. The security parameters are separated into three categories: mutual registration signals, premessage signals, and postmessage signals. Premessage signals are used to provide mutual authentication and exchange of the secret session key. Postmessage signals are used to provide document confirmation and integrity. Flow diagrams for secure facsimile operation on the PSTN in half-duplex mode and on ISDN and PSTN V.34 duplex mode. The procedures are defined in Annex G of ITU-T Recommendation T.30.

3.8.3.2 Procedures Using the RSA Security System

The procedures for the RSA security system are defined in Annex H of ITU-T Recommendation T.30. The RSA security system uses one pair of keys (public and secret) for authentication and message integrity and a separate pair of keys (encipherment public key and encipherment secret key) for document encryption.

For the exchange of security parameters at the protocol level, eight new signals are defined: digital extended request (DER); digital extended signal (DES); digital extended command (DEC); digital turnaround request (DTR); digital not acknowledge (DNK); transmitter not ready (TNR); transmitter ready? (TR); and present signature signal (PPS-PSS). DTR is sent by the calling terminal in the place of DTC to indicate that polling or turnaround is desired. The DNK signal indicates that the previous command has not been satisfactorily received and should be retransmitted. TNR is used to indicate that the transmitter is not yet ready to transmit. The PPS-PSS signal indicates the end of page and that a digital signature follows.

The use of "supertag" and "tag" octets for the classification of security parameters is included. The tags and their corresponding values are used to encode security parameters within the FIF of facsimile signals. Two supertags were created for security, one for the registration mode and another for the secure transmission mode. In each supertag, the various security parameters are separated by tags. Twenty-eight tags can be introduced by the two security supertags. The order of tags within the security supertags is not fixed.

In addition to several tags for the exchange of keys and identities, there are two tags reserved for security services and security mechanisms. The security services parameter indicates whether mutual authentication, security without confidentiality, or security with confidentiality will be used. The security mechanisms parameter indicates which hash function and encryption algorithm will be used in the facsimile communication. Currently, five optional encryption algorithms are registered for use with the RSA key management system, including secure and fast encryption routine with a key of 64 bits (SAFER-K64), fast data encipherment algorithm (FEAL), a block cipher developed by RSA Data Security (RC5), and the international data encryption algorithm (IDEA), and HFX40. Other optional algorithms could be registered in the future.

The hash function to be used can be either secured hash algorithm (SHA) or message digest 5 (MD5). MD5 is fifth in a series of message digest algorithms developed by RSA Data Security for use with its RSA algorithm. MD5 takes a message of arbitrary length and produces a 128-bit hash of the message. The length of the hashing result for SHA is 160 bits. By comparison, the hashing result for the HFX40-I algorithm is equivalent to 80 bits (24 decimal digits) in length. The choice between SHA and MD-5 is negotiated in the protocol. In the future, other optional hash functions may be added.

The registration mode permits the sender and the receiver to register and store the public keys of the other party prior to secure facsimile transmission. Two parties wishing to communicate can register their public keys with another user in two steps. First, the sender and the receiver each hash their identity and public key(s), and the hash results are exchanged out of band (directly, by mail or by phone, for example) and stored in the terminals. Then the identities and public keys of the two parties are exchanged and stored in the registration mode of the T.30 protocol. The validity of the identity and public key(s) of the other party is assessed by hashing these values and comparing them with the hash result that was exchanged out-of-band.

An optional security page follows the last page of the transmitted document. The security page contains the following parameters:

- Security page indicator;
- Identity of the sender;
- Public key of the sender;
- Identity of the recipient;
- Random number created by the recipient;

- Time stamp;
- Length of the document;
- Digital signature of the entity in brackets;
- Certificate of the public key of the sender;
- Security-page-type-identification.

The format of the security page uses a sequence of supertags, tags, and parameters similar to the tag sequences within the DER, DES, DEC, and DTR signals.

One other significant issue is the key length. It was agreed to limit the session key length for the RSA secure facsimile system to 40 bits. The amendment also adds a redundancy mechanism that repeats the 40-bit session key to fit the length of the various encipherment algorithms whose key lengths are required to be longer than 40 bits.

3.8.4 Encryption Policy

The successful development of secure facsimile standards and products depends to a large extent on the resolution of conflicting interests of governments and industry. Hardware and software manufacturers want strong encryption to safeguard their customers' data against attack thereby selling more products. Governments want to promote electronic commerce with strong encryption, but they are primarily concerned that law enforcement will be unable to intercept and decode encrypted messages. Therefore, governments want to mandate some form of key escrow, where the government or an escrow agent would keep a copy of everyone's private keys.

Encryption policy is further complicated by the varying regulations of different governments. That means facsimile manufacturers must create different versions of their equipment for use in different countries. Even with international secure facsimile standards, sending a secure facsimile from one country to another may be difficult or impossible depending on the key lengths and algorithms permitted in different countries. Although there has much discussion among industry and government leaders, encryption policy is far from stable.

3.8.5 Secure Facsimile Products

The possible implementations of secure facsimile can be broken down into the following five categories:

- Standalone facsimile machine with encryption capability built in.
- Self-contained encryption module connected between a facsimile terminal and the telephone network.
- Encryption-enabled facsimile applications that run on a personal computer with a facsimile modem. In this case, encryption is performed only in the software, and no custom hardware is needed. The lack of hardware complicates the use of smart cards for person-to-person security.
- Encryption toolkits for standalone facsimile machines or facsimile modem software.
- Encryption application programming interfaces (APIs).

There have been several products on the market that enable secure facsimile transmission and provide some or all of the following security services: mutual authentication, message confidentiality, message integrity, confirmation of receipt, support for smart cards, and storage of encrypted facsimile messages for personal retrieval. With few exceptions, these products are proprietary and do not conform to any national or international security standards. One exception is the Wordcraft toolkit, which provides secure facsimile services based on the HKM/HFX cryptosystem.

References

[1] Huffman, D. A., "A Method for the Construction of Minimum Redundancy Codes," *Proc. IRE*, Vol. 40, Sept. 1952, pp. 1098–1101.

[2] ITU-T Recommendation T.6, *Facsimile Coding Schemes and Coding Control Functions for Group 4 Facsimile Apparatus*, Vol. VII—Fascicle VII.3 pp. 48–57.

[3] *Error Control Option for Group 3 Facsimile Equipment*, National Communications System Tech. Info. Bulletin, 87-4, Jan. 1987.

[4] ISO 8859-1, *Information Processing—8-Bit Single Byte Coded Graphic Character Sets—Part 1: Latin Alphabet No. 5.*

[5] ITU-T Recommendation T.52, *Coded Character Sets for Telematic Services.*

[6] Mitchell, J. L., and W. B. Pennebaker, *JBEG Still Image Data Compression Standard*, Van Nostrand Reinhold, 1993.

[7] Rosenfeld, A., and A. Kak, *Digital Picture Processing*, 2nd ed., Vol. 1, Academic Press, p. 154.

[8] ISO DIS 10918-1/ITU-T T.81, *Digital Compression of Continuous-Tone Still Images, Part I: Requirements and Guidelines.*

[9] ITU-T Recommendation T.42 (1996), *Continuous-Tone Color Representation Method for Facsimile.*

[10] ITU-T Recommendation T.82 (1993), ISO/IEC 11544:1993, *Information Technology—Coded Representation of Picture and Audio Information—Progressive Bilevel Image Compression* (commonly referred to as the "JBIG standard").

4

Group 4 Facsimile

Long before adoption of the CCITT recommendation for Group 3 fax and the start of the Group 4 study by the CCITT, TR-29 and some other standards groups were working on concepts for a fax system that would send without modems on digital networks. Modems were expensive at the time. In 1977, when the first three-card sets of 9,600-bps modem cards were ordered for fax, they cost the manufacturers almost $1,000 per set. Although the interface to the public telephone network was analog, digital channels had already become part of the PSTN. Much earlier, the need for more voice circuits had led to development of analog carrier multiplex systems between central offices, providing 4 to 12 times as many voice channels. When the analog systems reached their limit, conversion to digital channels provided 24 voice channels.

Each digital voice channel was 64 Kbps full-duplex, giving the potential of sending high-speed digital fax on a voice channel, without modems, almost seven times as fast as the then new 9.6-Kbps high-speed modems. In addition to digital technology between central offices and on long distance circuits, the telephone companies had developed end-to-end digital voice technology and were planning to replace regular analog telephone service with ISDN service. The Japanese installed at least one fax network with direct interface of the fax machines to the 64-Kbps voice channels. Many large corporations had private data networks that were used intermittently for data traffic and would have free time for digital fax transmission.

Before the U.S. Bell System breakup, AT&T was dominant in the telephone business. With the few other telephone companies, AT&T had planned for ISDN to be introduced in an orderly fashion, with good assurance that its schedule would be met. That plan had pushed the start of a program to develop the Group 4 digital fax standard. It was thought that the digital channels needed would be available and waiting to be used by the time the first Group 4 fax equipment was available. A further assumption was that, in concert with the changeover of networks, Group 4 would begin to replace Group 3. After the breakup, however, with many new telephone providers involved, the PSTN switch to ISDN became much more complex, and its completion targets were not met. New targets were set, then missed many times. Subsequent estimates projected ISDN completion somewhere around 1994.

The planned market for Group 4 was quite different from the person-to-person communication that Group 3 covers so well. Group 4 was designed for computer-controlled network communication with multiaddress, store-and-forward, and electronic mail (message) systems. Group 4 was expected to usher in a new era in the facsimile business when ISDN became universally available. Some of the new capabilities would be:

- Acting as an input/output (I/O) device for remote computers;
- Accessing the public data network for the exchange of business mail;
- Printing documents stored in a computer on a remote Group 4 fax unit;
- Entering graphic data into a remote computer by a Group 4 scanner, where the operator would have the ability to add textual information or change the graphic image;
- Sending fax documents to a computer for storage in a fax database that then could automatically send the documents to Group 4 fax units when requested;
- Faxing documents to a central computer with sufficient processing power for low-error optical character recognition (OCR) conversion to machine readable text.

In the same 1980 session of the CCITT that adopted Group 3 fax recommendations, a decision was made that caused apprehension for many facsimile engineers. Study Group 14, the facsimile group, was disbanded, and its work was transferred to Study Group 8, now called Characteristics of Telematic Systems. The new standard for teletex also was adopted at that session. Teletex was designed as a high-speed, high-technology replacement for Telex™. With teletex, transmission speed jumped to 2.4 Kbps, making standardized enhanced text communication available for equipment such as word processors. The higher speeds needed by fax of up to 64 Kbps (and possibly higher) were not needed by teletex. Facsimile was not taken seriously by teletex engineers because they considered it to be an inefficient way to send text, requiring 10 times as much data to be sent for the same message. Fax was a minor optional add-on feature to teletex that had no need for more than 240-dpi printing.

It was soon decided that the Group 4 recommendations would utilize the Open System Interconnection (OSI) seven-layer model for digital networks. The protocols were expected to enable worldwide linkage of computers and terminals from different vendors, electronic-mail systems, Group 4 fax units, and other application processes. That would result in competitive products and lower prices, giving users a choice of vendors and a variety of communication equipment, with manufacturers and vendors having improved productivity and a broader market base. Communication between Group 4 fax units or with other devices (application processes) was broken down into seven subtasks called layers. Each layer groups certain related aspects of data communication (Figure 4.1).

A standard physical digital interface provides connection to the packet and circuit services of public data networks and the seven-layer model for Group 4 fax. Each layer can be thought of as an envelope with the written information on the outside indicating where and how to deliver the contents. The actual data transmitted are in the innermost envelope. The seven envelopes used at the transmitting end are stripped off one at a time for each layer at the receiving end. The control information on the outside of each envelope must be sent in addition to the data, adding to the total number of bits sent. Figure 4.2 shows how Group 4 fax would interface with other Telematic services.

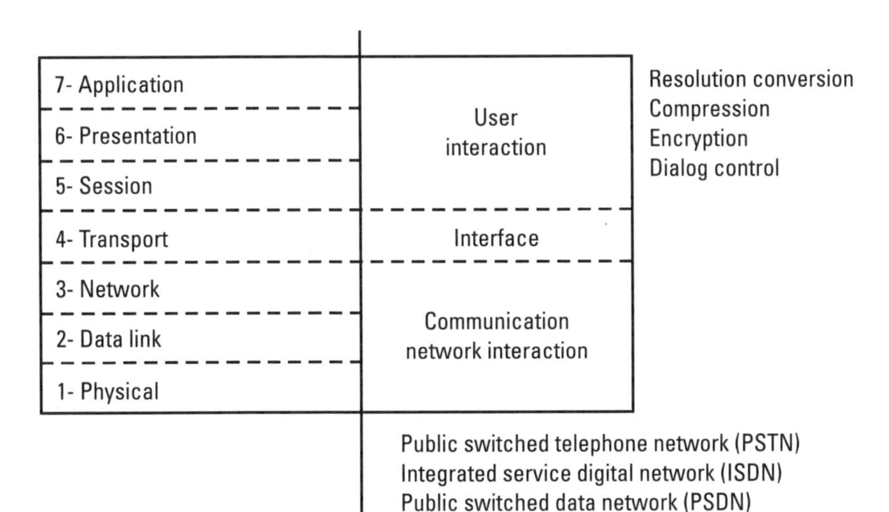

Figure 4.1 The OSI model.

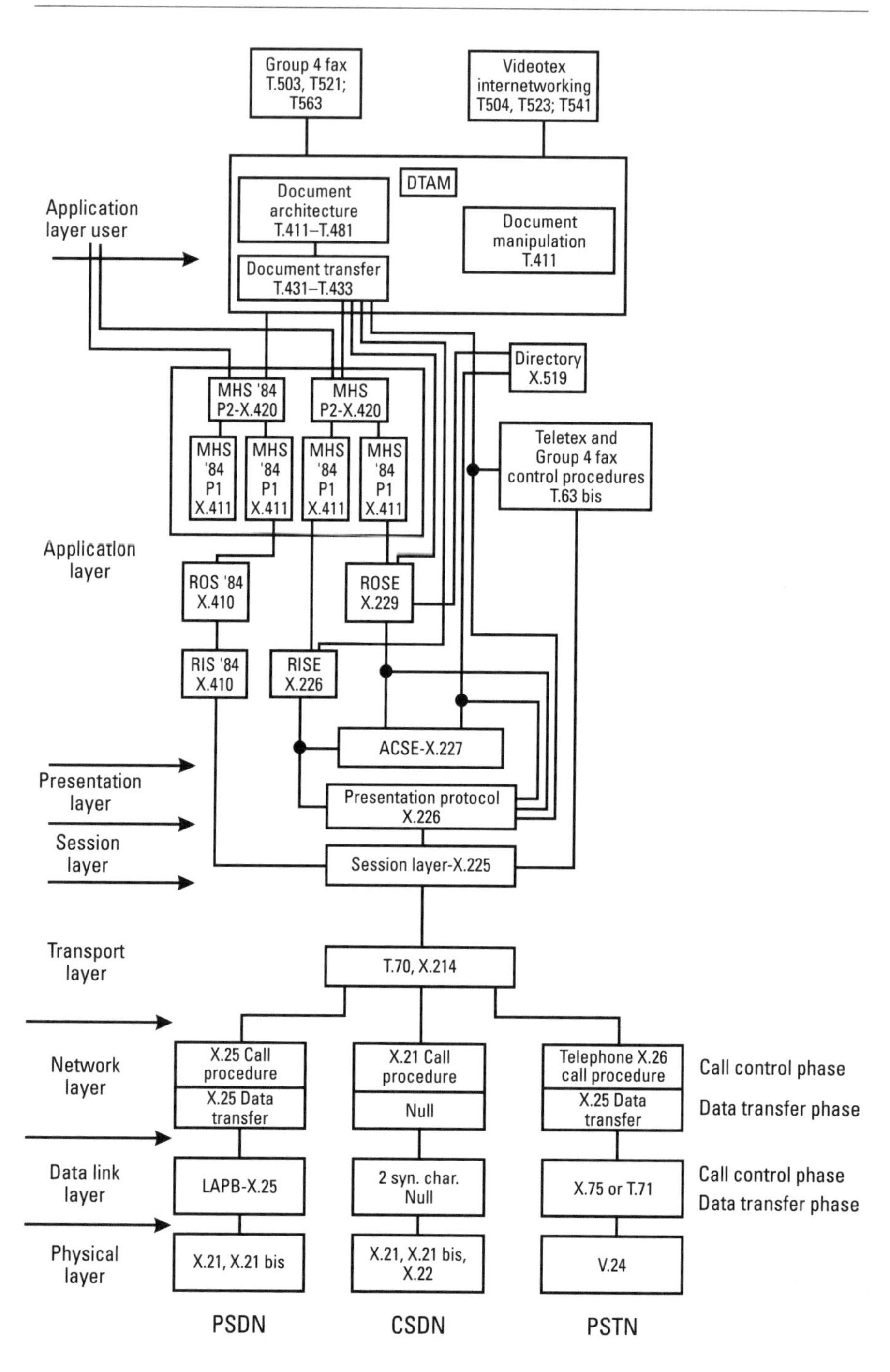

Figure 4.2 Group 4 facsimile recommendations in the OSI model.

The Group 4 recommendations were designed for digital networks before the OSI seven-layer model was completed, however, so Group 4 was not completely compatible with them.

Another Study Group 8 decision was that it would be necessary for Group 4 fax to send messages to and receive messages from teletex. This new mission had great impact on the work already underway. Much of the facsimile work effort switched to methods of making the two types of equipment compatible. Fax supporters knew most Group 4 users would not be willing to pay to add teletex, and teletex supporters knew their users would not be willing to pay to add fax. Eventually, a compromise was worked out that set up three classes of Group 4 fax machines, as defined by Recommendations T.6, T.503, and T.521.

Class 1 fax units would send and receive facsimile documents at 200 pels/inch. It was the lowest cost Group 4 because it did not need to interwork with teletex. Early Group 4 machines were permitted to use 4 pels/mm (196 pels/inch) instead of 200 pels/inch, allowing initial use of scan and print heads already in production for Group 3 fax. The CCITT later made communication with Group 3 units possible by an optional dual-mode Group 4 Class 1 fax. Dual-mode Group 4 fax could use a standard Group 3 fax modem, G3 C (64-Kbps capability), or both. The CCITT decided that the slight page-size distortions caused when interworking between Group 3 fax metric-based pels per millimeter and similar Group 4 pels per inch usually could be ignored.

Class 2 fax units would scan, communicate, and print facsimile-coded documents at 300 pels/inch. Conversion capability to 200 pels/inch for sending and receiving was required for compatibility with Class 1 units. Class 2 was further required to receive teletex and mixed-mode documents. To take advantage of the higher efficiency of character coding used by teletex, the mixed-mode provision would allow communication for text portions of a page in teletex mode and with the graphics sent in fax mode. For sending a logo or a signature, a unique escape code character received by the fax machine would be converted into a predetermined graphics representation. No provision was made for teletex machines to receive graphics sent from Group 4 fax machines.

Class 3 fax units would have Class 2 capabilities plus a keyboard to generate and send teletex and mixed-mode documents.

Any class Group 4 unit could use 200, 300, or 400 pels/inch. Most Class 1 fax units support 200 and 400 pels/inch by using a 400 pels/inch print head. There was no need for Group 4 to use the teletex 240 dpi, because the American Standard Code for Information Interchange (ASCII)–like characters received would be converted to one used by Group 4. Although there was much discussion of a provision for operation of Group 4 on the PSTN, a modem to make that possible was not selected.

Desirable features initially intended only for Group 4 were added to Group 3 recommendations, where they were invoked with less complexity (see Chapter 3). Group 4 Class 2 and Class 3 fax machines were never built, and in 1996, the CCITT dropped the teletex recommendations for Group 4 fax, eliminating both Class 2 and Class 3. The CCITT (now ITU-T) adopted the first versions of the Group 4 recommendations in 1984, but work on Group 4 recommendations now has almost stopped. That puts the future of Group 4 in doubt. In the total world population of fax machines, Group 4 comprises less than 0.05%, mainly in Japan, with that percentage getting lower each year. Few, if any, teletex machines exist in most countries.

4.1 The Group 4 Standard

In performance, Groups 3 and 4 are very similar. Both use the same techniques to process the picture signal before converting it to digital. The analog video output from the scanner is sampled and converted to a binary digital (black-white) signal. That is done by either a simple fixed-level slicing circuit or one that dynamically senses the background level. The Group 4 facsimile scanner and printer must be capable of storing at least one page of compressed data, but even low-cost Group 3 machines store a few pages, and some store 60 pages or more. Unlike Group 4, early Group 3 fax units did not include a DTE/DCE interface because the modem was built in. Group 4 uses full HDLC frame structure instead of an abbreviated version adapted for Group 3. Each Group 4 terminal has its own unique identification.

The initial image coding scheme developed for Group 4 fax, ITU-T Recommendation T.6, is a higher efficiency version of T.4, the two-dimensional MMR scheme of Group 3 fax, and uses the same encoding tables. T.6 use is limited to error-free channels, but it is also used for G3 fax with ECM. During the 1994–1997 time period, JBIG developed additional coding schemes for use by both Group 4 and Group 3 fax. T.85 is for bilevel (black-white) image compression. T.83 and T.84 are for continuous tone (gray scale) image compression. In addition, a continuous tone color presentation coding scheme developed by JPEG was approved by the ITU-T as Recommendation T.42 in 1996. See Chapter 3 for a description of these coding modes.

Group 4 page sizes and dots per inch (dpi) are given in Table 4.1. For point of reference, the center line is perpendicular to the scan lines and includes the point determined by the number of pels per line divided by 2. The raster point is the upper left corner of the ISO A4 page and occupies position (1,1). It

Table 4.1
Pel Transmission Densities

	ISO A4	North American	ISO B4	ISO A3	Japanese Legal	Japanese Letter
Resolution (pels/inch)						
200	1728/2339	1728/2200	2048/2780	2432/3307	2018/2866	1728/2024
240	2074/2806	2074/2640	2458/3335	2918/3969	2458/3439	2074/2428
300	2592/3508	2592/3300	3072/4169	3648/4961	3072/4299	2592/3035
400	3456/4677	3456/4400	4096/5559	4864/6614	4096/5732	3456/4047
Scan line length (mm) (P)	219.46	219.46	260.10	308.86	260.10	219.46
Paper width (mm) (Q)	210	215.9	250	297	257	182
P – Q	9.46	3.56	10.10	11.86	3.10	37.46
Nominal paper (mm)	297	297.4	353	420	364	257

is used as the starting point for character margins and positions. A guaranteed reproduction area of 196.6 by 277.23 mm applies for the ISO A4 page.

A more streamlined model based on ITU-T T.90 Internet protocol (IP) technologies (see Chapter 9) has overtaken the seven-layer OSI model as the network infrastructure of choice. Group 4 was designed to use the ISDN network, but some units operate over private data networks.

4.2 Group 4 Communication Application Profile (ITU-T T.503)

The bulk transfer (BT) communication application profile used by Group 4 facsimile is BT0 for direct document transfer. In addition, the following functional units are required: association use control capability, token control, exception report, and reliable transfer mode 1. Because the communication support function is a direct mapping to the session service, it means no presentation layer services are used. That is known as the *transparent mode.* It also means that all document transfer and manipulation (DATAM) services are mapped onto the application protocol data units (APDUs). From there, the APDU will be mapped onto the session services, and in turn the session services will be mapped onto the session protocol data unit (PDU).

4.3 Group 4 Document Application Profile (ITU-T T.521)

A typical document application profile for Group 4 facsimile is as follows:

- Document architecture level: formatted document;
- Content architecture level: formatted raster graphics;
- Document profile level: mandatory;
- Interchange format class: office document interchange format (ODIF) class B.

Group 4 facsimile does not require a logical structure because, according to the criteria for formatted document architecture, the only requirement is for a document profile and a specific layout structure. The specific layout structure has only two hierarchical levels, document layout root and page.

5

Image Components of the Fax System

This chapter discusses scanners and printers, as they relate to Group 3 and 4 fax machines, and PC-fax. Scanners and printers, which originally were developed as parts of fax machines, were later adapted as peripherals for those personal computers that had graphic capabilities. The scanning portion of the fax machine was combined with other electronics to enable the computer to view information on paper, and the fax printer provided the bit-map graphic method of recording computer-generated information on paper. New designs in scanners and printers benefited both computers and fax machines. Initially, office copiers formed an image on a photoreceptor surface before printing it, but they did not scan the image. Later, some high-end copiers employed scanning and printing techniques as used in fax equipment, ensuring uniform focus over the entire field and eliminating geometric distortions. These techniques also allowed digital signal processing to improve copy quality and image storage for collated printing, with separate bins for each copy of the document.

Today's fax machine has changed from an analog unit that needs high-precision mechanical scanning and printing techniques plus a high-precision frequency standard to a mass-produced specialized digital computer that incorporates a scanner and a printer with modem chips and fax chips. Once the analog signal representing a pixel of the page being sent is converted into digital format in the fax transmitter, it is handled seamlessly until it reaches the printing mechanism of the fax receiver. Transmission errors, if any, are automatically corrected by ECM. It is now possible to have a computer fax system without a scanner or a printer.

The ITU-approved resolution options (metric and inch-based) for Group 3 fax are listed in Chapter 3. Resolution is the pel density (not to be confused with density as used in optics and photography), the number of pels per inch, now often called dots per inch (dpi). For convenience, this chapter uses only the inch-based resolutions, but the comments apply equally to the almost identical metric resolutions. The slight geometric distortion in sending between metric and inch-based fax units is so small that the ITU does not consider it relevant. It also should be noted that resolution in this sense is not a measure of sharpness and detail of the received fax image, as in the field of optics. Higher resolution fax units are expected to give a sharper fax image, but design details make a substantial difference in the quality of the image produced with the scanning and printing portions of a fax machine. The difference is particularly noticeable at the older 100-by-100-dpi low-resolution of Groups 1 and 2 or photographic fax recordings of various fax picture services. Today's "standard" 100-by-200-dpi Group 3 resolution is better for text. If faxes are received from a PC-fax unit, that quality may be adequate for business office use. A fax at 200 by 200 dpi generated from a PC program provides a very acceptable received image sharpness. The fonts are designed for optimum bit-map printing on a fax receiver. The images generated do not skew or look ragged, as scanned images do. Lines on forms are of uniform width and do not stair step. Faxes sent from a fax machine, however, have a random match between text edges and pixel position caused by the physical scanning, making the fax copy difficult or tiring to read. "Fine" (200 by 200 dpi) transmission may be set as default to reduce the effect. Received-copy-quality expectations have increased since the Group 3 standard was set in 1980.

An alternative higher resolution of 300 or 400 dpi provides a sharper fax copy, more like a printed page and more suitable for OCR. The number of 300-dpi-capability fax scanners is growing, but some still have 200 dpi as the highest resolution. Some fax machines provide an alternative 200 dpi horizontal by "400" dpi vertical resolution by retaining the 200-dpi scanning spot size and halving the paper feed rate. This ITU-approved option for Group 3 fax gives little improvement over 200-by-200 dpi unless the scanning and printing spots are much smaller than 1/200 inch.

A fax system with a well-designed 300-dpi scanner and printer may produce better quality recorded copy than another system with 400 dpi. At resolutions of 300 dpi or higher, the differences are less noticeable, but they still affect the appearance of received copy, including the sharpness and readability of small fonts. Fax copies received from ITU test charts BW01, CT01, and 4CP01 will show the differences as a quantifiable measurement (see Chapter 10).

5.1 Scanners for Fax

A fax scanner converts the markings on a sheet of paper into electrical signals in digital format for transmission to a remote point for viewing, recording, storage, or processing. It may be part of a fax machine or a computer peripheral for PC-fax. The original scanner in 1843 was a stylus wire touching a raised metal type to complete an electrical circuit as it moved across. Around 1900, optical scanning was first used for fax and soon became the dominant method (see Chapter 2). Today's scanner may have one or more photosensors, producing an analog signal with a maximum amplitude pulse from a white background if there is no black marking in the square being viewed. The pulse amplitude is minimum if the square contains all black marking. For partially white squares, the pulse amplitude is proportional to the percentage of white. Gray pixels have intermediate amplitudes. A 1-bit A/D converter changes the pulses into a digital (black-white) signal. Pulses above the threshold level are sent as full white pels, and pulses below the threshold are sent as full black pels.

Much of business information sent by fax is printed material or drawings consisting of black, gray, or color markings on a page with a background of white, gray, or light color. To provide better contrast of the received image and to prevent gray or colored backgrounds from printing as black, fax machines use an automatic background control (ABC) circuit that makes the off-white backgrounds seem white to the fax transmitter. The ABC action allows the just-below-the-threshold markings of gray to show as black in the received image, while the same density gray as a background prints as white. If the background of the page being sent is gray or colored instead of full white, gain of the ABC increases until its output from the weaker signal is the same voltage it would be for a white background. The A/D converter then receives the background pulses at the same amplitude regardless of the background color. This compensation usually works on backgrounds over a range from white color to a gray of about 0.30D (optical density). If a white page is faxed, gray markings are sent the same as black markings. If down the page, the background density becomes the same as markings in the top part, a white background is sent and darker gray markings are sent as black on white. With colored backgrounds, it is difficult to predict just how dark the color may be before the signal sent switches to black and obliterates all the information. If in doubt, a test in copy mode will indicate the results. The ABC action works best when the background color is uniform across the full document width. When the background of the page being sent changes from white to gray, the ABC gain increases slowly, and a few scanning lines may print as black. If the ABC gain were to increase rapidly, information would be lost from low-contrast documents.

When the background changes from gray to white, the ABC gain decreases rapidly to ensure that light markings are not wiped out by being above the threshold. The combination of time constants for the change in ABC gain usually preserves markings being sent from low-contrast areas in documents, allowing most business documents to be sent without manual adjustment.

The scanner plays a critical role in a fax system because it alone sees all the original data on the page being sent. By the time the data have been converted into the binary digital format needed for Group 3 or 4 fax transmission, much information has been discarded. Signal processing after that point can only guess at what might make the received image more useful, while signal processing before digitizing can be based on real information. If the scanner sends a poor-quality image, the receiving point will still have a poor-quality image, even after signal processing.

For sending photographs, the ABC can use a white border for background color in gray-scale mode. Signal processing is then used to convert the gray shadings into clusters of black dots, close together for simulation of dark gray areas or far apart for light gray simulation. That is similar to the screening process for printing photographs (see Section 3.4). Before the analog signal from the photosensor(s) is digitized, a number of signal processing corrections may be made to improve the received fax copy. There is a limit to how small a contrast can be recognized by the human eye. Information in low-contrast originals not visible to the eye can be revealed by expanding the contrast of the lightest gray to white and the darkest gray to black. Some information not visible in blurred originals can be revealed by using software with sharpening algorithms that increase the contrast with nearby pixels.

Self-contained scanners and associated software available for use with PCs offer more signal processing capabilities than usually are available in fax machines. These scanners can be used with PC-fax, with OCR software, or for entering imagery from hard copy into computer memory for desktop publishing and other programs. When scanners first became available for use with PCs, most were based on fax standards with 203 dpi. Now, most scanners for PCs offer higher optical resolutions of 300, 400, 600 dpi. Interpolated resolutions listed are often two to four times higher, up to 2,400 dpi or more. Interpolation reduces the staircase effect on the edge of a line by making each scanned step into two, three, or four smaller steps. That improves the appearance if the printer or the viewing screen has enough resolution, but it does not add new information. It makes no sense to send a fax at a higher resolution than either the fax printer has or the resolution that will be used in viewing on a computer screen. Many computers are set up to view either 640 or 800 pixels across the screen instead of the 1,728 sent for Group 3 fax. Zooming the received image and viewing only a portion of it on the screen can be used to reveal the full resolution of the image.

Scanners connected to PCs may first store the image with 64, 128, or 256 gray levels. The black-white threshold for two-level transmission then can be manually adjusted for desired results on the PC screen before faxing. While no adjustment may be needed for most black-white documents, for those with a variable background density or variable marking gray levels, the optimum threshold setting may be critical. For sending photographs by Group 3 or 4 fax using an electronic screening process, manual adjustment of contrast or background may achieve substantially better copy quality than a default setting. Image-sharpening PC programs also can improve the image. Some programs employ unsharp masking techniques with many options for increasing the contrast between nearby pixels. These techniques improve edge sharpness or make visible information that otherwise would not be seen. Automatic default settings are available, but careful manual adjustments often give better results. The problem with manual adjustments is that the combinations are almost limitless, and it takes an expert to avoid squandering time to achieve optimum settings.

5.1.1 Drum Scanner

Few fax machines in use today use drum scanners, but they still are used to scan documents into PCs for applications such as prepress printing, where superior image quality is important. Images stored from drum scanners are also useful in checking the performance limits of fax printers. The page being scanned is mounted on a drum, and a lens forms an image of the page in a manner similar to that of a camera. Instead of forming the image on film in the back of a camera, the objective lens images the page (the document) on the aperture plate, an opaque panel with a small hole (the aperture) in the center (Figure 5.1).

Behind the aperture, a phototube or other photosensor converts the light passing through into an electrical signal. The signal represents the brightness of the very small spot on the page directly in front of the optical system. To get a strong signal, this spot is brightly illuminated by the exciter lamp, whose filament is focused there. As the drum rotates, the image sweeps across the aperture. The scanning spot traces a fine helical line of scanning dots on the page, converting brightness information into a signal, one small spot at a time. The image of the page is in focus at the aperture, and it does not matter that much of the page is not in focus. Assume that the image moves over 0.005 inch for each drum rotation, tracing a helix about 8.5 by 200 by 11 inches (18,700 inches total) to read sequentially every spot on the page. The spot should be called a *pixel* when it contains gray-scale information and a *pel* when it has only binary black-white information, but the terms are sometimes

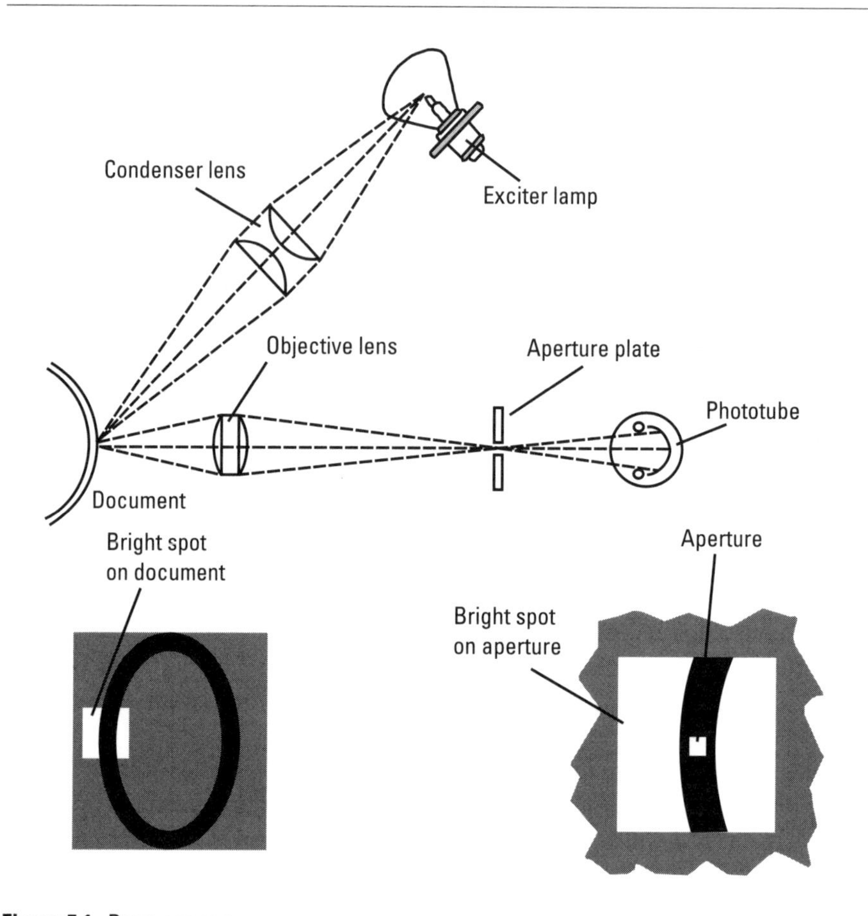

Figure 5.1 Drum scanner.

used interchangeably. Although drum scanners are no longer used by office fax machines for professional quality prepress scanning, the graphic arts trade still uses a well-designed drum scanner, which improves details in the highlights and shadows, increases overall sharpness, eliminates ghosting and most fogging between adjacent high contrast pixels, and has no geometric distortion.

5.1.2 Group 3 and 4 Scanners

Group 3 and 4 fax machines do not use drum scanners or recorders. Instead of a single scanning spot, they use a stationary strip (array) of 1,728 or more very small photosensors, one for each pixel across the page, with the whole width of the page image formed directly on the photosensor strip. The photosensor for the left edge of the page is read first. At each tick of a clock the adjacent photosensor is read, electronically scanning a line across the page instead of mechanically moving an image across one photosensor. The scanner reads sequentially as one scanning line, all of the 1,728 pels across the width of the page being sent. A stepper motor then moves the page down, and the next strip below is read. The whole page can be read in a few seconds. Wider sensor arrays of 2,048 or 2,432 pixels are used for the alternative wider page Group 3 fax machines. Higher resolutions with 300 or 400 pixels/inch use proportionally more photosensors.

5.1.2.1 CCD Chip Scanner

This scanning method with an optical system resembling a camera was used almost exclusively from the time it was invented, almost a hundred years ago, up through the first Group 3 and Group 4 fax machines. A single lens forms an image on a small CCD chip of 1,728 photosensors located where the 35-mm film would be in a camera (Figure 5.2).

As with a camera copying a whole page, the lens cannot be very close to the page. At least 30 cm (12 inches) is required between the scanning line and the CCD chip. That controls the minimum size of the fax machine, but a mirror can be added to fold the optical path to reduce the distance. The 8.5-inch

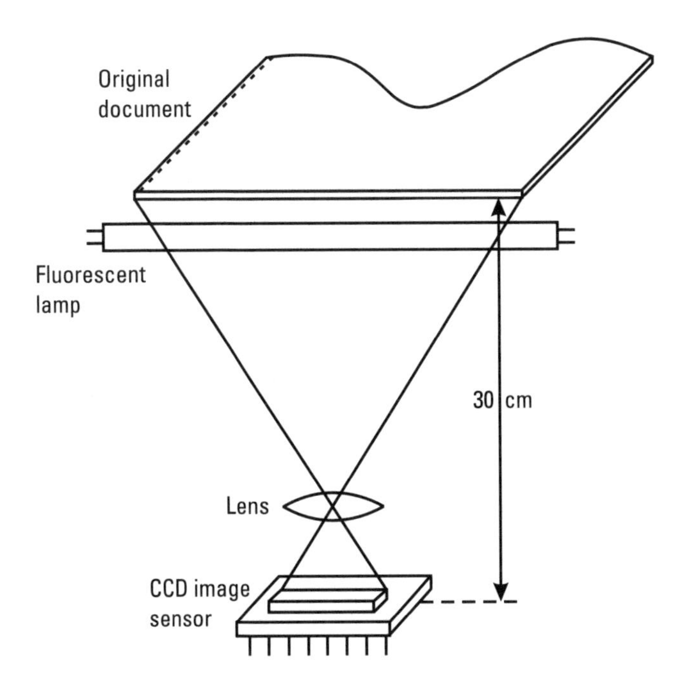

Figure 5.2 Camera type of CCD scanner.

page width matches the width of the 10:1 reduced-size image on the strip of 1,728 photosensors. The scanning line position on the page is determined by the part of the image falling onto the photosensitive strip of the chip. These sensors see only a thin 0.005-inch-high line across the page where the scanning line is illuminated as the paper moves past.

Each photosensor of the 1,728-element strip across the chip can be considered a light-sensitive resistor connected in parallel with a small capacitor. The capacitor starts with a full voltage charge. If the image spot is white, the photosensor has a low resistance, which drains the voltage off the capacitor during an exposure time. If the image spot is black, the photosensor has a high resistance and the capacitor is not discharged. A second 1,728-element strip near the photosensor strip is the CCD itself, an analog shift register (sometimes called a "bucket brigade"). This strip is divided with voltage gradient lines, forming wells matching the photosensor elements. Between the strips is a strip of transfer gates (switches). The gates are open during the exposure time for each scanning line. Then the gates all close momentarily. The current in-rush to recharge the capacitors forms charge packets that represent each photosensor brightness in the CCD wells. A network of electrodes that form these wells is connected to different phases of the clock that controls the electronic scan, moving the charge packets in series to the output terminal. Each packet voltage represents the brightness for a particular photosensor, starting at the left end of the scanning line. Exposure of the image on the photosensor strip continues for the time between scanning lines, providing good sensitivity by allowing the longest time possible for the voltage to leak off. After the 1,728th pixel is sampled, a stepper motor moves the page down 0.005 or 0.01 inch, and the process continues for the next scanning line. The photosensor capacitors are again charged and ready to accumulate a new line of data. The previous line in the analog shift register is transferred serially on clock pulses to the sensor's output, where each successive charge packet is an analog voltage. The output frequency may reach 20 MHz or more.

Decrease in output signal amplitude caused by falloff in brightness near the edge of the page can be partially compensated for by masking the light for the center of the line. Variations in sensitivity of the individual photosensor elements cause irregularities in the signal from the scan head. Without correction, vertical streaks would show in the received fax copy when a photograph is sent, and there would even be a small effect on the quality of text. An electronic shading circuit corrects each pixel for a uniform signal amplitude when a white strip is scanned. Digital memory for each pixel controls the correction. Such correction is important for scanners that generate 256 shades of gray scale. The analog picture signal is then changed to digital by a 1-bit per sample A/D converter. Software or a front panel control may allow the operator to change the

A/D threshold voltage, compensating for faded text or light markings on the page being sent. The output of the A/D converter is a digital signal, with each input pixel converted to a 1-bit pel (black or white). Binary digital processing is necessary for efficient coding of the picture signal. The black-white effect on the page being sent is similar to that produced by an office copier.

5.1.2.2 Contact Image Sensor Scanner

A major size reduction of Group 3 fax machines is an advantage when a contact image sensor (CIS) scanner is used instead of a camera type of scanner. Figure 5.3 shows how the compact CIS scan head mounts in a fax machine with an automatic document feeder.

All the scanner elements are contained in the CIS 8.5-inch-wide bar mounted to contact the page being scanned as it moves past. CIS scanning uses a strip of photosensors across the 8.5-inch page instead of a 0.8-inch line of photosensors across a CCD chip. The scan head contains lamps for illumination of the scan line, fiber optics, the sensor array, and video signal processing circuitry (Figure 5.4).

In a CIS scanner, a narrow 8.5-inch-wide fiber optics imaging bar (gradient–index rod–lens array) is mounted between the page being scanned and the photosensors. A narrow 8.5-inch strip across the document is focused with a 1:1 magnification onto the image sensor strip above it. Illumination is provided by one or two rows of light-emitting diodes (LED) with built-in lenses. Each photosensor element is the same size as the pixel it sees. The photosensor array can be made from silicon or cadmium sulfide–cadmium selenide. The discussion of the operation and circuitry given for the CCD chip scanner (see Section 5.1.2.1) also applies to CIS scanners. CIS scanners have the advantage of maintaining the exact dot position across the page and having the same sharpness of focus on the end of a line as in the center of the page being sent. In an alternative arrangement, a moving-scan-head design is commonly used in

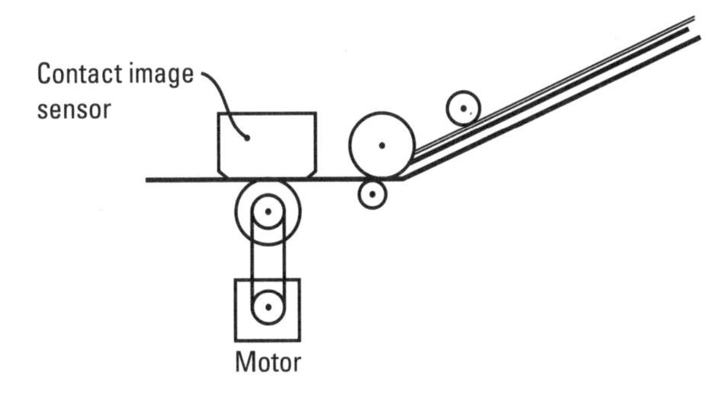

Figure 5.3 CIS scanner.

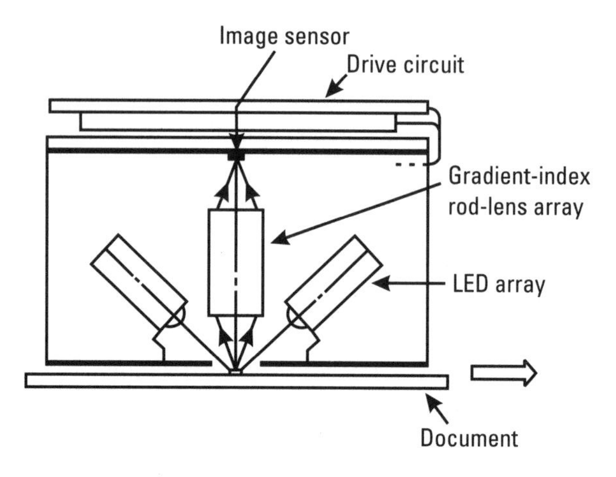

Figure 5.4 Contact image scan head.

office copiers. Only 5 cm (2 inches) is required between the sensor array and the page being scanned. The light source may be a fluorescent lamp mounted near the page being scanned. The page is held face down on a glass plate and the scan head reads through the glass as it moves down the page.

5.2 Printers for Fax

The printer, whether as a part of a Group 3 or Group 4 fax receiver or as a PC-fax computer peripheral, plays an important role in the quality of the output copy produced by a facsimile system. Both PC-fax and fax machines receive signals from a communication channel and convert them into black-white digital-format signals suitable for the printer. Pages sent from a fax scanner of poor design can be somewhat improved by signal processing at the fax receiver; however, signal processing before conversion to a binary signal at the fax transmitter can be far more effective. If the received image is a photograph, the signals are already in a format that will form simulated gray tones by clusters of black dots in an electronic screening process (see Section 3.4).

Since the last edition of this book, many new low-cost, plain-paper, nonimpact printers have become available for PC-fax. Some of these printers are part of standalone fax units, and some are part of a multipurpose machine—fax, printer, office copier, and scanner. The plain-paper recording methods described next have challenged thermal paper recording. They produce better paper handling and life of the recording. The copy looks and feels like a regular business document and does not have to be duplicated on an office copier before use or for permanent storage. The number of plain-paper Group 3 fax machine models is growing rapidly now that their costs are much lower.

5.2.1 Thermal Paper Recording

When thermal paper recording was introduced, it was a big improvement over earlier fax papers. Wet electrolytic recording paper, used since the first fax, had an unpleasant odor and produced low-contrast recording on paper that wrinkled as it dried, and sometimes the fumes discolored or removed paint from walls of the fax room! The alternative, dry burnoff recording paper, also had an odor and produced a black dust.

For many years, thermal recording directly onto heat-sensitive paper has been the low-cost method used by most fax receivers. The print head has a comb-shaped array of wires with a row of very small resistor-element spots across the recording paper width: 1,728 for 8.5 inches, 2,048 for 10 inches, and 2,432 for 11.9 inches (Figure 5.5).

The thermal recording paper touches the row of resistor spots (heating element). A pulse of current through a resistor heating element causes it to become hot enough to mark the paper in a spot about 0.005 inch in diameter. The spot temperature must be changed from nonmarking temperature to marking temperature and returned to nonmarking before the paper steps to the next recording line. The recording cycle may be 2.5 to 5 ms. In one design, 24V at a power level of 0.5W is applied to a marking resistor for about 0.6 ms. Marking temperature is about 200°F. Other head designs use 6V or 12V. Either thin-film or thick-film technology is used to make the print heads. Thin-film heads use a lower recording power and can be made with higher resolution. The head wires and integrated circuit elements are made in one compact head assembly. Thick-film print heads have enough resolution for Group 3 fax units plus good immunity to scratches and abrasive wear.

Direct thermal recording has several advantages:

- The fax equipment and the recording paper costs are low.
- The fax equipment size is small. A small-volume print head design with the print driver electronics in an integral package with the print head wires makes this possible.
- Print heads normally last the life of the fax machine without service calls or user maintenance.

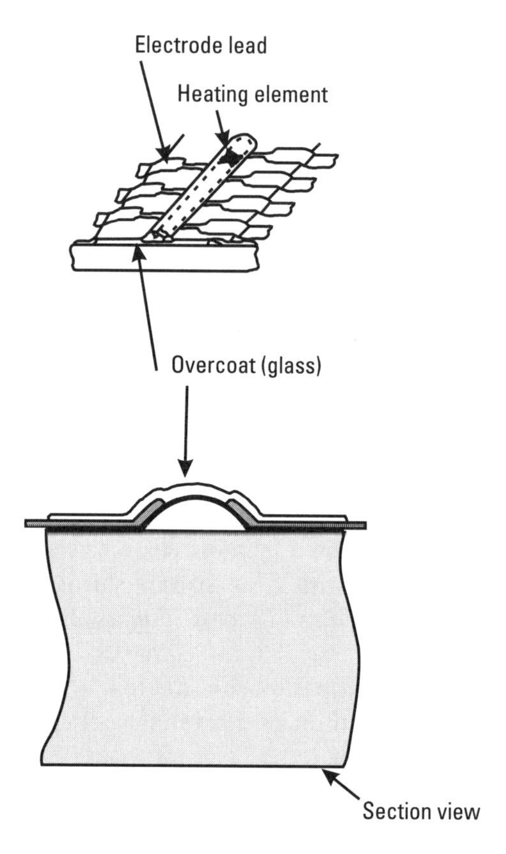

Figure 5.5 Thermal print head.

It also has several disadvantages:

- The contrast of the recording deteriorates rapidly if it is exposed to light or heat. (For archival records, an office copier print will circumvent that problem.)
- The paper is difficult to write on and does not look or handle like regular white bond paper.
- The replacement cost of the thermal head may be up to half the original price of the fax machine. The thermal print head is one of the most sensitive components of a fax machine. For that reason, the manufacturer's recommendations should be followed on which thermal recording papers will give good long-term results. To produce good copy quality, the sensitivity and paper thickness must match the printer requirements. Abrasive materials may cause premature failure due to

wear, and corrosive agents in the paper may be deposited on the print head.

- Although the following problem does not occur in most Group 3 fax machines, it may happen when a very small thermal fax machine is used for high-volume applications. The thermal printing process uses low power while reproducing white areas. However, when recording black all the way across the page, the print head alone may require 100W or more to heat all wires at once. Under normal circumstances, that happens intermittently, and the thermal mass of the print head assembly prevents overheating. Printing successive pages that contain large amounts of black area might overheat the print head, causing warping or burnout. To protect against that, a thermal sensor causes the print power to turn off until the head cools. A thermal fuse can be used as backup, in case the thermal sensor logic fails to perform properly. The design of the print head and its drivers determines how much black can be printed before the power shuts off.

5.2.2 Thermal-Transfer Recording

Thermal-transfer fax machines write on white paper with a print head basically the same as the one used in thermal fax recorders. Instead of hot spots of the recording head marking directly on heat sensitive paper, these spots heat a transfer film (similar to carbon paper) overlaying white paper. In early recorders, a roll of white paper and a matching roll of black thermal-transfer film (often called ribbon) were used, with the film placed between the thermal head and the white paper (Figure 5.6).

From the thermal-transfer film supply roll, black film moves with white paper from the plain paper supply roll past the recording head at the same rate, using the same quantities. The marking is made on white paper from a black pigment–impregnated wax coating (ink). Instead of impact transfer with carbon paper, heat from the hot spots across the recording line melts solid black ink on the film, causing black marks to be made on the white paper. The ink transfers to the white paper, leaving the transfer film with a negative record of all the pages recorded. The thermal transfer takeup roll of black film can be retained for backup purposes, but it is messy to handle. Caution is needed when the film roll is discarded if it contains confidential information. Most of the newer machines use a stack of cut-sheet white bond paper instead of a roll of paper.

Thermal-transfer recordings are much sharper than carbon copies. Recorded copy handles much the same as pages from an office copier, not like

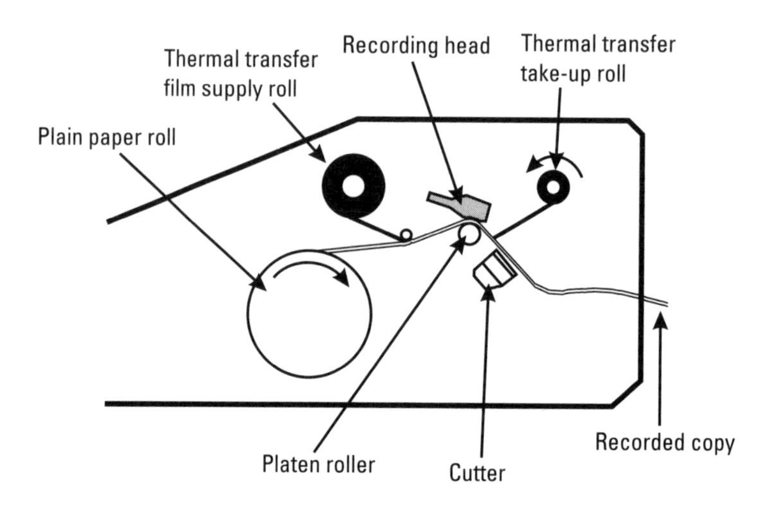

Figure 5.6 Thermal-transfer fax recording.

the slippery light gray thermal paper. Storage is not restricted by light or heat conditions, as is the case with thermal paper. Compared with direct thermal recording fax machines, the machine cost is higher and the size is larger. The cost of thermal-transfer recording materials varies from somewhat higher to about three times that of direct thermal recording. Due to the mechanical takeup reel, the overall reliability of the thermal-transfer fax machine may be less. Thermal-transfer recorders tend to be noisy when the black film that is stuck to the white paper at each recording dot peels off the paper as the film separates from it.

5.2.3 Laser Recording

Laser xerographic printers operating at 300 and 600 dpi are well established for office use in word processing, desktop publishing, and PC-fax systems. Similar laser printers, adapted to work at 400 dpi, are used as the recorders in almost all Group 4 fax machines. The quality of recorded copy is superior to that of thermal or thermal-transfer types; however, the laser printer's larger size and higher price have slowed it down from making substantial inroads into the Group 3 fax market. Recent lower printer prices and a printing cost per page of about 2 cents bode well for its future.

For recording, a small-diameter beam from a semiconductor laser is swept across a photosensitive semiconductor drum by a rotating polygon mirror, making one recording line per mirror face (Figure 5.7).

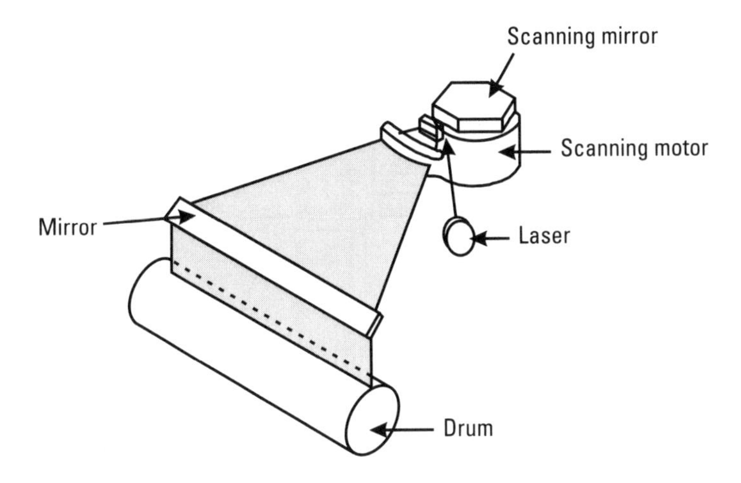

Figure 5.7 Laser fax xerographic recorder.

An imaging lens system keeps the spot in focus. The drum rotates the height of one recording line for each mirror-face sweep. Some designs use a photosensitive belt instead of a drum.

5.2.4 LED Recording

An LED contact-type linear array is used in some Group 3 fax recorders instead of a laser and rotating-mirror system. This simplifies the recorder and may increase its reliability by eliminating the rotating mirror (Figure 5.8).

The print head LEDs (300 or 400 dpi) are focused onto the photosensitive drum with a fiber optics rod–lens array. Before exposure to the LEDs, the drum has been given an electrostatic charge by a high-voltage corona unit. The charged area of the drum is enclosed in a light-tight chamber and is discharged at those points along the line exposed to light from the LEDs. The charge-pattern image on the drum corresponds to the black marks on the page being sent by the fax transmitter. As the image passes the development box, black toner powder is picked up to reproduce a mirror-image copy of the original. A sheet of paper is fed onto the drum surface to start in step with the page image. As the drum rotates, the sheet passes the transfer-corona station, and the toner image transfers from the drum to the paper. The image is heat or pressure fixed to the sheet to form a permanent copy of the page sent. Recording widths of 8.5 inches and wider are available. The focus and the spot size of the image are the same for all points (Figure 5.9).

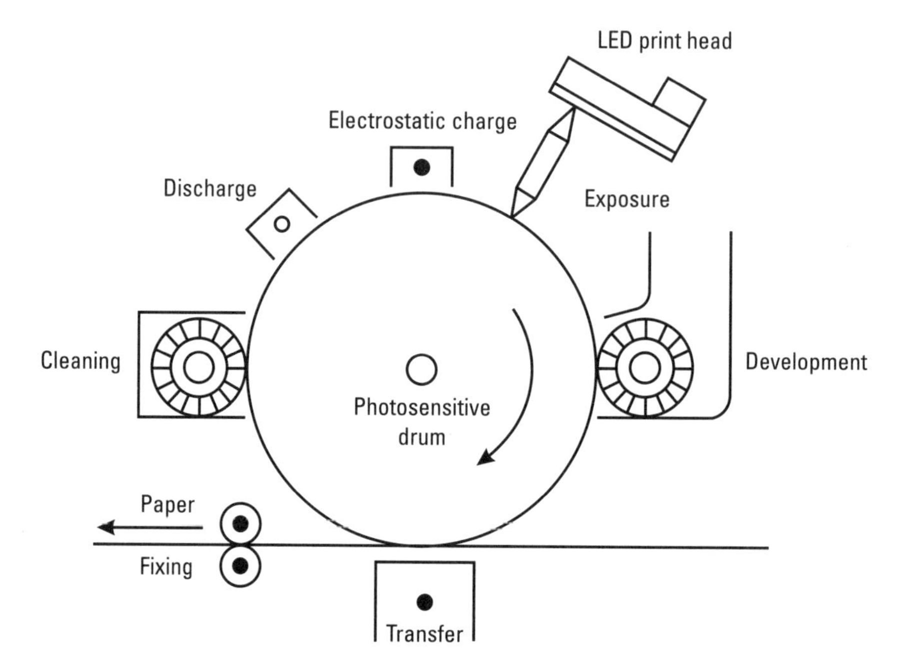

Figure 5.8 LED linear-array recorder.

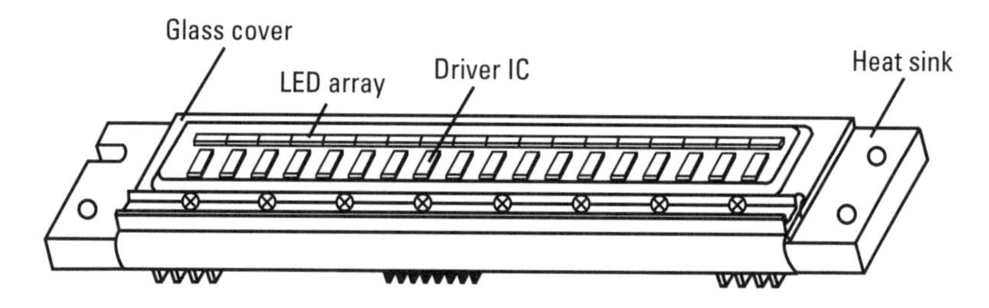

Figure 5.9 LED linear-array print head.

5.2.5 Ink-Jet Recording

Ink-jet recording is now rapidly replacing thermal recording for fax except in the low-cost retail channels. Ink-jet for fax recording began long before Group 1 fax but was dropped mainly because of messy cleaning needed to prevent clogging and handling the ink for refilling. Again, before the Group 2 standard was in place, a few manufacturers did put ink-jet fax machines on the market, but they never became popular and were withdrawn from sales.

One of the prime reasons for the current success of ink-jet recording is the incorporation of a replaceable cartridge that contains both the ink and the print jets. The cartridge was developed to produce a good computer printer at a lower cost than a laser printer. In the beginning, the price was still much higher than for thermal printers, and ink jet was not adopted for fax recording except by some PC-fax users. After the 300-dpi and 600-dpi ink-jet printers became popular with PC users, they became one of the lowest cost printers. The received facsimiles are sharp, and the plain paper avoids the curl and handling problems of thermal recording still used by most low-cost fax machines. The recording sharpness results in fax copy almost as good as that produced on a laser printer. The plain paper recording allows marking on the received copy with pen, pencil (erasable), or felt-tip marker. Ink fading is not a problem with documents, whether they are left out in the open or filed. The recording power required is low compared to laser, LED, and thermal recording. With the adoption of this type of printer for fax, a swing from thermal to ink jet is occurring, at least for low- to medium-volume users. The ability to record either metric-based or inch-based resolution on the same printer gives ink jet an advantage. The 300 dpi matches both Group 3 and Group 4. An ink-jet recording fax machine may record either 8 lines/mm, matching the original Group 3 standard, or 200 dpi, allowing the same fax machine to record both without size distortion. The major disadvantage is the cost of the replaceable ink-jet heads, bringing the cost per page to 10 or 15 cents.

6

Facsimile Digital Interfaces

6.1 Asynchronous Facsimile DCE/DTE Control Standards

The purpose of the asynchronous facsimile DCE/DTE control standards is to define a digital interface between data terminal equipment (DTE), such as a PC, and Group 3 fax standards, allowing the DTE to access a Group 3 fax machine or other fax data communication equipment (DCE) for the functions of scanning, printing, and communication on the PSTN. Other DCE includes PC-fax boards that plug into a PC and external fax adapters. Some fax boards have been furnished with the capability to use the high-speed Group 3 fax modems and ECM protocols for sending binary data between computers. Different vendors, however, used different interface techniques, making them incompatible with the PC software. Each fax modem vendor needed to develop its own applications software—a large job. The purpose of these standards is to define a common command language so vendors who write software for PCs can be sure their products will be compatible with multiple hardware vendors.

6.1.1 Background

A modem is a piece of DCE that allows DTE, such as a PC, to send digital signals over regular analog telephone channels (PSTN). The interface connection between the PC (DTE) and the modem (DCE) was standardized many years ago by the EIA as RS232 (EIA-232-F is the latest revision, released in September 1997) and by the CCITT as V.24, allowing PC manufacturers and modem manufacturers to build equipment that works together without having to customize the interface to match special characteristics between the devices. The connectors, circuit leads furnished, and electrical characteristics for each lead are specified. A shielded EIA-232 cable plugs into the RS232 port on the computer and into the EIA-232 port of the modem (Figure 6.1). EIA-574 describes the alternative 9-pin data connector commonly used on PCs. The cable carries a serial asynchronous bidirectional data stream. Line drivers are required at each end of the cable to allow the devices being connected to be up to 50 ft apart (special cables allow greater distances). A power supply furnishing +12V and −12V normally is used for the line drivers. Modern serial interface standards use ±5V.

When Group 3 digital fax standards were being developed, the set of modems needed was not available as a single device. The $V.27_{ter}$ 4.8-Kbps CCITT modem standard for use on the PSTN had just been adopted. The V.21 modem was available, but Group 3 needed only the return channel, for 300-bps handshake signaling. The high-speed V.29 9.6-Kbps modem standard was being developed for use on four-wire conditioned private telephone lines. Fortunately, CCITT SG 16 and the U.S. modem manufacturers assisted, and modem card sets with all the wanted functions were available before the Group 3 fax standards were adopted. The modem card sets were incorporated inside the fax machine without line drivers or the other bells and whistles of a complete modem. Those modem functions not easily available through the EIA-232 interface leads of a standard modem simplified the machine design, but there were no standards for the interface between the fax machine and the modem functions. The cost to the fax machine manufacturers of the first modem card sets (including 9.6-Kbps speed) was about 10% of what a 9.6-Kbps complete modem sold for at the time. Without that fortuitous combination, Group 3 fax may have been too expensive to succeed, and the modem manufacturers may have lost their first large high-speed modem market.

The use of the PC for business communication increased greatly when the PC-fax board became available. The PC was then able to send and receive files to Group 3 fax machines. The technical name for the hardware covered in these standards is *fax adapter.* That term covers both fax boards and the fax box used in this book to distinguish among the types of fax adapters. The PC-fax board plugs into a vacant slot in the PC, directly interfacing the PC processor bus (Figure 6.2). An alternative arrangement uses an external box containing a fax modem set that plugs into the PC through the EIA-232 connector

Figure 6.1 PC communications interface.

Figure 6.2 PC communications with a fax board.

(Figure 6.3). The asynchronous facsimile control standards apply directly only to this arrangement. However, because a PC processor bus provides a character serial bidirectional data stream, PC-fax boards installed inside PCs are now using the same protocols, particularly if they also include V.22$_{bis}$ or V.32 serial modem services. For sending ASCII files as fax pages, the PC generates a bit-mapped file as would be needed by a graphics printer. A character generator assembles, in memory, a row of characters across the page. The serial bit stream next generated is similar to one from a fax machine scanner. An encoder provides MH digital code words from a lookup table. The code words are sent over the PSTN by the modem. The sequence occurs in reverse at the receiver. Handshake commands and responses are generated, and control functions are performed.

With PC-fax boards, ITU-T Recommendations T.4 and T.30 provided the standards for making PC files compatible with Group 3 standards and sending files to fax machines worldwide. There was no standard for how and which functions were provided by the computer, however, and each manufacturer had its own method of emulating Group 3 fax. Lack of standardization forced much customization of software. Both the software programs designed to run the PC-fax operation and the application programs with fax drivers were affected. The features provided by different PC-fax adapters varied widely, and adding any new feature was expensive. The U.S. manufacturers of PC-fax boards prevailed on the TIA/TR-29 Facsimile Systems and Equipment Engineering Committee to set up a subcommittee, TR-29.2, Digital Facsimile Interfaces, under the chairmanship of Joseph Decuir, to develop the standards needed.

Equipment was manufactured that uses the Class 1 interface (TIA/EIA-578-A), and the Class 2 interface (TIA/EIA-592-A). These standards define an interface with an asynchronous serial bidirectional data stream between the PC (DTE) and a fax adapter (DCE) that contains a fax modem

Figure 6.3 PC communications with a fax box.

(Figure 6.4) and other parts needed to emulate Group 3 fax operation on the PSTN. The data stream carries the Hayes AT command set commonly used for computer modems, with extensions for commands issued by the PC to configure and control the PC-fax adapter, and responses of the PC-fax adapter. In 1996, the ITU-T adopted two recommendations based on that work. Service Class 1 is defined in ITU-T Recommendation T.31, while service Class 2 is defined in ITU-T Recommendation T.32. These standards restore the modem parts and protocols that were originally removed by Group 3 fax in the interest of cost and simplicity. The protocols can be used for interfaces to scanners, printers, local area networks (LANs), small computer serial interface (SCSI), American National Standards Institute (ANSI) X3.1311, and so on. Vendors that write software for PCs are now assured that their product will be compatible with multiple hardware vendors for PC-fax adapters and for other devices wanting to communicate with Group 3 fax machines. Protocols and procedures to adapt alternative communication schemes are beyond the scope of these standards.

A fax call is handled in the standard Group 3 protocol in five phases:

- Phase A: Dialing and answering a fax call;
- Phase B: Selecting modem speed and matching fax specifications;
- Phase C: Sending fax signals for the page being copied;
- Phase D: Confirmation of fax page received successfully;
- Phase E: Disconnect phone line after last page.

6.1.2 Service Class 1 (TIA/EIA-578-A and ITU-T T.31)

Service Class 1 is the simplest interface and is usually associated with a very small PC-fax adapter that would fit in a laptop or a portable PC. The Class 1

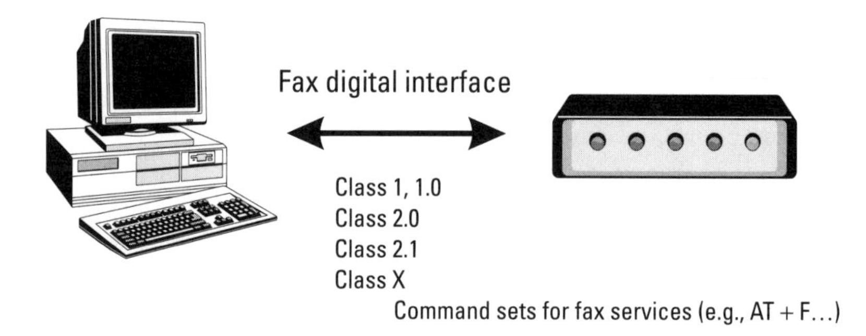

Figure 6.4 Facsimile digital interface.

facsimile DCE must provide the most basic level of services necessary to support Group 3 facsimile functionality. That means Class 1 must support CCITT T.30 procedures for document transmission and T.4 for representing images. The following services are required in Class 1: connection; waiting and silence detection; data transmission and reception; HDLC data framing, transparency, and error detection; and message generation. The PC-fax adapter provides interface with the PC, autodial functions, $V.27_{ter}$ (or optionally V.29 and V.17) modem functions for sending the fax data, V.22 with HDLC data formatting for T.30 commands and responses (handshaking), and PSTN interface through its LCU. It describes a set of services at the physical and data link layers: waiting, signaling, and HDLC data formatting. The DCE might be described as a tone generator, detector, and data pump (sends and receives 1s and 0s with HDLC framing without any knowledge of their meaning). The PC provides T.30 session management and the T.4 functions of MH and MR image coding and formatting of the data for each fax scanning line. These PC functions require substantial processing, so background operation of the facsimile function is difficult but has eased as computer processing power has risen and as OS multitasking has improved.

TIA/EIA-578-A was the base document in the ITU-T for developing an international service Class 1 recommendation. A number of modifications have been made to T.31 that did not exist in TIA/EIA-578-A. The most important change was support for the new V.34 and V.8 modem specifications. In addition, a new inactivity timer was added that allows a facsimile DCE to break away from an unsuccessful connection at any stage of the facsimile transfer. The EIA/TIA TR-29 committee is now in the process of updating the current TIA/EIA-578-A standard to account for the changes and modifications made in ITU-T T.31. When the standard is complete and has been approved, it will be renumbered as TIA/EIA-578-B. (It should be noted that Class 1 implementation is the market leader at this time.)

6.1.3 Service Class 2 (TIA/EIA-592-A and ITU-T T.32)

In addition to the Class 1 functions, Class 2 includes a set of services described in CCITT T.30. The PC-fax board makes and terminates calls, manages the communication session, and transports image data. The PC prepares and interprets image data in compressed form as specified in CCITT T.4. Commands are sent from the PC to the fax adapter while it is in a command state. The Hayes AT command set is used, including extensions for fax commands. Service Class 2 also includes facilities that establish a data link protocol between the DTE and the DCE to protect and recover from loss of data due to DTE system latency, multitasking, and so on. Optional features include copy quality checking, fault conversions, and control of scanners, printers, and local storage in fax machines.

TIA/EIA-592 was the base document in the ITU for developing an international service Class 2 recommendation. A number of modifications have been made to T.32 that did not exist in TIA/EIA-592. The most important change was support for the new V.34 and V.8 modem specifications. In addition, T.32 was modified to support new features added to T.30, including new resolutions, new commands, and new file transfer modes. Another significant change was designation of the FCLASS value. To distinguish T.32 compliant implementation from TIA/EIA-592-A implementations, the FCLASS value in T.32 has been changed to 2.1. The TIA/EIA TR-29 committee has updated the current TIA/EIA-592-A standard to account for the modifications made in ITU-T T.32. The standard has been approved and published as TIA/EIA-592-B.

6.1.4 Service Class 3

A service Class 3 standard was being developed by TIA/EIA TR-29 that, in addition to the features provided in Class 2, would provide a capability for the facsimile DCE to convert image data into CCITT T.4 compressed images for transmission and to reverse the conversion on reception. Tagged image file format (TIFF) and ASCII text are examples of file formats that were under study. Service Class 3 was never finished and has been dropped from further consideration.

6.1.5 Service Class 4

A service Class 4 standard was being developed by TIA/EIA TR-29 and was to take into account new work being done in the ITU-T on Recommendation V.80. Recommendation V.80 provides inband DCE control and synchronous data modes for asynchronous DCE. Because of the lack of progress and contributions on Class 4, it was suggested that modifications could be made to the Class 1 standard to account for some of the key areas that Class 4 was to address. Those areas include a new command to allow Class 1 commands to be embedded in the data stream, allowing DTE commands and DCE responses to be buffered and thereby providing relief from some of the critical DTE timing, and a method for the modem to automatically put itself in receive mode based on the last received DCS FIF. In all likelihood, Class 1 will be extended to meet the requirements that were being proposed for Class 4. It is anticipated that that will be accomplished through approval of an interim standard. This work will be carried out independent of the work on TIA/EIA-578-B.

6.1.6 S.100 Media Resource API

S.100 is an API for computer telephony (CT) applications, including facsimile. It was developed under the auspices of the Enterprise Computer Telephony Forum (ECTF). The ECTF, founded in 1995, is a nonprofit organization that comprises computer telephony suppliers, developers, system integrators, and users from the Americas, Europe, and Asia/Pacific. The ECTF facilitates the development, implementation, and acceptance of CT solutions; discusses, develops, and tests interoperability techniques; incorporates and augments existing industry standards; and publishes CT implementation agreements.

The work of the ECTF is carried out by an Architectural Working Group, whose mission is to define a high-level CT framework to ensure all ECTF working groups are working synergistically toward the goal of developing industry agreements for CT convergence and interoperability. Individual working groups define the architecture for their specific areas within the overall ECTF architecture framework. The Computer Telephony Services Platform Working Group is of interest to facsimile. Its mission is to increase the ability of computer telephony applications to share media resources and interoperate with existing call control architectures. The objective of this working group is to define a robust, scalable, and networkable CT services platform architecture. This architecture will be expressed as a series of implementation agreements that will enable a broad range of CT components and applications to interoperate successfully. One such implementation agreement is the ECTF S.100 API specification, which has applications for facsimile. It uses one of the profiles in ITU-T Recommendation T.611, which addresses programming communication interface (PCI) APPLI/COM for Facsimile Group 3, Facsimile Group 4, Teletex, Telex™, e-mail, and File Transfer Services. It also contains the first open industry APIs for low-level control of resources such as fax boards. (More information about the ECTF and the S.100 specification is available on the ECTF Web page, www.ectf.org.)

6.1.7 Group 3 Facsimile Apparatus Control (PN1906)

PCs and other DTE now have the ability to interface with scanners, printers, and modems. Group 3 fax machines provide those functions, but there is no standard for the DTE to use those features. Under the auspices of TIA/EIA Project Number 1906 (PN 1906), a standard digital interface that allows DTE access to those functions was developed. The scanner may provide the PC with imagery in various formats such as uncompressed bitmaps, MH, TIFF, pack bits, and GIFF formats. Formats printed on the fax machine may include TIFF, Epson, ASCII, PostScript™, or HP Laserjet PCL™, plus the standard facsimile formats. The PC should be able to control the fax machine and to provide or share storage for documents and phone numbers. Interworking will be provided for service Classes 1 and 2. As a result of the work accomplished under PN 1906, TIA/EIA Interim Standard (IS) IS-650 was published. IS-650, *Multifunction Peripheral Interface Standard (MFPI) Level 1*, defines an interface between a host computer and a multifunction peripheral (MFP). Such an interface is called the MFP interface (MFPI) (Figure 6.5). A typical MFP has a scanner, printer, fax-data-voice modem, operator console, and standalone controller (SAC). One or more of the components may be omitted from the MFP. IS-650 organizes the subsystems as components on a channel. The channel is a resource management component that manages the subsystems from a supervisory level. The standard assumes a simple channel, such as bidirectional parallel (IEEE-1284) or serial (TIA/EIA-232-F). Other channels, such as SCSI, are also amenable for use by this standard. This standard is limited to a single host on a given channel.

IEEE-1284, *IEEE Standard Signaling Method for a Bi-directional Parallel Peripheral Interface for Personal Computers*, defines a signaling method for asynchronous, fully interlocked, bidirectional parallel communications between hosts and printers or other peripherals. This standard was created because there existed no defined standard for bidirectional parallel communications between personal computers and printing peripherals.

Although TIA/EIA IS-650 defines a signaling protocol for use over the TIA/EIA-232 interface, it is not clear if there is any real level of implementation. In the absence of standards or if existing standards are not implemented, de facto approaches take hold. Consequently, fax units from different manufacturers are, in general, not compatible over the digital interface.

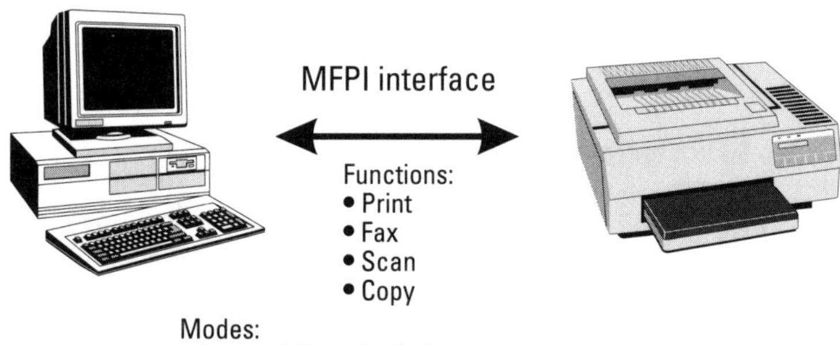

Figure 6.5 Multifunction peripheral interface (IS-650).

6.2 Programming Communication Interface for ISDN

In 1992, the ITU-T (then the CCITT) approved Recommendation T.611, *Programming Communication Interface (PCI) APPLI/COM for Facsimile Group 3, Facsimile Group 4, Teletex, Telex, E-mail, and File Transfer Services.* T.611 was subsequently revised in November 1994. T.611 (1992) defined a PCI called APPLI/COM, which provided unified access to Telefax Group 3, Telefax Group 4, Telex, and Teletex services. T.611 (1994) revises the 1992 version for the purposes of clarity and the inclusion of e-mail services in general, message handling services (MHS) as described in the ITU-T X.400-series recommendations in particular, and file transfer services. Special considerations were taken to ensure backward compatibility with the T.611 (1992) version.

During the 1993–1996 study period, ITU-T Study Group 8 (SG8) addressed the issue of PCIs under Question 1/8, 20/8, and 21/8. Work focused on the development of PCIs for the ISDN. An important aspect for the success of the ISDN is the availability of user applications (in particular, applications based on PCs) making use of ISDN services. The T.200 series recommendations that were developed defined a standard PCI, allowing applications to access and manage services provided by the ISDN. It provides mechanisms to support the most common protocols used for communications between ISDN applications. Contributions were received from Germany and France on PCI, and a series of draft recommendations for PCI for terminal equipment connected to the ISDN were developed as described in Sections 6.2.1 through 6.2.9.

6.2.1 Recommendation T.210: General Architecture

This section of the multipart recommendation gives a general description of the PCI; in particular, it provides an overview of the contents of each part of the standard.

6.2.2 Recommendation T.220: Basic Services

This section of the multipart recommendation provides a technical overview and defines the basic functions supported by the ISDN-PCI. It defines the PCI architecture and includes a detailed definition of the PCI messages and parameters used for the administration and connection control. It explains how to make use of these messages and parameters via a generic exchange mechanism.

6.2.3 Recommendation T.230: User Plane Protocol Architecture

This section of the multipart recommendation provides for general aspects for the management and access to the user plane protocol supported by the PCI. In particular, it includes the user plane protocol architecture and the detailed description of the protocol selection mechanism.

6.2.4 Recommendation T.231: Layer-1 Protocols

This section of the multipart recommendation details the procedures, messages, and parameters used to access the ISDN-PCIs user plane protocols that provide a layer-1 communication service.

6.2.5 Recommendation T.232: Layer-2 Protocols

This section of the multipart recommendation details the procedures, messages, and parameters to access the ISDN-PCIs user plane protocols that provide a layer-2 communication service.

6.2.6 Recommendation T.233: Layer-3 Protocols

This section of the multipart recommendation details the procedures, messages, and parameters to access the ISDN-PCIs user plane protocols that provide a layer-3 communication service.

6.2.7 Recommendation T.241: DOS Exchange Mechanism

This section of the multipart recommendation defines all the details of the operating system binding for an MS-DOS™ environment. (A general presentation of the binding mechanism can be found in the ISDN PCI Part 2.) It also describes the procedures, messages, and parameters to access the ISDN-PCIs user plane protocols that provide a layer-3 communication service.

6.2.8 Recommendation T.242: Windows Exchange Mechanism

This section of the multipart recommendation defines all the details of the operating system binding for a Windows™ environment. (A general presentation of the binding mechanism can be found in the ISDN PCI Part 2.)

6.2.9 Recommendation T.243: UNIX Exchange Mechanism

This section of the multipart recommendation defines all the details of the operating system binding for a UNIX™ environment. (A general presentation of the binding mechanism can be found in the ISDN PCI Part 2.)

6.2.10 SG8 and WTSC Resolution Regarding ISDN PCIs

It should be noted that Q1, Q20, and Q21 could not agree whether SG8 should approve the T.200 series. The reason was that a competing interface known as communication application programming interface (CAPI) became very successful in Europe during the time that the PCI recommendations were being completed. It was decided to bring the matter to the SG8 Plenary for adjudication. The Plenary considered two options: (1) forward the recommendations to the World Telecommunication Standardization Conference (WTSC-96) for approval and (2) terminate the work, freeze the documents, and consider this outcome as a reference paper in case work on the subject resumed. After a lengthy discussion, it was determined that the T.200 series recommendations would be submitted to the WTSC-96 for resolution. WTSC-96 did approve the recommendations. However, they determined that there would be only one recommendation, that is, T.200 would contain all nine parts previously identified as T.210, T.220, T.230, T.231, T.232, T.233, T.241, T.242, and T.243. Furthermore, it was not envisioned that any further work would be done in this area during the next study period. Therefore, Q1, Q20, and Q21 were deleted.

6.3 EIA-232 Interface

Figure 6.6, a functional block diagram of the typical facsimile machine, shows that the only I/O connection is a modem to interface with the PSTN. Table 6.1 lists the three possible modes of operation for a conventional facsimile system: transmit, receive, and copy. Many facsimile manufacturers are offering an optional digital interface to connect directly with a local computer, digital network, cryptographic device, local scanner, or external modem. The digital interface invariably is accomplished in accordance with the EIA-232 standard. ANSI/TIA/EIA-232-F, *Interface Between Data Terminal Equipment and Data Circuit-Terminating Equipment Employing Serial Binary Data Interchange*, is applicable to the interconnection of DTE and data circuit-terminating equipment employing serial binary data interchange. Figure 6.6 illustrates the digital configuration with the EIA-232 interface, and Table 6.1

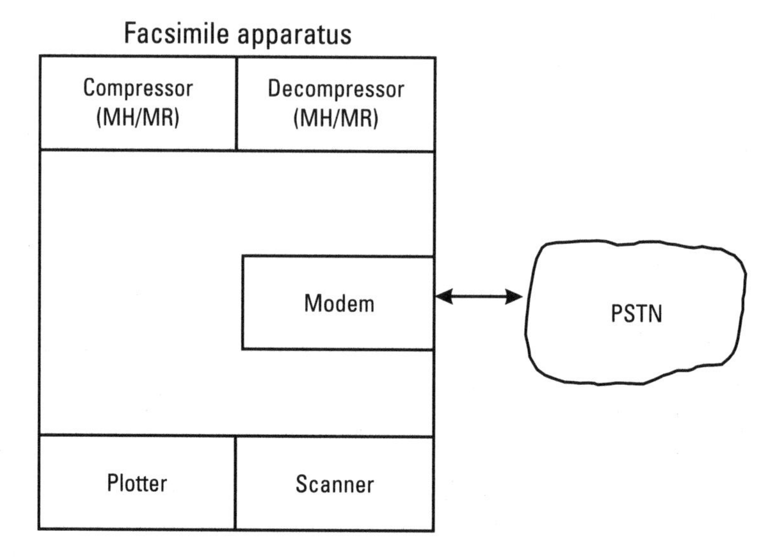

Figure 6.6 Typical facsimile terminal.

Table 6.1
Functions of Facsimile Machines With EIA-232

	Function	Source	Destination	Input	Output
Without EIA-232	Fax transmit	Scanner	Modem	Image	MH/MR
	Fax receive	Modem	Plotter	MH/MR	—
	Copy	Scanner	Plotter	Image	Image
EIA-232	Plot	EIA-232,	Plotter	MH/MR/image	Image
	Scan	Scanner	EIA-232	Image	—
	Transmit	EIA-232	Modem	MH/MR/image	MH/MR
	Receive	Modem	EIA-232	MH/MR	—
	Compress	EIA-232	EIA-232	Image	MH/MR
	Decompress	EIA-232	EIA-232	MH/MR	—
EIA-232 plus character generator	Print	EIA-232	Plotter	ASCII	Image
	Remote print	EIA-232	Modem	ASCII	MH/MR
	Create image	EIA-232	EIA-232	ASCII	MH/MR/image

lists the six different modes of operation that can be accomplished by using the digital interface. In the transmit mode, imagery that originates in a computer is fed to the fax unit for transmission. In the receive mode, pictures received by the fax unit are fed directly to a computer, LAN, or cryptographic machine.

Another typical option frequently associated with the EIA-232 mode is the inclusion of a character generator, shown in Figure 6.7 and listed in Table 6.1. In the remote print mode, the fax unit accepts ASCII character data from the local digital source, the character generator converts the data to raster format, and the MH/MR compressor prepares the signal for conventional fax transmission over the PSTN. The advantage of the remote print mode as well as the transmit mode is that the input image is never scanned, so the output image does not have the degradation associated with the scanning process.

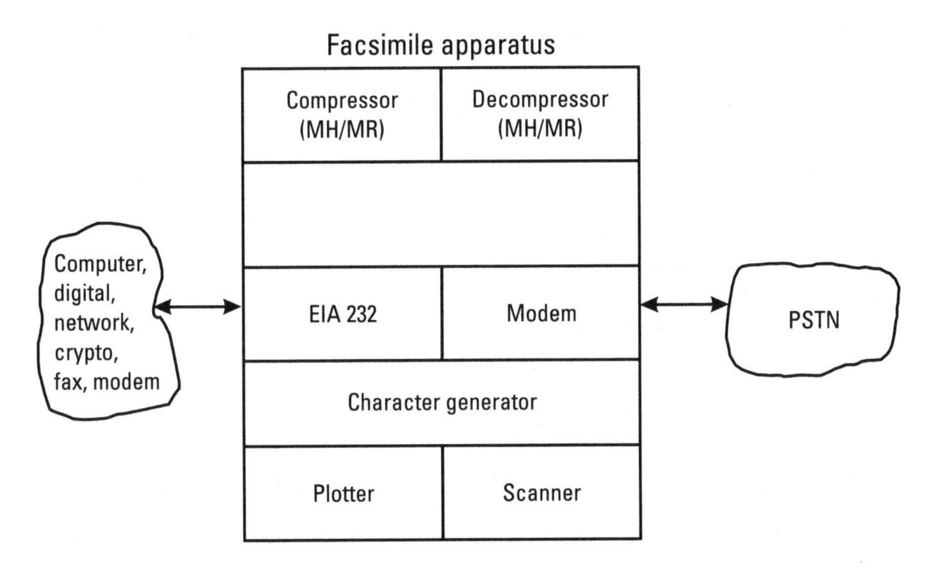

Figure 6.7 Facsimile terminal with EIA-232 interface.

7

PC-Fax

For convenience, *PC* as used in this book, includes all computers. *PC-fax* covers those PCs, including multifunction communication devices, capable of sending and/or receiving Group 3 fax signals, regardless of the computer operating system in use. PC-fax (sometimes known as computer-based fax, or CBF) allows PCs and fax machines to exchange documents using the Group 3 fax protocols discussed in other sections of this book. Users can fax images and text to the more than 100 million Group 3 fax machines and to even greater numbers of PC-fax units. Most fax machines are connected to the PSTN or the ISDN and are commonly available 24 hours a day, like a regular telephone. PC-fax units, however, usually are not enabled for automatic fax reception unless they are connected through a LAN server.

The first U.S. fax cards for PCs were designed by Henry Magnuski and appeared in the early 1980s. To produce the cards, Magnuski formed a company named Gammalink (now a division of Dialogic). Success followed, other companies entered the competition to add fax capabilities to PCs, and today PC-fax units outnumber fax machines. A portable computer with fax capability can even communicate from an automobile or other vehicle using a cellular phone system. PC-fax e-mail on the Internet provides a way of sending faxes without the cost of long-distance phone rates. Binary file transfer of computer data is also possible between PCs using Group 3 fax protocols and modems.

It might seem preferable to exchange computer-prepared documents between PCs, bypassing fax protocols. That might allow higher resolution or higher speeds, but the received copy may be formatted differently or have different fonts or font sizes (more like a retyped document than a copy). As a text file, it also would risk modification of the document by the recipient in any manner wanted, as well as attaching a stored image of a signature, like a rubber stamp. For legal purposes, PC document exchange lacks the recognized authenticity of a faxed document.

7.1 How PCs Fax

Most PCs purchased today have the capability to emulate a Group 3 fax machine, using built-in modems and the Group 3 fax software furnished. Generating a file that will be faxed uses the same process as that for creating a PC document or image on the computer screen. A PC-fax will send by Group 3 fax protocol files generated by word processing, spreadsheet, graphic, or other programs. Instead of scanning an image or text on a sheet of paper, PC-fax scans the image generated by the PC and stored in its memory. That eliminates the many signal processing steps made by a fax machine, including compensation for:

- Differences in brightness of the image caused by variations in the paper color and the marking density of the image;
- Sensitivity variations between the multiple photosensors;
- Unevenness in illumination of the page being scanned;
- Deficiencies in the scanning optics;
- Raggedness of line edges produced by the scanning process itself.

The better received quality of text faxed from a PC-fax rather than a Group 3 fax machine results from the use of a character generator to place each dot of the image matrix for a character in the optimum position for the typeface used (similar to the display on a PC screen). When a fax machine scans a character, the dot matrix formed depends on the exact character position related to the scan lines and sampling points along the line. Figure 7.1 shows a computer printout of a Group 3 scanned image. Notice that the character *a* has a different dot pattern each time it is printed. When graphics are sent, PC-fax transmission also produces better received quality.

7.2 PC-Fax Transmission

The PC-fax modem sends at the highest speed suitable for the telephone line and both fax units. A computer-generated file can be transmitted by fax using the printing function of the program in use. Windows™ and Macintosh™

Figure 7.1 Scanning distortion of character shapes.

operating systems have a fax transmission option in the Print pull-down menu, and its selection causes the modem to switch to fax mode. Alternative software programs may provide additional or different features. Menu-driven screens guide the user through the steps for sending. Selection of the telephone number of the receiving fax machine is either from a stored directory or by direct entry. Computers with other operating systems such as UNIX™ and MS-DOS™ also have this capability, but supplemental software may be needed.

Hard copy is not used in the transmission of computer-generated faxes, but the document can be saved and printed for a paper record. The automatic personalized cover page feature of the software program adds an extra page. Some users prefer instead to add a short header on the first page. Delayed transmission can be selected to send at the time of lowest telephone line charges, but that defeats the prime advantage of facsimile—immediate delivery. The rapid delivery of fax permits multiple-turnaround exchange of business records between distant points in one day. That is valuable in contract and other legal negotiations and in dissemination of time-critical information. Rapid delivery is also valuable in conducting business across multiple time zones. Consider an end-of-workday request from New York for an urgent task needed to be done in Tokyo. The work could be done overnight New York time with the resulting report waiting in the New York office at the start of work the next morning. The extra direct cost per page for sending in prime time may be pennies or less, while the rapid fax delivery could make the difference between success and failure of a project.

The earliest PCs that offered fax capabilities required a $2,000 add-on unit with its own fax modem. The add-on unit became a virtual printer that was selected by the PC to furnish a known signal format for processing. When the computer file was converted to Group 3 format for sending, there was usually a delay while a new file was generated and stored in memory. ASCII format files could be sent directly, but text created by word processing had to be converted to ASCII by stripping out control characters before calling the receiving fax unit. Each PC-fax design needed a specially written software program, and font selection was limited. Although later methods could send imagery in addition to text and simplified the step-by-step procedure, a separate fax modem still was required.

When it was realized that the PC data modem could be programmed to provide fax capability, it became standard in most PCs being sold. Software already controlled the modem for different specifications, and adding fax cost practically nothing. To further lower costs, many PCs used software and the host processor to perform fax operations. The downside of that method was that it often included only the lowest efficiency image compression (MH), a maximum modem speed of 4.8 Kbps, and the original 1980 fax resolutions. Timing errors, especially with background operation of the fax, often caused much longer transmission time or perhaps complete failure. Background operation should allow the sending or receiving of facsimile copies without interruption of other work being done by the computer; otherwise, the computer is just an expensive fax machine. Total cost to the user was likely to be not lower but considerably higher, because of increased transmission time. With heavy or even moderate fax traffic, the initial cost saving was dissipated. Nevertheless, many infrequent fax users on a low budget still preferred the lower purchase price. Fortunately, PC designs evolved, and fax performance from new PCs is much better. With today's much higher speed processors, improved software, and other refinements, many problems have been minimized or eliminated. Typical PC-fax units sold now may have a 14.4- or 33.6-Kbps modem and MMR image compression.

Performance differences, such as transmission time per page, between various PC-fax designs may still be substantial. For best performance, the PC-fax should incorporate a well-designed fax board with its own microprocessor to handle fax activities, freeing the host PC to do regular computing during fax activity. Software or an integrated circuit chip can be used to furnish image compression with MH, MR, or MMR and higher resolutions of 300 or 400 dpi. Conversion to a fax signal is done during sending, without the delay of storing the image in a fax format file before starting to send. The separate microprocessor chip handles the task fast enough to send it directly. Timing errors are eliminated, and full background operation should cause no problems.

To send from a physical document, a scanner must be connected to the PC-fax. The page to be sent is scanned and its image viewed on the PC screen. Adjustments to improve the image quality can be made before it is sent as a fax. A selection of signal processing options can improve the image to compensate for defects such as light-density markings and fuzzy or overinked printing. Fine tuning the black-white threshold setting can make a dramatic difference in resolution of small print, thin lines, or the information shown from low-contrast documents. Contrast and gamma adjustments also can make a big difference when photographs are sent. (See Section 5.1 for more information.)

7.3 Flow Control

To understand the need for flow control, examine first the operation of pre–Group 3 analog fax machines that used real-time scanning and printing functions. The spot being printed at the receiver was synchronized to be in the same position on the page as the one at the transmitter (minus the fixed transit time of the signal). A frequency standard in each machine provided the reference needed to generate identical spot speeds in both units. Group 3 fax eliminated the frequency standard but introduced a new requirement that was not handled properly by some of the low-cost PC-fax designs. Instead of sending a signal for each pixel, the scanning spot position information is coded in the signal sent (not locked to a frequency standard). The recording operation occurs later than the scanning operation, and the time difference between them varies while the page is being sent. The time taken to scan a fax line may be shorter than the time needed to record a fax line. The older Group 3 fax units may have a slow recording rate and practically no memory for storing an image while waiting for the recording operation to finish. Without protocol intervention; especially when receiving from a fast PC-fax, part of the image would be lost, and the transmission would fail. To prevent such loss, the time between the starts of successive transmitter scan lines may be artificially increased to be no shorter than the time needed for printing the lines. The page is then printed as it is received, without data loss.

To accomplish that, the fax protocol uses a method called bit stuffing. When all the fax information for one line has been sent and it is too soon to start sending the next fax line, dummy bits are sent instead of coded bits. Dummy bits send no fax information but are needed to slow down the fax transmitting unit. They are thrown away at the fax receiver, having served their purpose by holding the sending speed to just below the highest speed that the fax receiver can handle. In the initial handshake, the receiving unit notifies the transmitter of its minimum scan line time (MSLT), which can be as long as 40 ms for older mechanical fax machines. It might seem simpler to merely stop signals from the sending modem instead of using dummy bits, but for the receiving modem to work properly, it must continuously lock into the fixed symbol rate sent. If the modem signal were turned off between fax scan lines to prevent sending fax data too fast, the transmission would fail.

When sending between PC-faxes, there is no waiting to print each fax line, and the coded image information can be sent from fax transmitter memory to receiver memory as fast as the modem can send over the telephone line. Then speed limitations of mechanical scanning or printing should not slow things down, but they may and it is cost related. One method of sending from a PC-fax unit ignores the MSLT signal from the fax receiver. Each scanning line is sent as if going to the slowest old Group 3 fax machine, taking at least 40 ms per scan line, and the time for even a blank page at fine resolution is then 0.04 s/line times 2,200 lines/page = 88 sec, even with the fastest modem available. Even though MSLT is 0 ms, no calculations are made to determine whether fill bits are needed or how many. Equipment cost is lowered by ignoring the MSLT handshake information, at the expense of a much longer transmission time and its greater cost. An occasional user may be satisfied, but for others that may be an expensive way to send.

7.4 PC-Fax Reception

During receiving, the fax image information is stored in memory as a file on a computer disk (MSLT 0), even if printing occurs while the fax signals are being received. That allows completion of the fax receiving session long before the printing is finished. A page in storage typically takes 50–60K of memory, but 100K or more may be required. Storage may be on a floppy disk (about 20 pages) or on the hard disk. Many fax machines have storage capabilities similar to PC-fax, but almost none has a screen for viewing the received image. The stored image can be viewed on the computer screen (soft copy) as it is received or on page completion, but a printed image is easier to read. Although computer-generated characters can be read without strain when displayed in 80 columns across the screen, received faxes present a much more difficult task. Where each dot for the internally generated computer font matches the optimum position, the random match between the scanning dot and the edge of a character causes much edge raggedness, making text difficult to read. With VGA 640 dot resolution display, typed fax characters are almost impossible to read unless the image size is expanded to where only a small portion of the page can be viewed at one time. Even with higher screen resolutions, when a full 8.5-by-11-inch received fax page is viewed on the PC monitor, text readability is almost impossible because only half the screen width is utilized to fit the fax page image to the screen height. By matching the fax line width to the screen width, the text may be readable, but for the full length of the received fax page to be viewed, the image must be stepped down the screen, three partial-page frames, and the image is still ragged. Even super-VGA resolution shows less than half the number of dots (800) across the screen as Group 3 standard resolution (1,728 dots). Although new PCs have available 1,024, 1,280, or more dots across the screen, many PCs are still set up with the old standard VGA 640 dot display.

When a Group 3 fax is printed at fine resolution, 200 by 200 dpi, copy is sharp enough for most applications. The PC-fax printer may provide the 300-dpi or 400-dpi fax copy that some PC-fax board users demand. Having a high-resolution printer does not ensure high-quality printing of received faxes. Early PC-fax software designs converted the 200-dpi received fax signal into 150 dpi for a laser printer rather than to 300 dpi, thus producing images not as sharp as those received on a standard fax machine.

Security against unwanted viewing of the received document is better on a PC-fax than on a Group 3 fax machine, where the pages are stacked in a tray, accessible for any passerby to pick up and read. Some fax machines and PC-fax units have a mailbox option, where the document is kept in memory until accessed by means of a password. To prevent unauthorized interception or printing, the fax signal can be encrypted (see Section 3.8). An intercepted signal cannot be decoded without the proper encryption key. Finding the proper key by trial and error is almost impossible if a powerful encryption system is used.

7.5 PC-Fax Compared With Group 3 Fax

7.5.1 Advantages

PC-fax has some advantages over Group 3 fax.

- Computer-generated documents can be sent directly from the computer without first printing and then scanning them. The copy quality at the receiving fax machine is much better because the edges of the text and other markings are matched to the fax scanning rather than being random. A given typed character always has the same optimized matrix of dots as produced by the sending computer.
- Received fax copy can be viewed on a screen without incurring printing costs, which can be more than the telephone line charges.

- Received fax documents can go directly to computer memory and not be visible for unintended viewing on the PC screen. Unauthorized viewing or printing of the fax can be further denied to computer users without the proper password, providing even better security.

7.5.2 Disadvantages

The disadvantages of PC-fax compared with Group 3 fax are as follows.

- For even simple fax transmissions, users must operate a keyboard and have some computer knowledge, whereas they can send via Group 3 fax machines by inserting a document into a tray and pressing a button or two.
- Existing documents cannot be sent unless a scanner is connected to the PC.
- For older PCs that use a DOS operating system, there may be memory allocation problems that prevent transmissions from being made as expected. The computer may not have enough memory to load the PC-fax program or to run regular computer application programs after the software has been loaded. Even PC-fax programs that require only a small amount of memory after loading may take a large amount of memory while being loaded. Because modern word processing and other programs use a large portion of the basic 640-Kb low memory where the fax software is also loaded, the PC cannot load both programs. That can happen even if the PC has many megabytes of unused RAM in extended memory.

7.6 Fax/Voice Line Switching

Most of the newer PC-fax and fax machines have the ability to determine whether the incoming call is fax, voice, or data. The fax function answers a fax call; the data modem answers a data call; and the telephone rings if a voice call is received. The CNG tone is a 1,100-Hz 0.5 sec "beep" sent every 3 sec by automatic calling fax units. CNG is not needed to start the fax receiver, but it identifies a fax call and causes the fax to answer, starting the handshake procedure. Older fax units may not send CNG on calls dialed manually, and automatic detection of a fax call will fail unless an alternative method, such as detection of a quiet interval, causes the fax to answer. CNG is not needed for manual sending operation, because the caller starts sending fax when the CED answer signal is heard. If a person answers instead, the caller can say that a fax is being sent and then wait for the CED tone from the receiving fax before starting transmission. The beep tone, heard by the person answering a wrong-number call indicates that a machine is calling, rather than silence, which might indicate a harassment call.

7.7 Printer Fax Adapters

Some companies offer a Group 3 fax adapter that connects between a computer and its printer and performs the same functions as a PC-fax board, but the PC does not need to be on while receiving. Connection is made directly to the telephone line, and memory in the adapter stores incoming faxes if the printer is in use. The memory is also utilized for sending faxes, and conversion of files to Group 3 fax format is done in the adapter, which may provide performance similar to the high-level PC-fax boards. Alternatively, during sending between the same type of adapter units, it may be possible to send character data from the PC (instead of Group 3 fax signals) to the receiving unit adapter and print with 300-dpi resolution.

7.8 PC-Fax Broadcasting

Broadcast in radio and TV means sending a single stream of signals by radio waves for reception by multiple receivers at different locations. In the 1920s, fax broadcasting started with a single stream of signals sent by radio waves to multiple receivers. Much later, fax broadcasting started over a dedicated phone line network with transmission of a single stream of signals to multiple receivers (e.g., AP Wirephoto™ network and the U.S. Air Force Weatherfax network). These applications were described in Chapter 2.

Multiple-line fax broadcast services that started before Group 3 fax offer simultaneous transmission of documents to hundreds of destinations at the same time. A separate PSTN telephone line connects to each user. Delivery of the document to the service provider is by fax, PC data transmission, or physical delivery.

When some customers wanted to send simultaneous transmissions with their own equipment, PC-fax boards were developed for sending on four PSTN lines. Today, a single PC card slot can handle 24 PSTN channels, or with two PC-fax cards, 48 channels in a single PC. To match possible differing characteristics of channels and fax receivers, separate buffer memories allow individual handshaking and transmission protocols. A T-1 channel telephone line interface is used instead of 24 individual telephone lines. Fractional T-1 interfaces are available for fewer than 24 lines. Multiple documents can be preassembled in memory and sent in one session to maximize efficiency and lower phone costs. If one of the telephone lines being used for broadcast goes out of service, the task is automatically allocated to the remaining lines and a "service needed" flag is alerted. A multiline receive-and-print feature allows receiving directly to hard disk over four or more lines simultaneously and printing from memory, a type of fax broadcast often used on LAN networks. At one time, an alternative method was offered by companies such as 3M, which sold rather expensive fax broadcast units with the ability to send faxes on 48 or so telephone lines simultaneously. Some Group 3 fax machines and some PC-fax units have sequential broadcast ability; that is, they automatically call one recipient at a time on the PSTN and send the fax pages from memory.

8

Enhanced Services Using Group 3 Protocols

8.1 Audiographic Conferencing

Teleconferencing standardization has evolved along two paths. One approach, begun in the EIA/TIA TR-29 committee under Neil Starkey, concentrated on audiographic conferencing. That effort led to the T.120 series of recommendations on multimedia conferencing developed by the ITU-T. The other approach, also developed by the ITU-T, focused on videoconferencing. The culmination of the videoconferencing work was the H.32x series of recommendations, which can operate over a variety of networks. Eventually, the two sets of standards were tied together, that is, the T.120 series can operate alone or in the data channel of the H.32x series over the same set of networks.

8.1.1 T.120 Series of Recommendations

The T.120 series of recommendations collectively define a multipoint communication service for use in multimedia conferencing environments. The T.120 series contains a set of communication protocols and services that support real-time multipoint data communications. The recommendations can be used in many different ways. For example, T.120 can be used as the protocol of the H.320 terminal data channel. It also can be used as the primary protocol to manage a multipoint audioconference.

The architecture relies on a multilayered approach with defined protocols and service definitions between layers (Figure 8.1). Each layer presumes the

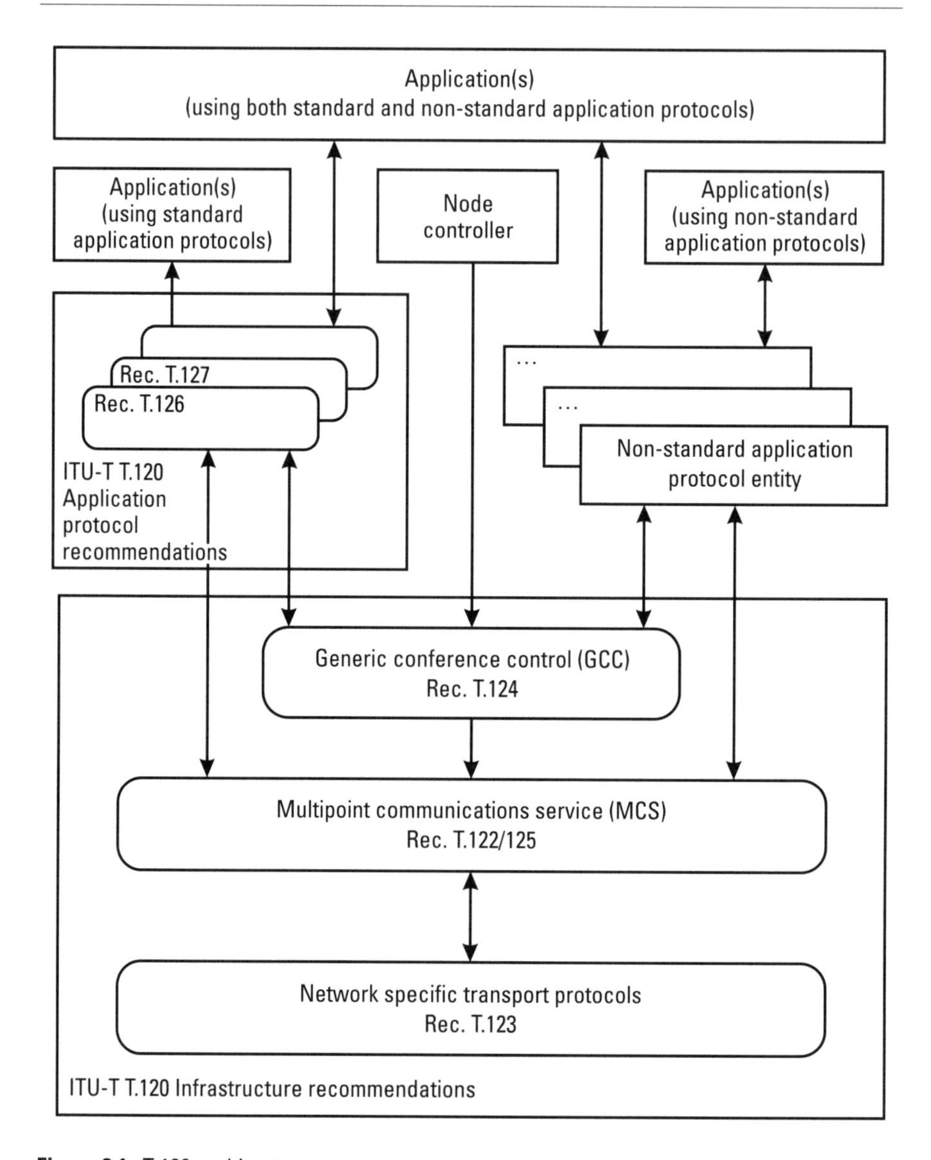

Figure 8.1 T.120 architecture.

existence of all layers below. The lower layers (T.122, T.123, T.124 and T.125) specify an application dependent mechanism for providing multipoint data communications services to applications. The upper layers (T.126 and T.127) define protocols for specific conferencing applications, such as screen sharing, still image transfer, and binary file transfer. Support for facsimile images compressed according to Recommendation T.4 or T.6 is included.

8.1.1.1 T.120: Data Protocols for Multimedia Conferencing

Recommendation T.120 introduces the T.120 series of recommendations. It describes the T.120 system model that provides an architecture for multipoint data communication in a multimedia conferencing environment. This recommendation provides facilities to establish and manage interactive communications (conferences) involving two or more participants on and between a variety of different networks. It provides a comprehensive data communication service for those participants, independent of the underlying network. Within a conference, it allows communications to be established between any combination of conference participants. It also provides support for applications and their associated protocols, defining startup mechanisms, and capability exchange procedures.

8.1.1.2 T.124: Generic Conference Control

Traditionally, telephony services have been constrained to point-to-point operation. To support group activities involving physically separated participants, there is a requirement to join together more than two locations. The term *multipoint communication* describes the interconnection of multiple terminals. Typically a special network element, known as a multipoint control unit, or bridge, provides that function. The term *conference* typically refers to a group of geographically dispersed nodes that are electronically joined and that are capable of exchanging audiographic and audiovisual information across various communication networks.

Generic conference control (GCC) provides a set of services for setting up and managing the multipoint conference. It provides access control and arbitration of capabilities. Applications use GCC facilities to coordinate the use of channels and tokens. GCC facilities can be used to query a multipoint control unit or a multiport terminal node to find a desired conference. Multiple applications can be running on any given node and can be dynamically launched, used, and shut down during a conference. Using mechanisms provided by GCC, applications create conferences, join conferences, and invite others to conferences. As endpoints join and leave conferences, the information base in GCC is revised and the information used to notify all endpoints automatically.

8.1.1.3 T.123: Network-Specific Data Protocol Stacks for Multimedia Conferencing

The T.120 protocols are designed to operate over a wide range of networks and to provide communication between endpoints on a mixture of networks. The T.120 suite provides for operation over the following networks: the ISDN, as defined in the I-series recommendations; circuit-switched digital networks

(CSDNs); packet-switched data networks (PSDNs) using Recommendation X.25; and the PSTN. The differences in T.120 operation for the various networks are confined to the lowest layers, as detailed in Recommendation T.123, which defines the network-specific transport stacks for each supported network. Generally, existing link layer protocols appropriate to each network are selected and then mapped into a common interface layer, thus defining a transport profile for a given network.

8.1.1.4 T.126: Multipoint Still-Image and Annotation Protocol

Recommendation T.126 defines the protocol to be used by a broad set of user applications that require interoperable graphical information exchange in a multivendor environment. It can be employed by user applications that require simple whiteboarding, annotated image exchange, and hard copy image exchange, as well as for more advanced functions such as remote computer application piloting and screen sharing. Point-to-point still-image transfer cutting across application borders includes JPEG, JBIG, T.4, and T.6. The protocol manages the conferencewide synchronization of multiplane or multiview graphical workspaces. An extensible set of bitmap, pointer, and parametric drawing primitives can be directed to the workspaces. Advanced options such as keyboard and pointing device signaling to support computer application remote piloting and screen sharing also are defined.

8.1.1.5 T.127: Multipoint Binary File Transfer Protocol

Recommendation T.127 defines a protocol to support interchange of binary files within an interactive conferencing or group working environment. It does not impose any restrictions on the content of the files to be transferred. It provides the functionality to allow interworking between applications requiring a basic general-purpose file transfer capability. Mechanisms are provided in the multipoint binary file transfer (MBFT) protocol that facilitate both distribution and retrieval of files. A basic file transfer application conforming to Recommendation T.127 may offer the ability simply to broadcast one file at a time or to broadcast multiple files simultaneously. It also provides for private distribution of files to a selected subset of the conference and conductor control of file distribution.

8.1.2 H.32x Series for Videoconferencing

The ITU-T has developed a number of standards for videoconferencing over existing and emerging networks. Table 8.1 lists the key recommendations that have been developed for videoconferencing. The year that each of the families was first put into service is also shown in Table 8.1; work is continuing on all

Table 8.1
ITU-T Videoconferencing Recommendations

	H.320	H.324	H.323	H.322	H.321
Network	N-ISDN	PSTN (low bit rate)	Nonguaranteed QoS* packet-switched networks (Ethernet)	Guaranteed QoS packet-switched networks	Broadband ISDN, ATM,† LAN
First approved	1990	1996	1996	1995	1995
Video	H.261, H.263	H.261, H.263	H.261, H.263	H.261, H.263	H.261, H.263
Audio	G.711, G.722, G.728	G.723	G.711, G.722, G.728, G.723, G.729	G.711, G.722, G.728	G.711, G.722, G.728
Data	T.120	T.120	T.120	T.120	T.120
Multiplex	H.221	H.223	H.225.0	H.221	H.221
Signaling	H.230, H.242	H.245	H.245	H.230, H.242	H.230

*Quality of service
†Asynchronous transfer mode

the families to improve performance and functionality. The most mature family of recommendations is H.320, designed for operation over narrowband ISDN (N-ISDN) at rates of 64–1,920 Kbps. H.320 is the basis in many ways of the videoconferencing standards that followed, with appropriate modifications to accommodate the characteristics of the other networks and to take advantage of advancing technology.

Recommendation H.324 defines a videoconferencing terminal that communicates speech, data, and video signals over the PSTN. H.324 functionality is similar to that of H.320, but new standards are included for speech coding, video coding, control, and multiplex to accommodate the narrower bandwidth of the PSTN. The remaining families of recommendations defined in Table 8.1 adapt H.320 and H.324 to other networks. H.321 defines videoconferencing for transmission via the broadband ISDN/asynchronous transfer mode (B-ISDN/ATM) networks for transmission rates greater than 2,000 Kbps. H.322 adapts the H.320 standard for those LAN networks with guaranteed quality of service (QoS), such as ISO Ethernet.

H.323, the newest family of standards shown in Table 8.1, is designed to provide videoconferencing over LANs that do not provide guaranteed QoS. These networks dominate corporate desktops and include packet-switched networks such as Ethernet, Fast Ethernet, Token Ring technologies, and, equally important, the Internet. Note that many of the recommendations are common to all networks. For example, H.261 for video coding is mandatory for all services listed in the table, and although H.263 was added later for H.324 terminals, it is now optional for all the other services. This greatly enhances interoperability among networks. A great deal of commonality also exists in audio, multiplex, and control. The T.120 series of recommendations is used by all services.

8.1.2.1 H.320

Recommendation H.320 is the overview that specifies a multimedia terminal for transmission over the ISDN network. The five standards that define the H.320 terminal are listed in Table 8.2.

H.261: Video Codec

Recommendation H.261 specifies the video coding algorithm, picture format, and FEC technique. It uses a two-dimensional DCT. The use of motion compensation is optional.

Table 8.2
H.320 Recommendations

Designation	Title	Purpose
H.320	Narrowband visual telephone systems and terminal equipment	Defines requirements and provides a general system description
H.261	Video codec for audiovisual services at P × 64 Kbps	Specifies the video coding algorithm, picture format, and FEC technique
H.221	Frame structure for a 64–1,920 Kbps channel in audiovisual teleservices	Defines frame structure for synchronizing single or multiple B or H0 channels or a single H11 or H12 channel, containing video, audio, and data
H.242	System for establishing communication between audiovisual terminals using digital channels up to 2 Mbps	Defines the protocol and procedures using BAS* codes for point-to-point communications
H.230	Frame synchronous control and indication signals for audiovisual systems	Defines the C&I† signals related to video, audio, and maintenance for point-to-point and simple multipoint. Signaling is accomplished by means of BAS codes.

*Bit-rate allocation signal
†Control and indication

H.221: Frame Structure

Recommendation H.221 provides for subdivision of an overall transmission channel of 64–1,920 Kbps (in multiples of 64 Kbps) into lower rates suitable for audio, video, and data. The overall transmission channel is constructed by synchronizing and ordering transmissions over from one to six B connections, from one to five H0 connections, or an H11 or H12 connection. In addition to audio, video, and data, the channel is used for inband signaling, containing a frame alignment signal (FAS) and a bit-rate allocation signal (BAS). The 8-bit FAS code is used to frame (or synchronize) the 80 bytes of information in a B channel (and the 480 bytes in an H0 channel). The 8-bit BAS code provides a means to send and receive capabilities and commands.

H.242: Protocol and Procedures

Recommendation H.242 defines the protocol and point-to-point procedures involving BAS codes, including:

- Initial capability exchange and mode initialization;
- Dynamic mode switching and recovery from fault conditions;
- Procedures for activation and deactivation of audio, video, and data channels;
- Procedures for operation of terminals in restricted networks.

H.230: Control and Indication Signals

Control and indication (C&I) signals include information for the proper functioning of the system and are signaled via BAS codes. They facilitate maintenance and testing, for example, loopback commands. They also are used for simple multipoint where T.120 and multilayer protocol (MLP) are not available.

Audio

The H.320 recommendations refer to three other ITU recommendations that define audio coding: G.711, G.722, and G.728. The G.711 and G.722 recommendations have been used extensively in Px64 systems operating at high bit rates (e.g., 1.544 Mbps and 768 Kbps), where the 64 Kbps for audio is a relatively small percentage of the transmission bit rate. However, for a 2B connection, the channel bit rate is 128 Kbps, where 64 Kbps audio would require too large a fraction of the channel capacity. The newer G.728 recommendation operating at 16 Kbps alleviates that crowding.

Data

The H.320 recommendations provide for the transmission of data along with video and audio information. Four types of data channels are specified: low-speed data (LSD), high-speed data (HSD), MLP, and high-speed MLP (H-MLP) channels. These data channels operate up to 64 Kbps and at multiples of 64 Kbps for the high-speed versions. For the channels to be useful, it is necessary to develop communication and application protocols to define data flow. The MLP channels are used by the T.120 communication protocol stack. Both MLP and H-MLP can be activated simultaneously, but typically not at the same time as HSD and LSD.

The data channel is optional. Options for data applications include:

- T.120 series for point-to-point and multipoint audiographic conferencing, including database access, still-image transfer and annotation, application sharing, and real-time file transfer;
- T.84 still picture interchange file format (SPIFF) point-to-point still-image transfer, including JPEG, JBIG, T.4, and T.6;
- T.434 point-to-point telematic file transfer (BFT);
- H.224 for real-time control of simplex applications, including far-end camera control;
- Unspecified user data from external data ports.

These data applications can reside in an external computer or other dedicated device attached to the terminal, or they can be integrated into the terminal itself. Each data application makes use of an underlying data protocol for link layer transport. Other terminals in the H.32x family provide for similar applications.

8.1.2.2 H.324

Recommendation H.324 describes a terminal for low-bit-rate multimedia communication, utilizing a V.34 modem operating over the PSTN. H.324 terminals can carry real-time voice, data, video, or any combination, including video telephony. H.324 terminals can be integrated into personal computers or implemented in standalone devices such as video telephones. Support for each media type (voice, data, video) is optional, but if supported, the ability to use a specified common mode of operation is required, so that all terminals supporting that media type can interwork. H.324 allows more than one channel of each type to be in use. Other recommendations in the H.324 series include H.223 multiplex, H.245 control, H.263 video codec, and G.723 audio codec.

H.324 makes use of the logical channel-signaling procedures of Recommendation H.245, in which the content of each logical channel is described when the channel is opened. Procedures are provided for expression of receiver and transmitter capabilities, so that transmissions are limited to what receivers can decode and so that receivers can request a particular desired mode from transmitters. Because the procedures of H.245 also are used by Recommendation H.323 for nonguaranteed bandwidth LANs, interworking with these systems should be practical.

As with H.320 systems, H.324 terminals can be used in multipoint configurations through multipoint control units and can interwork with H.320 terminals on the ISDN through a gateway. H.324 implementations are not required to have all functional elements, but they must support the V.34 modem, H.223 multiplex, and H.245 system control protocol. H.324 terminals offering audio communication must support the G.723 audio codec. H.324 terminals offering video communication must support the H.263 and H.261 video codecs. H.324 terminals offering real-time audiographic conferencing should support the T.120 protocol suite.

G.723: Speech Coder Transmitting at 5.3/6.3 Kbps

The G.723 speech coder can be used for a wide range of audio signals but is optimized for speech. The system has two mandatory bit rates, 5.3 Kbps and 6.3 Kbps. The coder is based on the general structure of the multipulse-maximum likelihood quantizer (MP-MLQ) speech coder. The MP-MLQ excitation will be used for the high-rate version of the coder. The algebraic codebook excitation linear prediction (ACELP) excitation is used for the low-rate version. The coder is designed to provide a quality essentially equivalent to that of a "typical" toll call. For clear speech or with background speech, the 6.3-Kbps mode provides speech quality equivalent to the 32-Kbps G.726 coder.

H.245: Control Protocol for Multimedia Communications

The control channel carries end-to-end control messages governing operation of the H.324 system, including capabilities exchange, opening and closing of logical channels, mode preference requests, multiplex table entry transmission, flow control messages, and general commands and indications. H.245 messages fall into four categories: request, response, command, and indication. Request messages require a specific action by the receiver, including an immediate response. Response messages respond to a corresponding request. Command messages require a specific action but do not require a response. Indication messages are informative only and do not require any action or response.

H.223: Multiplexing Protocol

The H.223 multiplexer is a two-layer structure; the adaptation layer (AL) provides the interface between the individual virtual channels (video, speech, data, control) and the multiplex (MUX) layer. The AL supports the underlying MUX layer and the next higher application layer. Variable-length packets are generated by the MUX layer for transmission. This approach provides a form of dynamic multiplexing that can adapt quickly to the needed bandwidth for video, voice, and data. This functionality was a significant improvement over the multiplexing used in H.320, which either does not adapt or adapts very slowly.

8.1.2.3 H.323

Recommendation H.323, the newest member of the family of videoconferencing standards, is designed to provide videoconferencing over LANs that do not provide guaranteed QoS. H.323 terminals and equipment can carry real-time voice, data, video, or any combination. Support for voice is mandatory, while data and video are optional, but if supported, the ability to use a specified common mode of operation is required. The LAN over which H.323 terminals communicate can be a single segment, or ring, or it can be multiple segments with complex topologies. Examples of these networks include Ethernet, Fast Ethernet, and Token Ring technologies. Multiple LAN segments (including the Internet) are supported; for that reason, H.323 technology is expected to dominate videoconferencing. H.323 shares many elements from both H.320 and H.324 in structure, modularity, and audio and video codec recommendations.

Recommendation H.323 enables real-time, two-way communications between two H.323 terminals. Communications consist of control, indications, voice, video, and data between two locations on a LAN. Multipoint communication among more than two terminals is also supported (with and without using a multipoint control unit). Terminals defined in this recommendation can be integrated into PCs or workstations, or they can be standalone units.

Systems and terminal equipment complying with Recommendation H.323 are able to interwork with each other and with those complying with Recommendations H.320 (ISDN), H.322 (GQOS-LAN), and H.321 (ATM). The H.323 gateway unit provides an interconnection between the H.323 LAN and the wide area network (WAN), which can be N-ISDN, B-ISDN, or both. A gateway can be used to connect an H.320 terminal to an H.323 LAN. The gateway can be connected via N-ISDN or B-ISDN to other gateways and

LANs to provide communication between H.323 terminals not on the same LAN. An H.323 terminal communicates with another H.323 terminal on the same LAN directly and without involving the gateway. The gateway can be omitted if communication with terminals not on the same LAN is not required. Interworking with videophones on the PSTN (H.324 series of recommendations) also is possible.

H.323 supports point-to-point and multipoint data conferencing through the T.120 series of recommendations. Version 2 of H.323, approved in 1998, improved the integration of H.323 with T.120 by requiring endpoints that support both T.120 and H.323 to lead the call with H.323. T.120 is an optional part of the H.323 conference. When supported, it enables collaboration through applications such as whiteboards, application sharing, facsimile, and file transfer.

8.1.2.4 H.321/H.322

Recommendation H.322 covers the technical requirements for visual telephone services in those situations where the transmission path includes one or more LANs, each of which is configured and managed to provide guaranteed QoS such that no additional protection or recovery mechanisms beyond those mandated by Recommendation H.320 need be provided in the terminals. Recommendation H.322 does not encompass ATM LANs because they are in the scope of Recommendation H.321. Systems and terminal equipment complying with Recommendation H.322 are able to interwork with each other and with those complying with Recommendations H.320, H.321, and H.323. An H.322 terminal communicates with another H.322 terminal on the same LAN directly. As explained in Section 8.1.2.3, a gateway can be connected to other gateways and LANs to provide communication between H.322 or H.323 terminals not on the same LAN.

8.2 Simultaneous Voice and Facsimile

This section examines the technical developments that made simultaneous voice and facsimile (SVF) possible. Two techniques for splitting the communications channel were proposed, an analog technique and a digital technique. Each technique has advantages and drawbacks. In both methods, the speech signal is compressed, necessitating the selection of an appropriate compression method. The work resulted in a new ITU-T Recommendation, T.39, *Application Profiles for Simultaneous Voice and Facsimile Terminals.*

8.2.1 Communications Channel Sharing Technology

Simultaneous voice and data/facsimile capabilities depend on using in-channel signaling to change modalities. Calls can start as voice calls and add a data/facsimile channel during the call, start as a data call and add voice later, or start as an SVF call and switch to voice only. The techniques for this in-channel signaling, which can take place prior to the establishment of modem communication, are standardized in Recommendation V.8 *bis*.

V.8 *bis* signaling can be used to identify the type of modem for the required communication mode, as well as details of the communication mode. V.8 *bis* permits a list of supported communication modes, as well as software applications, to be exchanged between terminals in the initial interchange. The capabilities exchange feature is similar to the one that takes place in T.30 facsimile, except that it occurs before the applications are communicating. The procedure ensures that the chosen communication mode is possible and avoids attempts to establish incompatible modes of operation.

A capabilities exchange can be performed either automatically at call setup under the control of either the calling or answering station or at any time during the telephone connection. When it occurs during the connection, the communication link may be configured either to return to voice tele- phony mode or to enter immediately into one of the common modes of communication when the information exchange is completed.

When a capabilities exchange takes place in telephony, the interruption in voice communications is short (less than approximately 2 sec) and is designed to be as unobtrusive as possible. Because V.8 *bis* permits automatic switching between voice and fax, users do not have to coordinate their transference of voice and data.

8.2.2 Use of T.30

In normal Group 3 facsimile operation, the connection is established by the modems, and the communication of the T.30 protocol begins at low speed (300 bps). The capabilities exchange occurs during the low-speed operation. Then, if line conditions permit, modems "train" to operate at the speed selected during the capabilities exchange. This procedure is not possible in SVF operation, because the PSTN connection has already been established with a voice call, or the modem connection has been established by using V.8 before the start of the facsimile protocol. Therefore, T.30 Annex C was selected for use in the SVF environment. Annex C originally was designed for use in the ISDN environment, where call establishment also takes place before initiating T.30.

8.2.3 Operational Modes

The following operational modes are possible in an SVF-capable facsimile terminal:

- Initiation as SVF call;
- Switch to SVF call during voice call;
- SVF terminal to standard facsimile terminal;
- Voice-only call.

8.2.3.1 Initiation as SVF Call

When the user initiates the call, the calling equipment detects a dial tone and dials the desired number. To clearly indicate to normal telephone users that they are inadvertently connected to facsimile equipment, CNG is transmitted during the time that the attempt to connect is being made. When the called equipment detects a ring, it answers the call. The called equipment than transmits the menu of available modes (modem, facsimile, data). In response, the calling terminal, using V.8 *bis*, makes a selection from the menu. The selected SVF mode (V.61, V.70, or H.324) is then initiated. All this takes place before the modem connection is made. After establishment of the modem connection, the T.30 duplex procedures are initiated (or data channel, in the case of H.324).

8.2.3.2 Switching to SVF During a Voice Call

There are two ways to switch to SVF capability during a voice call. The first method assumes prior knowledge that the receiver has SVF capability. The second method assumes no prior knowledge of SVF capability.

- *Prior knowledge of SVF capability.* While a voice call is in progress, the user indicates to the equipment that a facsimile connection is desired. The user is aware that the other party has SVF-capable facsimile equipment. The SVF equipment sends an indication of its SVF capability. When the called equipment acknowledges the availability of that mode, it is initiated and both enter into T.30 Annex C duplex procedures.
- *No prior knowledge of SVF capability.* If the caller does not know whether the called terminal has SVF capability, the calling equipment transmits a request for the called equipment's SVF capabilities. If the called equipment is SVF capable, the call proceeds as described for

the first method. Otherwise, the voice call proceeds as before without facsimile.

8.2.3.3 Switch to Voice Only During SVF Call

When the transmission of the fax is completed, the facsimile connection can be terminated automatically by the equipment or manually by the user. The facsimile equipment sends a DCN signal to terminate the SVF mode.

8.2.3.4 SVF Terminal to Standard Facsimile Terminal

The SVF terminal transmits the V.8 *bis* signal indicating that it has SVF capability. Because the called terminal does not have the SVF capability, no response is provided to the calling terminal. Not hearing a response to its request for an SVF session, the calling terminal defaults to normal T.30 procedures.

8.2.3.5 Voice-Only Call

In a voice call, no V.8 *bis* signal is sent and a normal voice call takes place, even though SVF capable equipment may be connected to the line.

8.2.4 Analog Simultaneous Voice and Facsimile

Analog simultaneous voice and facsimile (ASVF) uses the simultaneous voice and data (SVD) modem defined in ITU-T Recommendation V.61. V.61 modems operate at voice plus data signaling rates of 4,800 bps, with optional automatic switching to data-only signaling rates of up to 14,400 bps. The modem can transmit any type of audio signal.

Facsimile signals are sent over the data channel because it is full duplex with 4,800-bps rates. If audio is not present, the modem reverts to V.32 *bis* for data transmissions and to V.17 for facsimile transmissions.

A proposal has been made for a V.34 version of the analog simultaneous voice plus data (ASVD) modem. This version is called V.34Q. It is similar in operation to the V.61 modem except that the speech is encoded prior to its mixing. That does not present a problem from the SVF standpoint because it does not affect the data channel. The proposal for a V.34Q modem has not been accepted for standardization.

8.2.5 Digital Simultaneous Voice and Facsimile

Digital simultaneous voice and facsimile (DSVF) is far more complex than ASVD. Unlike ASVD, a suite of recommendations is needed to describe

DSVD. The basic modem used in DSVD is the V.34 modem. Multiplexing is used to derive two or more channels, one of which can be used for voice. The overall system description is in Recommendation V.70. The terminal control procedures are described in Recommendation V.75, and the multiplexing method is described in Recommendation V.76. V.42-based link access protocol M (LAPM) multiplexing is used in a device-independent mode. The speech coder for the voice channel is defined in Recommendation G.729 Annex A.

The DSVD modem is designed to provide communication over analog channels with capabilities similar to those for terminals in the H.324 series of recommendations. Unfortunately the DSVD standard differs from the other H.324 series recommendations in the selection of multiplexing and voice compression algorithms, which will present some obstacles to interworking.

The following items make up a DSVD system:

- Supervisory and control function;
- Data processing function;
- Voice processing function;
- Control entity;
- Multiplex function;
- Modem.

The supervisory and control function is responsible for the control of the DSVD system. The data processing function is responsible for converting the user data into a format suitable for handling by the multiplex function, (e.g., conversion of start-stop framed data to a format suitable for synchronous transmission). The data channel(s) can be synchronous or asynchronous, and the protocols and applications supported can be indicated and negotiated using the DSVD control entity.

The voice processing function performs the digital encoding and decoding of the speech signal according to Recommendation G.729 Annex A. The voice processing function also includes voice activity detection. A noise generator may be provided to generate noise during idle periods to match circuit conditions and save bandwidth. Optional speech coders can be negotiated using the control entity.

The DSVD control entity is used for overall systems management. It establishes and releases the channels, exchanges capability and parameter information, converts the control information into the H.245 message format, and transfers data, voice, and control information to and from the multiplex function.

DSVD terminals use the V.76 multiplexing capability and can combine one or more voice channels with one or more data channels. A data channel can be used for facsimile to provide DSVF functionality. An optional out-of-band control channel also can be configured. Transmission is bidirectional in all channels. The multiplexer also provides frame delimiting and error protection.

The multiplexed bit stream is transmitted over analog circuits using a V.34 modem. In the future, newly defined high-speed modems such as the PCM modems also may be used. V.8 *bis* is used for mode negotiation and selection at call setup and during the telephone connection.

A system control entity also can be included for functions such as terminal configuration, assignment of channel priority, and capability selection decisions. Although the DSVD/DSVF configurations are more complex than ASVD/ASVF, the increased functionality, upward compatibility, and design commonalties with the H.324 series of multimedia recommendations provide increased flexibility.

8.2.6 Audio Coding for SVF

8.2.6.1 Digital SVF

Because any bandwidth assigned to support voice reduces the bandwidth available for facsimile transmission, it is important to choose an efficient algorithm for the voice coder. In addition, the voice coder has to share computational resources with the facsimile encoder/decoder. The codec used in H.323, defined in G.729, would have been a prime candidate, but it was found to have a greater complexity than desirable. A less complex version, G.729 Annex A, was selected. This codec provides some measure of implementation commonality with G.729.

There is the possibility in V.70-based implementations to negotiate a greater bandwidth for the audio channel if better speech transmission quality is desired. It should be noted that both G.729 Annex A and G.729 are optimized for speech signals. If other types of audio transmission were found to be desirable, codecs optimized for program material other than speech could be used.

8.2.6.2 Analog SVF

One of the advantages claimed for ASVD is that the audio signal is not digitally encoded because the entire procedure takes place in the analog domain. The audio channel in the ASVD modem, therefore, is equally suitable for program material other than voice. The V.61 ASVD modem currently standardized has a maximum data rate of 14.4 Kbps and is not considered viable in competition with V.70 DSVD modems. The independence of signal types was one of the strongest arguments fielded for adoption of the V.34Q proposal. It was countered by the ability of V.70 to negotiate different encodings. An optional codec suitable for general program material has not yet been standardized.

8.2.7 Applications of SVF

8.2.7.1 Point-to-Point Communication

SVF point-to-point communication can be applied as follows:

- During a voice telephone call, one or more pages of a document relevant to the conversation can be transmitted.
- After a facsimile transmission has been initiated, the sender can discuss a matter with the called party.

8.2.7.2 Conferencing

SVF conferencing enables these applications:

- With PC facsimile, a document can be displayed to both parties simultaneously.
- Multiple parties can receive and display facsimile in a conference if a voice conference bridge is used (call setup must be in voice mode).

8.2.8 Interworking of SVF With Other Transmission Media

While the most obvious application for SVF is direct connectivity between two voice and fax applications, a number of interesting scenarios are possible when SVF is used in the H.324 environment. Because T.39 is H.324 enabled, SVF can make use of all the services that H.324 specifies, including the T.120 suite of conferencing services.

T.39 depends on V.8 *bis* for negotiation and session establishment. Thus, only applications that are V.8 *bis* enabled can support SVF. Through the use of the T.125, a voice/facsimile conference can be set up. Because both T.39 and T.127 support binary file transfer, any type of binary file can be distributed during a voice conference.

SVF was conceived to provide voice and facsimile capability for cases in which only PSTN connectivity was available. The rapid evolution of the Internet is making simultaneous voice and data/facsimile available to all Internet users. With the development of methods to transmit facsimile over the Internet, it may be that widespread use of SVF will not take place. Still, SVF does not depend on availability of an ISP and therefore may find some limited application.

8.2.9 Application of SVF to Future Modem Capabilities

The currently standardized SVF capabilities are based on V.32 *bis* and V.34 modems. New modem types under development include V.xdsl, which is designed to operate at data rates higher than 33.6 Kbps as long as the path from the information source to the serving central office is totally digital. The V.90 modems can operate at a maximum data rate of 56 Kbps, although FCC regulations limit the rate to 53 Kbps. That rate is obtainable only for data transfer in the direction to the modem, making it an unlikely candidate for SVF operation.

Digital subscriber line (DSL) modems are also designed to work over copper loops but at rates in the megabit-per-second region, depending on the DSL type. Unlike modems, DSLs require special equipment in the central office. The high data rates make DSL a more reasonable candidate for line sharing between voice and facsimile/data.

A third recent development is bonding, a procedure analogous to the capability in ISDN of tying the two B channels together to obtain a single 128-Kbps circuit. In the analog modem world, it requires two telephone lines and two modems. By using V.90 modems, a theoretical throughput of 112 Kbps can be obtained. It could be argued that that is not an advantage over a separate voice and data line. However, because compressed voice requires only 8 Kbps, a data or facsimile transmission rate of 100 Kbps could be obtained using SVD/SVF techniques.

8.3 Binary File Transfer Using Facsimile Protocols

The Group 3 facsimile recommendations originally were geared toward facsimile terminals, or paper-to-paper transmission. Today, facsimile terminals are being integrated with computers and networks. In addition, the facsimile function is now being implemented in PCs, commonly known as PC-fax. These new products are represented by both add-in PC facsimile boards and by remote facsimile devices connected to the PC through its parallel or serial port. Along with the standard facsimile functions, these new devices offer additional capabilities inherent in network and personal computer environments. Examples of these new features are higher resolution scanners and printers, color, large memories for buffering and storage, scheduling and distribution, and confidential reception. Binary file transfer (BFT) was developed to enhance the use of those new features by providing a general method to transmit files error free from one computer to another.

Although file transmission can be accomplished using standard data modems, there are good reasons for incorporating this feature in PC-fax. For example, facsimile transfer and file transfer can be accomplished in one telephone call. A PC with a PC-fax board uses the same board for fax and for data transfer between computers. The Group 3 handshake, starting when the called computer answers a call, is automatic. It is not necessary for the caller to have any knowledge of the capabilities of the PC being called except that it has BFT capability. Nothing has to be set by the caller to match the called machine, not bits per character, parity, or error correction used. The Group 3 fax ECM is a standard part of the BFT protocol. The protocols of Group 3 fax are extended to provide for the additional BFT function without affecting the Group 3 compatibility of existing Group 3 fax machines. The coding rules of BFT are technically aligned with ISO Standard 8751, file transfer, access, and management (FTAM) coding according to ITU-T Recommendation X.209 [1]. A virtual file store describes a file system that is independent of its underlying physical hardware or operating system. Using BFT, a telematic service can transfer data files between apparatuses for eventual storage on the real file store of the destination.

BFT was approved first as a U.S. standard [2] (EIA/TIA 614) and then as ITU-T Recommendation T.434 [3]. In the United States, the standard was prepared by TIA TR-29.1, Subcommittee on Binary File Transfer (Phil Bogosian, Chair). Recommendation T.434 was first published in 1992 and revised in 1996.

8.3.1 T.434: BFT Format for Telematic Services

Recommendation T.434 describes the semantics and syntax necessary to represent a data file to transfer it through the protocols of various telematic services, such as facsimile Group 3 and Group 4, document transfer and manipulation (DTAM) normal mode and message handling. A binary file message consists of a sequence of attributes that include the file data itself. BFT file attributes are listed in Table 8.3. All the attributes are optional except for protocol version, which is defaultable. The attributes are described using abstract syntax notation (ASN.1).

BFT files are coded according to ASN.1 rules, which translate a binary file and its attributes into a binary range. ASN.1 specifies a set of basic encoding rules that are used to derive the specification of the transfer syntax for values of types defined, using the notation specified in ISO 8824. These basic encoding rules are also for decoding the transfer syntax to identify the data values being

Table 8.3
BFT Attributes

BFT Attributes
Protocol version
Filename
Permitted actions
Contents type
Storage account
Date and time of creation
Date and time of last modification
Date and time of last read access
Identity of creator
Identity of last modifier
Identity of last reader
File size
Future file size
Access control
Legal qualifications
Private use
Structure
Application reference
Machine
Operating system
Recipient
Character set
Compression
Environment
Pathname
User visible string
Data file content

transferred. The ASN.1 transfer syntax notation is very similar to that used by FTAM. Using ASN.1, the encoding of a data value consists of four components, which appear in the following order: (1) identifier octet, (2) length octets, (3) contents octets, and (4) end-of-contents octets. The length of the contents octets can be specified by either the length octets or the end-of-contents octets.

- *Identifier octet.* The identifier octet encodes the ASN.1 tag (class and number) of the type of the data value. This tag is used to identify the data value in the context in which it is coded. The value of the tag can be determined explicitly or implicitly from the production rules being applied.
- *Length octets.* The length octet encodes the length of the contents octets. Two forms of length octets are specified: (1) the definite form, which consists of one or more octets and represents the number of octets in the contents octets; and (2) the indefinite form, which indicates that the contents octets are terminated by end-of-contents octets and consists of a single octet with a value of 80 hex.
- *Contents octets.* The contents octets consist of zero and one or more octets; they encode the data value as specified in ISO 8824 and ISO 8825.
- *End-of-contents octets.* The end-of-contents octets are present if the length is encoded as a single octet with a value of 80 hex; otherwise, they are not present. The end-of-contents octets consist of two zero octets.

8.3.2 BFT Versus E-mail

BFT provides the user of Group 3 equipment with the means to exchange files of any kind with additional information included in a file description and be automatically processed at the receiving side. The file description is a structured document that contains information regarding the file (e.g., file name, content types). It is aimed mainly at being processed automatically at the receiving side. The file description is transmitted ahead of the data file itself and concatenated with it.

The rapid improvement of e-mail functionality on the Internet may decrease the use of BFT on the PSTN or at least slow its growth. At this writing, there are still problems with attaching files to e-mail messages: not all methods of encoding files are compatible, and there are limitations on the file sizes. The biggest difference between e-mail and BFT, however, is that the BFT file is a structured document that is aimed to be automatically processed at the receiving side. E-mail is not set up to do that, at least at this time. With Internet facsimile, BFT is available on the Internet as part of Group 3 (discussed in Chapter 9).

References

[1] ITU-T Recommendation X.209, *Specification of Basic Encoding Rules for Abstract Notation One (ASN.1).*

[2] EIA/TIA-614, *Binary File Transfer Format for Group 3 Facsimile.*

[3] ITU-T T.434, *Binary File Transfer Format for Group 3 Facsimile.*

9

Transmission of Group 3 Facsimile

The facsimile protocol, as defined by Recommendation T.30, is a real-time protocol. Its operation depends on a constant interaction between the sending terminal and the receiving terminal. While there are a number of standard defaults, most transmissions currently use options that have to be negotiated at the beginning of the transmission and that may be renegotiated between pages. The highest currently available transmission rate defined in the standard on the (analog) PSTN is 33.6 Kbps. On digital networks, such as ISDN, a rate of 64 Kbps or higher is possible. Because T.30 defines a protocol, transmission can take place at any rate the digital network can support, but it may have to be negotiated as a nonstandard facility option.

Procedures also have been standardized for facsimile transmission over certain data networks, such as the X.25 packet network. There are also procedures for store-and-forward (non-real-time) facsimile transmission and carriage of facsimile information in electronic mail services such as X.400 and the simple mail transfer protocol (SMTP). At this writing, procedures are being standardized for facsimile transmission over the Internet. The basic Internet transport protocols are user datagram protocol (UDP) and transport control protocol/Internet protocol (TCP/IP).

Facsimile terminals are of two types: standalone dedicated terminals (fax machines) and facsimile protocols implemented on a PC and using the PC facsimile/data modem for facsimile transmission. Both types implement the same standard protocols. A facsimile/data modem implemented as a plug-in board is often called a facsimile board. An extension of this concept is the facsimile server, which contains one or more boards in a PC or workstation. Facsimile servers normally do not support data protocols, are more robust, and may support 4 to 12 lines per board. They typically are used on a LAN. PC-based facsimile normally is used to transmit a document created on the PC. The procedure does not require the creation of a printed copy, and the output from PC facsimile is of higher quality for the same resolution, because it does not contain distortions attributable to a scanner. For the transmission of existing documents, a dedicated facsimile terminal is more convenient. One reason for the success of facsimile as a communications medium is the ease with which it can be accomplished when facsimile terminals are used. The only skills required of the operator are the ability to insert the paper, dial a telephone number, and press the start button. At the receiving end, all that is required is removal of the transmitted document.

The facsimile protocols originally were designed to operate in real time over the PSTN. Regardless of the transmission medium, successful operation of the facsimile protocol still depends on maintaining the connection between the two terminals. Many command-response sequences occur, and several timers are set during the procedure. If an appropriate response to a command is not received in time, the procedure may revert to a previous point in the protocol (called a retry), or the call may be disconnected. This requirement of the facsimile protocol presents a special challenge for transmission over some networks, such as cellular networks and the Internet.

Another important feature of facsimile communication is the available confirmation of sender and receiver identities. The sender knows immediately that connection has been successful when the recipient's terminal ID (phone number or name) appears on the sending terminal's LCD screen. The sender's ID is also printed on the top of each page, in the caller identification line. Those two features are actually part of the protocol rather than content. Many facsimile terminals also accumulate a log of all transactions. The date and time of transmission also are usually part of the identification procedure. Because of the identification procedure, documents transmitted by facsimile have some legal status.

9.1 PSTN (Modems)

At the facsimile transmitter, a facsimile modem (from *mo*dulation/*dem*odulation device) accepts digital information, codes it into a complex format, modulates it to form an analog signal, and delivers it to the telephone line. At the facsimile receiver, the modem decodes the analog signal and converts it back to a duplicate of the digital signal fed to the modem at the transmitter. Recommendation T.4 specifies modulation and demodulation schemes, with the necessary portions of the modem built into the Group 3 facsimile machines. A full modem requires an EIA-232 interface, but there is none between the Group 3 facsimile digital signal and its internal modems.

The mandatory modem requirements for Group 3 facsimile are Recommendation V.27$_{ter}$, for sending facsimile message data at 4.8 and 2.4 Kbps, and V.21, channel 2 (return channel), for 300-bps handshake signaling, used primarily before and after each page. Due to its slow speed, the 300-bps rate does not require telephone line equalization. The receiving modem for the facsimile signal, however, has an automatic equalizer to compensate for the telephone line problems of amplitude distortion and envelope-delay distortion. This improves the accuracy of the delivered digital facsimile signal, resulting in a greatly reduced error rate.

Phase jitter and noise cause the received signal state to move about from its assigned amplitude-phase position. When the distortions are severe, the signal state may be in the wrong amplitude-phase position when sampled and be mistaken for an adjacent signal state, causing bit errors. One sample error can have up to six bit errors.

The receive modem has an automatic equalizer to compensate for the telephone line amplitude distortion and envelope-delay distortion, improving the accuracy of the delivered digital facsimile signal. Optional modems are V.29 (9.6 and 7.2 Kbps), V.17 with trellis coding for improved error immunity (14.4, 12, 9.6, and 7.2 Kbps), and recently V.34 (up to 33.6 Kbps).

Group 3 facsimile modems automatically test the transmission characteristics of the telephone connection. Before sending data, the modem sends a training signal of fixed format known to the receiver. The highest rate available in both facsimile machines is tried first. If that speed would give too many errors, the transmitter tries the next lower speed. If that fails, the modem rate again steps down to the next lower speed. For example, for facsimile machines that have V.29 modems, the rates tried are 9.6, then 7.2, switching to V.27$_{ter}$ for 4.8 and 2.4 Kbps. Facsimile machines with a top rate of 4.8 Kbps start the handshake at that speed. The system ensures transmission at the highest rate consistent with the quality of the phone line connection. When V.29 operates at 7.2 Kbps, it samples 3-bit segments of digital facsimile signal. That gives eight analog signal states on the telephone line for the same 2,400-baud rate with the same telephone line bandwidth as for 9.6 Kbps. At 7.2 Kbps, the modem receiver circuitry makes a signal state decision on only 8 states instead of 16, and a larger phase/amplitude error is tolerated, lowering the error rate on some telephone channels.

A reduction in handshake time results from use of a short train of 143 ms at all times except for the initial training, when a 1.4-sec training signal is used. The V.29 modem uses a long training signal each time it is used in handshaking before starting to send new information. The new facsimile machines that use V.17 still have V.29, V.27_{ter}, and V.21, channel 2, to maintain compatibility with the existing Group 3 facsimile machines.

Because international circuits now carry a heavy load of facsimile traffic, equipment that detects a facsimile call has been added. The V.29 or V.27_{ter} facsimile signal is then demodulated before reaching the international portion and sent on a 32- or 40-Kbps channel, saving bandwidth. At the other end of the channel, the facsimile signal is restored to normal for the analog PSTN channels.

9.1.1 Standard Operation: V.27_{ter} (4.8 and 2.4 Kbps)

At 4.8 Kbps, the modulation rate (or baud rate) is 1,600 baud, or 1.6 KB. The data stream to be transmitted is divided into groups of three consecutive bits (tribits). Each tribit is encoded as a phase change relative to the phase of the preceding signal tribit element (Table 9.1). At the receiver, the tribits are decoded and the bits reassembled in correct order.

At 2.4 Kbps, the modulation rate is 1,200 baud, or 1.2 KB. The data stream to be transmitted is divided into groups of two bits (dibits). Each dibit is encoded as a phase change relative to the phase of the immediately preceding signal element (Table 9.2). At the receiver, the dibits are decoded and reassembled in the correct order.

Table 9.1
4.8-Kbps Phase Encodings

Tribit Values	Phase Change
0 0 1	0 degrees
0 0 0	45 degrees
0 1 0	90 degrees
0 1 1	135 degrees
1 1 1	180 degrees
1 1 0	225 degrees
1 0 0	270 degrees
1 0 1	315 degrees

Table 9.2
2.4-Kbps Phase Encodings

Dibit Values	Phase Change
00	0 degrees
01	90 degrees
11	180 degrees
10	270 degrees

9.1.2 Optional Operation A: V.29 (9.6 and 7.2 Kbps)

For optional higher speed operation, as may be possible on high-quality circuits, V.29 at 9.6 or 7.2 Kbps can be used. This optional mode has the following characteristics:

- Primary operation at 9.6 Kbps, with a fallback to 7.2 Kbps;
- Combined amplitude and phase modulation with synchronous mode of operation;
- Inclusion of an automatic adaptive equalizer.

At 9.6 Kbps, the modulation rate is 2.4 KB. The scrambled data stream to be transmitted is divided into groups of four consecutive data bits (quadbits). The first bit (Q1) in time of each quadbit is used to determine the signal element amplitude to be transmitted. The second (Q2), third (Q3), and fourth (Q4) bits are encoded as phase changes relative to the phase of the immediately preceding element (Table 9.3). The relative amplitude of the transmitted signal element is determined by the first bit (Q1) of the quadbit and the absolute phase of the signal element (Table 9.4). The absolute phase is initially established by the training signal.

At the fallback rate of 7.2 Kbps, the modulation rate is 2.4 KB. The scrambled data stream to be transmitted is divided into tribits. The first data bit determines Q2 of Table 9.3, and the second and third data bits determine Q3 and Q4, respectively.

Table 9.3
9.6-Kbps Phase Encodings

Q2	Q3	Q4	Phase Change
0	0	1	0 degrees
0	0	0	45 degrees
0	1	0	90 degrees
0	1	1	135 degrees
1	1	1	180 degrees
1	1	0	225 degrees
1	0	0	270 degrees
1	0	1	315 degrees

Table 9.4
Relative Amplitudes

Absolute Phase (Degrees)	Q1	Relative Signal Element Amplitude
0, 90, 180, 270	0	3
	1	5
45, 135, 225, 315	0	$\sqrt{2}$
	1	$3\sqrt{2}$

9.1.3 Optional Operation B: V.17 (14.4, 12, 9.6, and 7.2 Kbps)

For higher speed operation on telephone circuits that can support the higher speed, V.17 can be used. This optional mode has the following characteristics:

- Provision for half-duplex synchronous operation;
- Quadrature amplitude modulation with synchronous line transmission at 2.4 KB;
- Inclusion of data scramblers, adaptive equalizers, and eight-state trellis coding;

- Two sequences for training and synchronization: long train and resync.

At all speeds, the modulation rate is 2.4 KB. At 14.4 Kbps, the scrambled data stream to be transmitted is divided into groups of six consecutive data bits and mapped onto a signal space of 128 elements. At 12 Kbps, the scrambled data stream is divided into groups of five consecutive data bits and mapped onto a signal space of 64 elements. At 9.6 Kbps, the data are divided into groups of four consecutive data bits and mapped onto a signal space of 32 elements. At 7.2 Kbps, the data are divided into groups of three consecutive data bits and mapped onto a signal space of 16 elements. The V.17 modem is more robust than V.29, due to its trellis coding method; for example, V.17 can operate at 9.6 Kbps on certain telephone lines while V.29 cannot.

9.1.4 V.34/V.8

Higher data rates are theoretically possible and highly desirable for transmission of continuous tone color. The V.34 modem standard, approved in 1994, provides speeds to 33.6 Kbps. Use of the V.34 modem is described in Annex F of Recommendation T.30. This modem also uses trellis coding and a number of symbol rates from 2,400 to 3,429 symbols/s. It requires the use of ECM and runs in half-duplex for facsimile. An unusual feature of this modem is the use of Recommendation V.8 for the modem startup procedures. V.8 provides the means to determine the best mode of operation *before* the initiation of the modem handshake. The initial V.34/V.8 standard had some problems relative to facsimile use. The evidence for that can be found in the large number of modifications to V.34 contained in Annex F of Recommendation T.30. At this writing, V.34 has not had wide use in facsimile.

9.2 ISDN

Group 3 operation over the ISDN is described in Annex C of Recommendation T.30. This option, commonly known as Group 3C, originally was designed to be used on the ISDN, but the protocols optionally can be used on digital networks other than the ISDN. In fact, the protocols also can be used on the PSTN using modulation schemes. The procedures and signals used are based on those defined in the main body of Recommendation T.30 as well as Annex A, because error correction is an integral part of the protocol. The protocol was designed to operate in duplex mode to take advantage of the capability of the ISDN. It also can operate in half-duplex mode. In either case, error correction is required. The procedure is very similar to that defined in Annex A, but it is modified to accommodate the duplex mode. The error correction method used on the ISDN is outlined next.

As in Annex A, the error correction method is based on a page-selective repeat ARQ technique. An HDLC frame structure is utilized for all facsimile message procedures. The transmitting terminal divides the message into a number of concatenated frames as defined in Annex A of Recommendation T.4 and transmits it as a number of pages or partial pages. The transmitting terminal uses a frame size of 256 octets as indicated in the DCS command, and the receiving terminal must be able to receive a frame of that size. Optionally, in operation over analog networks, a frame size of 64 octets may be indicated by the transmitting terminal.

In the duplex mode of operation, the transmitting terminal transmits subsequent partial pages without waiting for a response to the preceding partial page. If corrections are required, they are sent at the end of the next partial-page transmission. Unacknowledged commands from previous pages or partial pages are retransmitted prior to any corrections. For half-duplex, all corrections are sent and acknowledged before a subsequent partial page is sent. When the previous message has not been satisfactorily received, the receiving terminal transmits a partial page request (PPR) response to indicate that the frames specified in the associated facsimile information field must be retransmitted. The PPR signal contains the page and block numbers as well as the required frame numbers. When a PPR signal is received, the transmitting terminal retransmits the requested frames specified in the PPR information field.

The maximum number of attempts to correct a page is left up to the transmitter. If the transmitter decides that too many attempts have been made, it sends the DCN signal. If the receiver is unable to continue to receive new information, it sends receive not ready (RNR) continuously until it is ready to receive new information. During that time, the transmitter sends any outstanding correction frames and any unacknowledged commands. If there are no outstanding corrections, then it continuously transmits any unacknowledged commands until it receives a response other than RNR. The transmitter sends no new information until all previously transmitted pages have been acknowledged as received correctly.

The format of the initial identification is a repeated sequence of XID + DIS, or XID + NSF + DIS, or XID + NSF + CSI + DIS sent three times concatenated, followed by 256 flags. The sequence is transmitted until a valid response is received from the calling terminal subject to a maximum time of 5 sec. The exchange identification (XID) procedure indicates that the called terminal has Group 3C capabilities. This signal is defined in Recommendation T.90. (The format of the XID frame is defined in Annex F/T.90.)

9.3 Packetized Facsimile

In addition to the point-to-point facsimile transmission methods, two other methods for facsimile transmission have been defined: packetized facsimile and store-and-forward facsimile. Packetized facsimile is a method developed to allow transmission of facsimile data over X.25 packet networks. Standards for packetized facsimile transmission were approved in 1992. Packetized facsimile operates in real time and thus provides for end-to-end negotiation between terminals. It supports all standardized options and nonstandard facilities other than nonstandard modulation schemes. The following standards define the operation of packetized facsimile:

- Recommendation X.5: Facsimile Packet Assembly/Disassembly Facility (FPAD) in A Public Data Network;
- Recommendation X.38: Group 3 Facsimile Equipment/DCE Interface for Group 3 Facsimile Equipment Accessing the FPAD in a Public Data Network Situated in the Same Country;
- Recommendation X.39: Procedures for the Exchange of Control Information and User Data Between an FPAD Facility and Packet-mode DTE or Another FPAD.

Because FPADs normally are network based, facsimile transmissions can occur over any X.25 network.

To access an FPAD, the facsimile terminal uses the public telephone network. Connection to the FPAD can be manual or automatic. One automatic method is the insertion of a bump-in-cord (BIC) device between the facsimile terminal and the wall jack. The BIC is preprogrammed with the FPAD's number and any necessary authorization codes. The BIC also can be programmed to allow facsimile transmissions only to certain destinations. When an FPAD answers the call from the calling facsimile terminal, the FPAD determines the destination address; using dual-tone multiple frequency (DTMF), it establishes a modem connection with the called FPAD. The called FPAD then establishes a modem connection with the called facsimile terminal. When the called facsimile terminal answers, the FPAD transmits the calling facsimile terminal's address and establishes a virtual connection between the facsimile terminals. The called facsimile terminal then transmits its 2,100-Hz answer tone. The FPAD procedures preserve almost all standard and optional facsimile features.

9.4 Store-and-Forward Facsimile

Another standardized facsimile communication method is called store-and-forward facsimile. Because the store-and-forward service is not capable of real-time operation, negotiation between sender and receiver is not possible; the result may be low QoS. For instance, if a high-resolution image is sent and the receiving terminal supports only standard resolution, the image will have to be converted to ensure delivery. The conversion to the lower resolutions may not satisfy the requirements of the sender, who might have decided not to transmit the image if afforded the opportunity to negotiate capabilities. The following recommendations define the operation of store-and-forward facsimile.

- F.162: Service and operational requirements of store-and-forward facsimile service;
- F.163: Operational requirements of the interconnection of facsimile store-and-forward units.

Standardized store-and-forward facsimile is also known as COMFAX and was designed to take advantage of an existing messaging (e-mail) infrastructure, such as X.400 based message systems.

The store-and-forward procedure requires two-stage dialing similar to that in FPADs. From that point, however, a complete facsimile transaction takes place between the facsimile terminal and store-and-forward unit (SFU) in the network. The received facsimile message remains in its compressed form and is encapsulated in a Group 3 message body part as defined in the X.400 recommendations. The body part also carries information about the characteristics of the facsimile encoded message. The message header carries the destination information. The receiving SFU establishes a connection with the called facsimile terminal, uses the transmitted characteristics to negotiate the connection, and transmits the message. The SFU may generate a cover page.

Two types of delivery information can be returned: delivery to the SFU (level 1) and delivery to the facsimile terminal (level 2). The sender is not required to be notified of successful delivery, since that is an option, but must be notified of delivery failure. The SFU may implement conversion facilities if the receiving facsimile terminal does not support options used to encode the facsimile message. Address lists can be stored in an SFU, allowing a single message to be sent to multiple recipients. Some services allow for the submission of address lists via a cover page.

A number of international carriers support COMFAX-like services, but their use has been limited. With the introduction of standardized Internet facsimile, COMFAX services are not likely to grow significantly.

9.5 Internet

9.5.1 Internet Protocols

The basic Internet transport protocols are user datagram protocol (UDP) and TCP/IP. The TCP/IP protocol provides a packetized data transport service between two endpoints. TCP/IP error control is provided by packet retransmission. UDP transmission is faster but does not provide any error control. In addition to the data transport service between the applications at Internet addresses, several specialized communications protocols can make use of TCP/IP. Of interest here are file transfer protocol (FTP), for file transport, and SMTP, for electronic mail. Transfer speeds depend on the speed available in the links that make up the connection between the two Internet addresses, error conditions such as lost packets, and the number of routers the packets have to traverse. Packets belonging to a particular message do not all necessarily follow the same paths through the network. Because Group 3 facsimile uses a real-time interactive protocol, it cannot operate directly over a connection that includes the Internet without the use of gateways. A specialized protocol that is also widely implemented is the hypertext transport protocol (HTTP), which is intended for use in World Wide Web (WWW) applications.

9.5.2 Mechanisms for Transmitting Facsimile Over the Internet

The mechanisms for transmitting facsimile between two facsimile terminals over the Internet are similar to facsimile transmission over any packet-type network: store-and-forward service and real-time connectivity.

9.5.2.1 Store-and-Forward Service

Electronic mail over the Internet is implemented as a store-and-forward service. Mail transfer is accomplished in three steps. The first step involves the connection of the user agent or messaging client with a network server or network node. The protocol used here is usually post office protocol (POP). POP is being superseded by a protocol with more capabilities: Internet mail access protocol (IMAP). IMAP has many additional capabilities, such as selective retrieval. The protocol used to transport mail over the Internet between servers is SMTP or extended SMTP. SMTP was originally designed to handle text messages only. Messages with binary content, such as images or computer files, have to be converted to an all-7-bit text format by procedures such as uuencode/uudecode or base64, which increase the size of the files being transported. With the addition of nontext messages, a method was needed to identify the message type. That has been provided by the multipurpose Internet mail extensions (MIME) protocol extensions to SMTP. For each new message type, a MIME content-type has to be agreed on and standardized.

A server consists of an application implemented in hardware that has connectivity to the telephone network on one side and the Internet on the other side. Servers that provide the connection to the Internet usually are furnished by ISPs. On the Internet side, ISPs implement some or all of the following protocols: FTP, HTTP, SMTP, TCP/IP, and UDP. Those protocols have been standardized by the Internet Engineering Task Force (IETF). Of the protocols listed, only SMTP makes provisions for the transmission of service messages such as notifications.

9.5.2.2 Real-Time (Session-Oriented) Internet Facsimile

Session-oriented facsimile is the currently agreed-on terminology for what has also been called *real-time facsimile* and *direct terminal-to-terminal facsimile.* Standard facsimile protocols designed for use over the PSTN, as well as other networks, are the basis for this mode. Poor-quality networks can be handled by reducing the modem speeds. ECM can be employed to maintain image quality, although over extremely poor connections that may result in no transmission whatsoever. Sometimes, there are difficulties over satellite connections with more than one hop.

The principal impairment encountered over the Internet is the long latency, which is not due to a single factor. It is due mainly to packet loss and router processing times. Packet losses of 30% or more are not uncommon. Error-free transmission is obtained by retransmission of the lost packets.

The T.30 protocol is extremely delay sensitive. It contains six timers, which are sensitive to certain actions or responses. Many connections also have to traverse 10 to 16 routers, each of which adds some delay. Facsimile terminals cannot connect to the Internet directly but have to use a server or gateway, which introduces some additional delay. If a response is delayed because of the latency of the Internet, the session will terminate. Several mechanisms, such as padding lines in T.4, can be employed to decrease these timing sensitivities.

9.5.3 Early Implementations

Several early implementations claimed to provide a "facsimile over the Internet" service. On closer examination, most did not actually provide terminal-to-terminal service. At the time this investigation was compiled, all implementations were store-and-forward-type services. The variations on Internet facsimile include several whose only connection with the Internet is the use of Internet e-mail. All required software supplied by the service provider to be installed on the PC. The types that have been identified are:

- The word processing document is converted to facsimile file, and the file is e-mailed from the desktop to a node, then faxed to a terminal.
- The word processing document is converted to facsimile file, and the file is e-mailed to another PC.
- The facsimile terminal transmits to node, and the facsimile is delivered by e-mail.
- The word processing document is converted to a facsimile file and submitted via Web page to base location. Then the facsimile is delivered via public or private network.

9.5.4 Standardization of Facsimile Transmission Over the Internet

Standardization of Internet facsimile communication methods is essential for their development. Two standards organizations have taken the challenge of defining and writing standards for Internet facsimile. As to be expected, Study Group 8 of ITU-T has initiated work from the facsimile side. From the Internet side, the IETF has started a new facsimile working group in the applications area. Both groups have expressed interest in cooperation, which will be essential to avoid incompatible standards.

9.5.4.1 ITU-T Study Group 8

Initially, the Internet facsimile standardization work in Study Group 8 was to be divided: Question 1 was to be responsible for developing protocols, while Question 4 was to be responsible for service definitions. Soon it was decided that all the work should be done in Question 4.

At the February 1997 meeting of Study Group 8, Question 4 drafted a set of objectives for the study of Internet facsimile, based in part on a U.S. contribution. The group agreed to investigate how facsimile terminals, in their current form, should communicate over the Internet while (1) directly connected to the Internet and (2) using the PSTN or ISDN for access to and from the Internet. Two modes of operation were to be provided for (1) a real-time, session-oriented mode and (2) an indirect mode using store-and-forward facilities. It was also agreed to look at whether new types of terminals might be defined for facsimile communication over the Internet. Three basic requirements were to be met: (1) Terminals must provide for the identification of terminal capabilities; (2) terminals must permit facsimile interworking between the installed base of terminals and any new terminals specifically designed for facsimile over the Internet; and (3) terminals must provide facilities for confirming the results of the facsimile transaction. These requirements would ensure that facsimile communication over the Internet will maintain the quality of service that users have grown to expect from Group 3 facsimile communication. The study group also agreed that it would be useful to establish communication with the IETF facsimile working group on issues related to Internet facsimile.

9.5.4.2 IETF Facsimile Working Group

The IETF facsimile working group established a charter and agreed to take steps for specifying a facsimile-related core messaging service over the Internet. It was to develop a shared set of terminology and definitions to ensure a common framework for participants having differing backgrounds in Internet protocols and facsimile telecommunication. The group was to review existing facsimile-related Internet data specifications and accept, modify, replace, or augment them with particular attention to their encapsulation, such as via MIME. The mechanisms for addressing and receipt notification for facsimile data carried via Internet mail were to be specified. The initial project for the IETF facsimile working group was to be the definition of the facsimile message format based on TIFF, with some facsimile-specific additions.

As may be seen from the above descriptions, the ITU-T tends to take an overall service-oriented approach, while the IETF prefers to tackle smaller chunks amenable to rapid resolution. Initially, facsimile knowledge in the IETF was limited, but with cooperation considerable progress was made in both groups. Both the IETF facsimile working group and the ITU-T Study Group 8 have progressed standards on Internet facsimile. It is notable that, while the two efforts started independently, the end result came out of a cooperative effort. The bulk of the cooperative effort was on the subject of store-and-forward facsimile; there was no effective input on real-time facsimile from the IETF.

9.5.5 Service Requirements for Internet Facsimile

Service requirements are the subject of standards documents from the ITU-T and the IETF. The ITU-T standard is Draft Recommendation F.185, *Internet*

Facsimile: Guidelines for the Support of the Communication of Facsimile Documents. The IETF standard is RFC 2805, *A Simple Mode of Facsimile Using Internet.* Draft Recommendation F.185 is more general in nature and covers both real-time and store-and-forward e-mail Internet fax (IFax). The IETF document covers only a small feature set of e-mail–based IFax called basic mode. Work on a more extensive set called full mode is in progress.

The interface between the PSTN and the Internet is called a gateway. From the PSTN to the Internet, a gateway usually is called an onramp; from the Internet to the PSTN, it is called an offramp. This terminology is Internet-centric. For real-time IFax, the onramp is called the emitting gateway and the offramp is the receiving gateway.

9.5.5.1 Definition of Store-and-Forward Facsimile

Store-and-forward IFax occurs when the sending and the receiving terminal are not in direct communication with each other. The transmission and reception take place via the store-and-forward mode on the Internet using Internet e-mail. In store-and-forward mode, the facsimile protocol "stops" at the gateway to the Internet. It is reestablished at the gateway leaving the Internet.

Two modes of store-and-forward facsimile are defined. In simple-mode store-and-forward, only the coded image is transmitted. In full-mode store-and-forward facsimile over the Internet, three requirements must be satisfied:

- The capabilities of the terminals are exchanged.
- An acknowledgment of receipt is exchanged between gateways and may be transferred from the receiving terminal to the sending terminal.
- The contents of standard messages used by the transmitting terminal are preserved.

The IETF term for store-and-forward IFax is *e-mail fax.*

Table 9.5 summarizes the implementation requirements for simple-mode store-and-forward facsimile.

9.5.5.2 Definition of Real-Time Internet Facsimile

Real-time IFax involves the transmission/reception of a facsimile in a single session such that confirmation of receipt is provided to the transmitting terminal prior to termination of the call. A real-time IFax communication over the Internet between two standard facsimile terminals involves preserving the contents and sequence of standard messages exchanged between the sending and receiving terminals. In addition, the duration of the communication between

Table 9.5
Implementation Requirements for Simple-Mode Store-and-Forward Facsimile

Sender	
Required	Send image data as a single MIME multipage TIFF profile-S file
	Provide notice in case of local transmission problems
	Provide return address of an Internet e-mail receiver that is MIME compliant
Strongly recommended	Include message-ID
	Use base64 encoding for image data
Optional	Use other TIFF profiles if it has prior knowledge that such profiles are supported by the receiver
	Provide notice on receipt of DSN or other notifications
Receiver	
Required	Be MIME compliant except that placing a MIME attachment in a file is not required and a received file may be printed rather than displayed
	Be capable of processing multiple MIME TIFF profile-S image files within a single message
	Provide notice in case of reception or processing problems
Offramp Gateway	
Required	Be SMTP compliant
	Provide delivery failure notification in the form of a DSN
	Be able to process PSTN/facsimile e-mail addresses
	Comply with the relevant ITU-T recommendations relating to facsimile transmission
	Attempt to relay authorized e-mail to the corresponding Group 3 facsimile terminals
	Ensure that local legal requirements relating to facsimile transmissions are met
Strongly recommended	Use DSN for delivery failure notification
	Use an approved mailbox access protocol when serving multiple users
Optional	Translate image data into a format acceptable by the receiving Group 3 facsimile terminal
	Use a mailbox access protocol when serving a single mail recipient
Mailstore	
Required	Be SMTP compliant
	Provide delivery failure notification in the form of a DSN

the two terminals should not take appreciably longer than would have been the case if the two terminals had been connected directly over the PSTN.

9.5.6 Store-and-Forward Internet Facsimile

The ITU-T Rec. T.37, *Procedures for the Transfer of Facsimile Data via Store-and-Forward on the Internet*, does not in itself define those procedures. It references a set of IETF documents called Request for Comment documents (RFCs) that define the procedures for facsimile communication over the Internet using e-mail. The current draft of T.37 standardizes only the use of simple mode; full mode is left for further study. The intention is that full mode should support, to the greatest extent possible, the standard and optional features of Group 3 facsimile. That would include, among others, delivery notification, capabilities exchange, color and other optional coding mechanisms, and file transfer. Because the Internet e-mail protocols do not as yet allow many of those functionalities, much work remains to be done. Figure 9.1 illustrates the reference model for store-and-forward IFax.

A critical part of the support for store-and-forward facsimile is the definition of the format in which the facsimile image is carried, the file format. The format is defined in RFC 2301, *File Format for Internet Facsimile.* The file format defined in RFC 2301 supports all the image formats standardized for use with Group 3 in Rec. T.30. The document describes the TIFF representation of image data specified by the ITU-T recommendations for bilevel and color facsimile. This file format specification is commonly known as TIFF-FX. It formally defines minimal, extended, and lossless JBIG modes (Profiles S, F, J) for black-and-white facsimile, and base JPEG, lossless JBIG, and mixed raster content modes (Profiles C, L, M) for color and gray-scale facsimile. The basic TIFF standard is proprietary to Adobe, but a free license has been granted for use with IFax. Simple-mode store-and-forward facsimile supports only the S profile.

The reason for developing an e-mail-based store-and-forward mode is to permit interworking among facsimile and e-mail users. Existing e-mail users should be able to send normal messages to lists of users, including facsimile-based recipients. E-mail recipients also should be able to reply to the received facsimile message and continue to include facsimile recipients. Existing e-mail software should work without modification and should not have to process new or different data structures beyond what is normal for Internet mail users. The store-and-forward facsimile mode, therefore, is based on existing IETF e-mail service standards to provide compatibility with existing e-mail service.

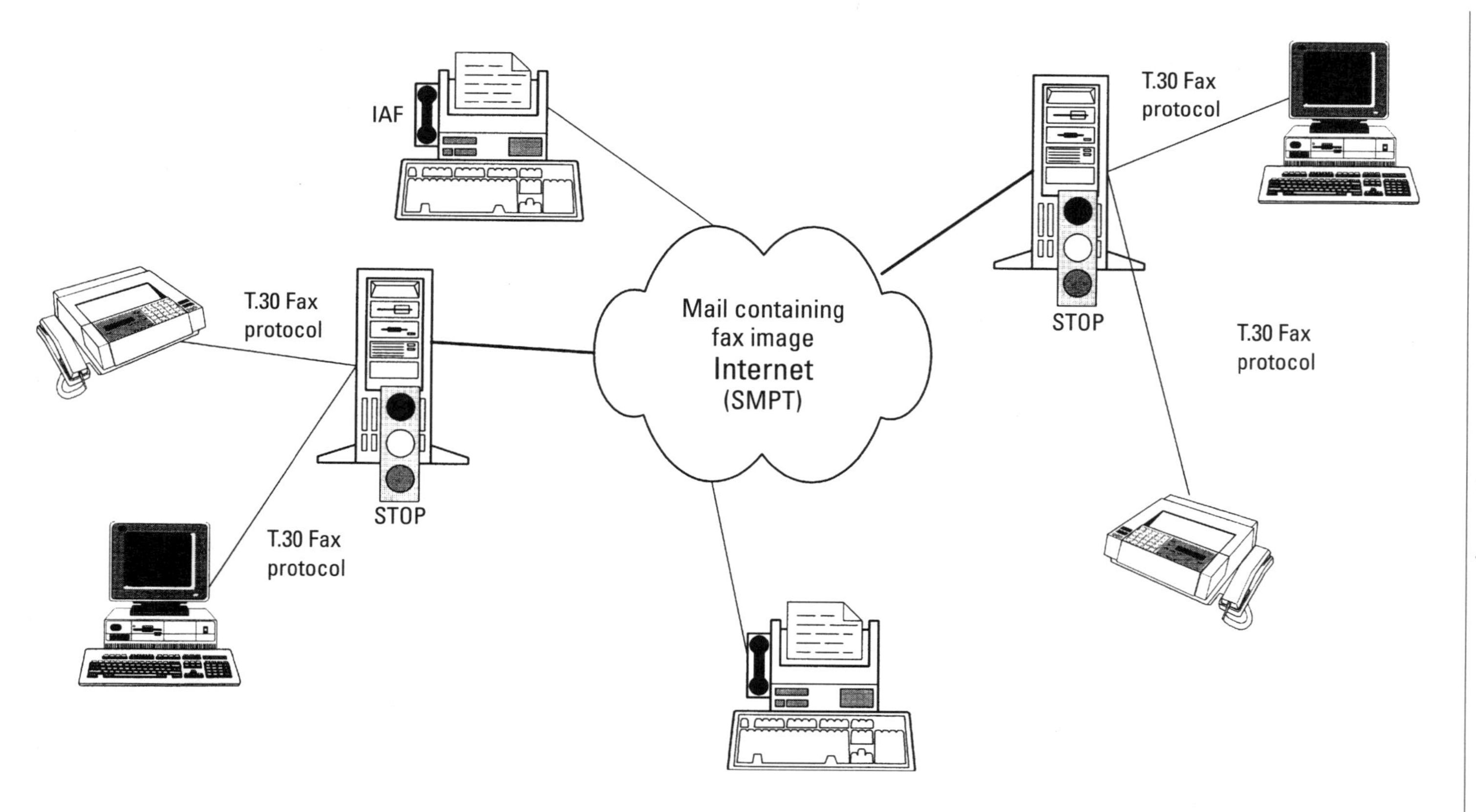

Figure 9.1 Store-and-forward IFax reference model.

Several interchange modes are possible under the RFC 2301 scenario:

- Internet mail–network printer;
- Internet mail–offramp gateway (forward to Group 3 facsimile);
- Network scanner–network printer;
- Network scanner–offramp gateway (forward to Group 3 facsimile);
- Network scanner–Internet mail.

9.5.6.1 Mail Protocols

The e-mail message containing the facsimile information in the form of a TIFF file is identified by means of a MIME type. The MIME registration is standardized in RFC 2302, *Tag Image File Format (TIFF)—Image/TIFF MIME Sub-type Registration.* This enables the receiving application to invoke the appropriate processing of the content. The receiving application may be a mail user agent, or it may be the gateway converting it back into the T.30 protocol for sending to a Group 3 facsimile machine. The standard protocol for transferring mail in the Internet between message transfer agents is SMTP. User agents can use POP3 to retrieve mail.

Positive delivery notifications generally are not available in the Internet. Delivery failure notifications usually are available and should be sent using delivery service notification (DSN) if the facsimile message transfer fails at any place in the delivery chain.

9.5.6.2 Internet Facsimile Addressing Issues

The addressing scheme of the Internet differs from that of facsimile, which is a telephony application. A conventional facsimile terminal has no facilities to enter an alphanumeric address in the traditional Internet user@domain format. The destination address information must be conveyed to the ISP, which in turn must convey it to an ISP located in the vicinity of the destination. The information could be part of the Internet address or it could be carried as part of the message.

The addressing requirements for various modes of operation are examined next.

Case 1: Traditional Facsimile to Traditional Facsimile

For traditional facsimile to traditional facsimile communication, the sender does not know anything about an Internet address or whether it might be a facsimile server with a direct connection to the Internet. The originator has to transmit the destination numeric facsimile address to the ISP. There are three possibilities:

- Manual double dialing with voice prompts;
- Using a dialing device;
- Using a cover page, which must be scanned by the ISP and interpreted by means of OCR.

The ISP then transfers the facsimile to a destination ISP, based on the facsimile address and any previous relationship to the ISP. The ISP dials out and delivers the facsimile.

Case 2: PC Facsimile to Traditional Facsimile

The PC facsimile can use a complete Internet address, but it has to know the domain name of a service that delivers facsimile messages, either from their own offramp or by arrangement with another operator.

Case 3: PC Facsimile to PC Facsimile

In reality, this is a standard e-mail-based delivery.

Case 4: Traditional Facsimile to PC Facsimile

This mode of operation uses the ISP behavior of Case 1. Offramp is also similar to Case 1, because the ISP operator generally does not know that the destination is a PC facsimile, except in special cases. If OCR is used and the sender happens to know the domain name of the destination PC facsimile, delivery could be like Case 3.

Case 5: Internet-Aware Facsimile

An Internet-aware facsimile (IAF) is a facsimile machine that can access the Internet directly. An IAF is equipped with a keyboard for entering Internet addresses. It also has built-in protocols so it can act as its own gateway.

Two new standards were written to describe the format of telephony addresses for transmissions over the Internet. RFC 2304, *Minimal FAX Address Format in Internet Mail*, specifies the format for addressing a facsimile machine connected to the PSTN over the Internet. The facsimile address is specified as a fully qualified E.164 address. An example of this is FAX = +3940226338. The format also allows a T.33 subaddress to be specified in the string, for example, FAX = +12027653000/T33S = 1387.

When the destination facsimile is an IAF or is reached by means of a facsimile server on a LAN, the format would be FAX = +3940226338@Internetaddress.com

9.5.7 Real-Time Internet Facsimile

The new Draft Recommendation T.38, *Procedures for Real-Time Group 3 Facsimile Communication Between Terminals Using IP Networks*, is nearing approval in the ITU-T. This standard describes how facsimile transmission takes place between Internet gateways. Communication between the facsimile machine and the gateway is by means of T.30. The T.38 procedures work in conjunction with Recommendation H.323. H.323 procedures are used for call control, protocol selection, conveyance of the called terminal's address, and call progress information. The protocol defined in T.38 specifies the messages and data exchanged between facsimile gateways connected via an IP network. Figure 9.2 illustrates the reference model for this real-time IFax.

The model in Figure 9.2 shows a traditional Group 3 facsimile terminal connected to a gateway, emitting a facsimile through an IP network to a receiving gateway, which makes a PSTN call to the receiving Group 3 facsimile equipment. Once the PSTN calls are established on both ends, the two Group 3 terminals are virtually linked. All T.30 session establishment and capabilities negotiation are carried out between the terminals. This is in contrast to the store-and-forward mode, in which the facsimile is terminated at the gateway.

An alternative scenario would be a connection to a facsimile-enabled device (e.g., a PC) that is directly connected to an IP network. In that case, a virtual receiving gateway is part of the device's facsimile-enabling software and/or hardware. In other environments, the roles could be reversed, or there might be two facsimile-enabled network devices, such as an IAF. The T.38 protocol thus operates directly between the emitting and receiving gateways.

9.5.7.1 Basic Principles

The transmission method in T.38 can be described as a demodulation/remodulation procedure (demod/remod). The facsimile transmission arrives at the gateway as a digital T.30 data stream modulated by the facsimile modem into an analog signal. At the gateway, the signal is demodulated. It is then packaged into data packets to be conveyed over the Internet by the emitting gateway. The receiving gateway reassembles the packets and modulates them into a modem signal, which continues on to the receiving facsimile terminal. Neither facsimile terminal is aware that part of the transmission was over the PSTN and part over the Internet.

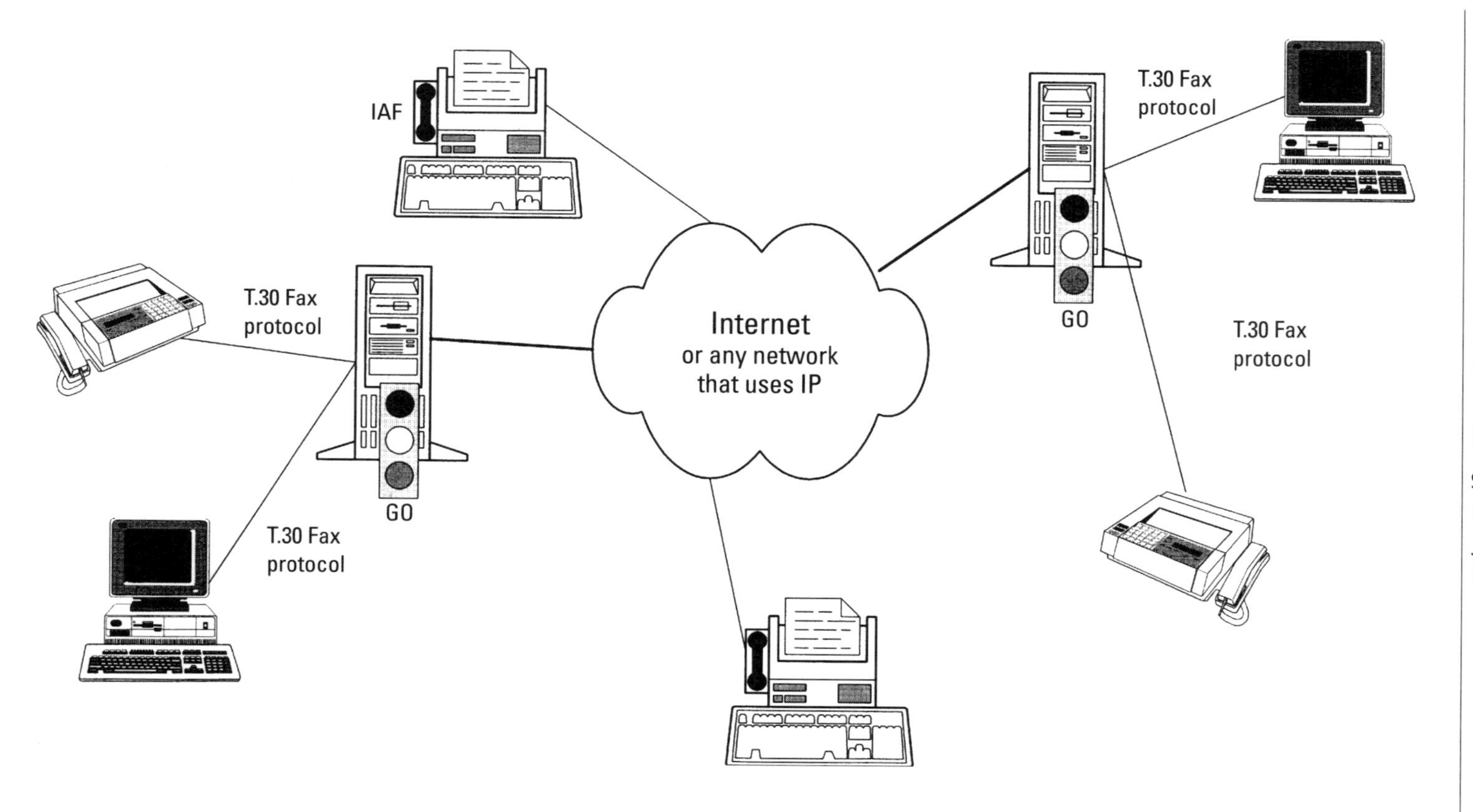

Figure 9.2 Real-time IFax reference model.

Some signals are not transferred between gateways but are generated or handled locally between the gateway and the facsimile machine. The gateways indicate the detection of the facsimile tonal signals in the data stream so that the other gateway can generate them. The emitting gateway may ignore signals associated with nonstandard facilities (NSF, NSC, and NSS), or it may take appropriate action and pass the information to the receiving gateway. The receiving gateway may do the same or take appropriate action, including passing the information to the receiving facsimile terminal.

9.5.7.2 Modem Rate Alignment

When the TCP protocol method is selected, each gateway independently provides rate negotiation with the facsimile modems, and training is carried out independently. It would be possible to have each end of the connection negotiate to a different data rate, which would be a problem. To prevent that from occurring, a set of messages is provided to align the modem speeds. Because of the lower delays experienced with using UDP, the modems can negotiate data rates without relaying on gateways.

9.5.7.3 Transmission Protocols

There are two candidates for transmission protocols over the Internet, TCP and UDP. The TCP protocol is error free, because it has built-in error correction by means of retransmission. UDP has no error control. The principal way in which errors manifest themselves in the Internet is by lost packets. Thus, in TCP, errors translate into delay. T.30 is sensitive to delay, because there are a number of timers active in the protocol, which, if they expire, would cause transmission to fail. UDP does not suffer from delay, but lost packets could have serious consequences. If a packet is lost in the negotiation phase, the transmission could fail. A packet lost in the image transmission phase could cause lost scan lines in the image, unless ECM is in effect. If UDP is to be used, then some form of error control should be available.

There are some additional considerations when TCP is used. Because delay is more probable, some mechanisms may be used to keep the transmitting or receiving terminal active. This mechanism, called "spoofing," also requires the gateways to track their progression throughout the T.30 process. Thus, a state machine is required. The tradeoff for the freedom from error correction is an increased complexity and slower throughput.

9.5.7.4 Real-Time Internet Facsimile Packet Structure

The basic packet carrying facsimile information is the Internet facsimile packet (IFP) and is the same for TCP and UDP. The way the packet is carried over the IP layer differs for TCP and UDP (Figure 9.3).

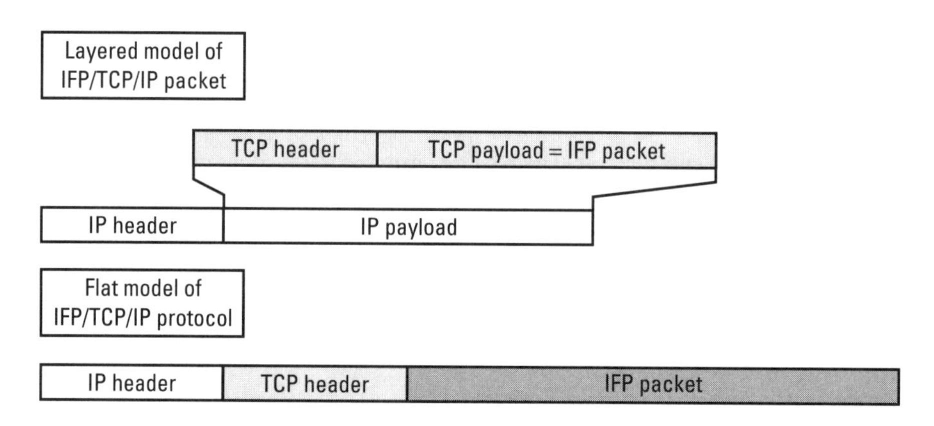

Figure 9.3 High-level IFP/TCP packet structure.

The IP payload consists of the IFP packet plus the TCP header. The complete packet thus contains the IP header, the TCP header and the IFP packet. In UDP provision has to be made for error correction. The topmost element, the UDP transport layer (UDPTL) payload, contains the IFP packet and the error control mechanism. This is preceded by the UDPTL header and forms the UDP payload. The UDP payload and the UDP header make up the IP payload (Figure 9.4).

The TCP packet structure is simpler, but that is offset by the increased delay and complexity needed in the gateway.

9.5.7.5 User Datagram Protocol Error Control

The UDP error control mechanism is a simple redundancy mechanism in which each IFP packet is transmitted twice but in a different IP packet. Each IP

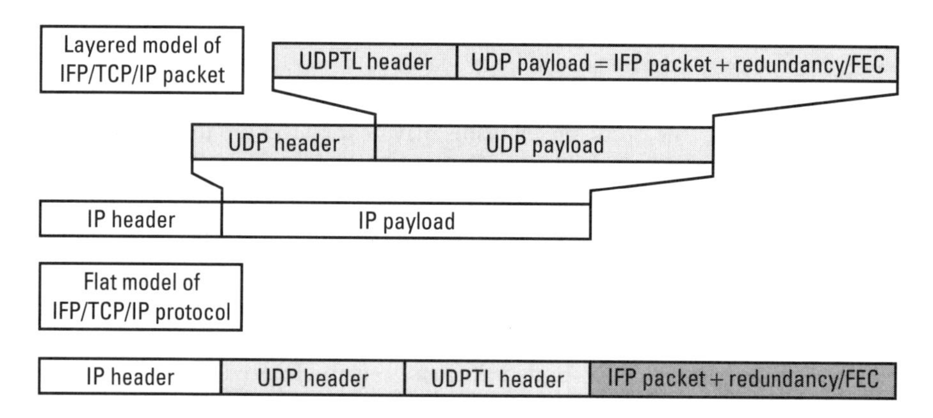

Figure 9.4 High-level UDPTL/IP packet structure.

packet contains the previous and the current IFP packets. If one IP packet is lost, it can be replaced by the redundancy information. Of course, if two packets in sequence are lost, this method does not work. That is an unlikely event, however, because the loss distribution over time is random. The effect on error control is a small increase in the required bandwidth. Resistance to error could be increased by including more than two IFP packets. A more sophisticated error correction method using FEC is available as an option. Preliminary experience with the simple error correction method is showing that this probably will not be required. Additional error protection is available if ECM is available between the two communicating facsimile terminals. The use of redundancy is optional. Gateway implementations can choose strategies to use in this case.

9.5.8 Future Work in Internet Facsimile

At this writing, much IFax standardization work remains to be done. The store-and-forward IFax work remaining includes full mode with capabilities negotiation, notifications, and image compression beyond MH and MR. Security is another issue that needs to be addressed. In real-time IFax, the major remaining issues are V.34 modem access and H.323 protocol negotiation.

10

Fax Test Charts and Images

Before the availability of standardized fax test charts, it was difficult to compare performance of fax machines because each company used its own test charts. Standardized test charts started in 1955, and additional charts were added as needed. The latest set of test charts was developed for the new higher performance Groups 3 and 4 fax equipment. This set has much higher image sharpness, expanded gray scale, and full color test images for testing both fax equipment performance limitations and transmission channel characteristics. Fax machine manufacturers can use the test charts to measure the level of performance achieved in a new design and to ensure the quality of production line fax machines. Users may find the charts helpful in comparing brands and models of fax machines.

10.1 Early Standardized Fax Test Charts

The test charts in this section once played an important role in testing fax equipment, but they have been replaced by newer test charts.

10.1.1 IEEE Test Chart 167

The first standardized fax test chart, issued in 1955, was developed by the IRE and continued in publication by the IEEE[1] (Figure 10.1).

---

1. The Institute of Electrical and Electronics Engineers (IEEE) was established in 1963 as a result of the merger of the IRE and the AIEE.

Figure 10.1 IEEE facsimile test chart 167A.

The original chart was assembled by author Ken McConnell, chair of the IRE Facsimile Committee, using favorite test patterns of committee members. He then worked with the Eastman Kodak committee member to achieve the desired appearance of the photographic prints. It was expected that after testing many patterns could be eliminated. Each one was vigorously defended, however, and all were kept. Although some of the patterns may appear to be redundant, each one is better than others for a particular image sharpness measurement or distortion evaluation. The Eastman Kodak secretary (otherwise known as the "fax lady"), whose picture appears in all issues of the chart, little realized that her likeness would be spread worldwide on more than tens of thousands of fax test charts. Each chart, as it was used, made many more fax copies until her image became universally recognized and associated with fax test patterns.

The chart was still widely used 35 years later for testing performance of fax, electronic imagery devices, office copiers, and other photographic systems. When more copies were needed in 1990, however, the master negative had degraded so much that it was no longer possible to reproduce the chart. The last printing was in 1987 (IEEE STD 167A-1987). After considering other alternatives, the IEEE replaced test chart 167A with separate charts for continuous-tone gray scale and black-white patterns.[2]

10.1.2 CCITT Chart No. 1

The first CCITT facsimile test chart was for photo facsimile transmission (Recommendation T.20, 1960). Figure 10.2 shows test chart No. 1, the 1964 and 1968 versions.

The recommendation states: "This chart has been designed for measuring the quality of both picture and black-and-white transmissions, and it enables the apparatus used and the communication channels to be judged by means of objective measurements, the results of which may be expressed in code."

The chart size is 110 by 250 mm (4.33 by 9.84 inches). Gray-scale performance is checked by a 15-step density pattern and the photograph. Black-white patterns measure single-line and multiple-line image sharpness up to 6 lines/mm (152 lines/inch). Other patterns show analog facsimile distortions caused by telephone lines and equipment design. The ITU has deleted T.20 and replaced it with Recommendation T.22.

10.1.3 CCITT Chart No. 2

In 1980, the CCITT issued Recommendation T.21 for test charts No. 2 and No. 3 for document facsimile transmission. The size of these charts is 210 by 297 mm (8.23 by 11.69 inches), the ISO A4 international paper size. The

2. IEEE Standards P167A.1 and P167A.2 have the same images as test charts No. 4 and No. 5 contained in Recommendation T.22.

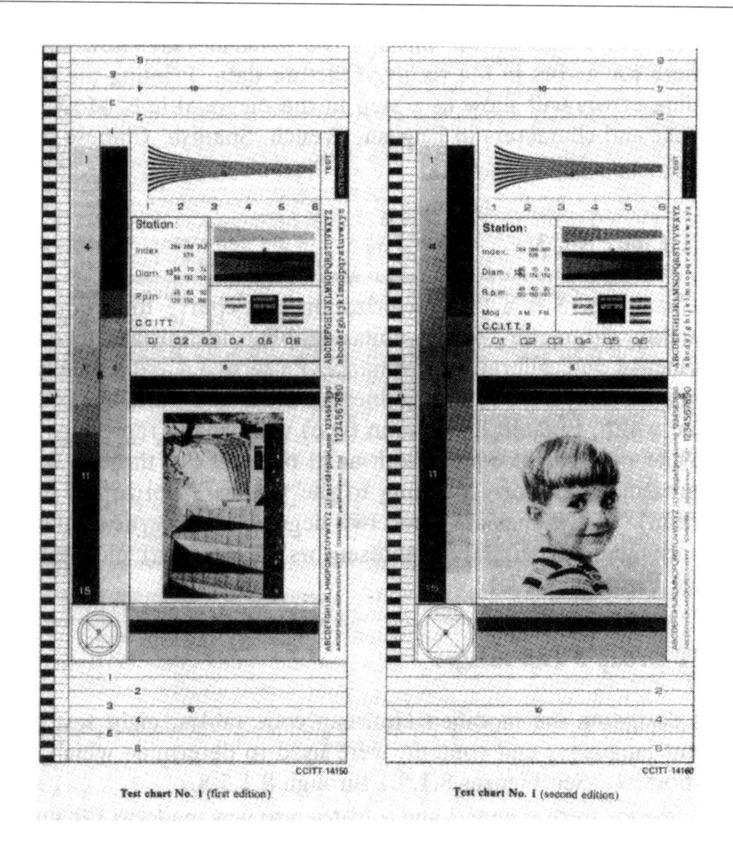

Figure 10.2 CCITT test chart No. 1: (a) first edition (1964); (b) second edition (1968).

recommendation states "that a standardized test chart to check the quality of document facsimile transmissions will have great advantages. Owing to the development of international document facsimile transmission services, a great variety of characters and symbols, including ideographic symbols, are involved and must be taken into consideration."

Test chart No. 2 was a "transmission test chart," with patterns for quantitative evaluation of distortion and character groups for evaluation of readability (Figure 10.3).

Test chart No. 2 contains test patterns for measuring lost margins and image sharpness. The diagonal line through the patterns may show steps or white streaks if there are errors in the received picture data. Printing the previous line for MR coding errors will show as a step in the diagonal line. Most of the chart consists of text and characters in English, French, Spanish, Chinese, Arabic, and Russian. The ITU has deleted T.21 and replaced it with Recommendation T.22.

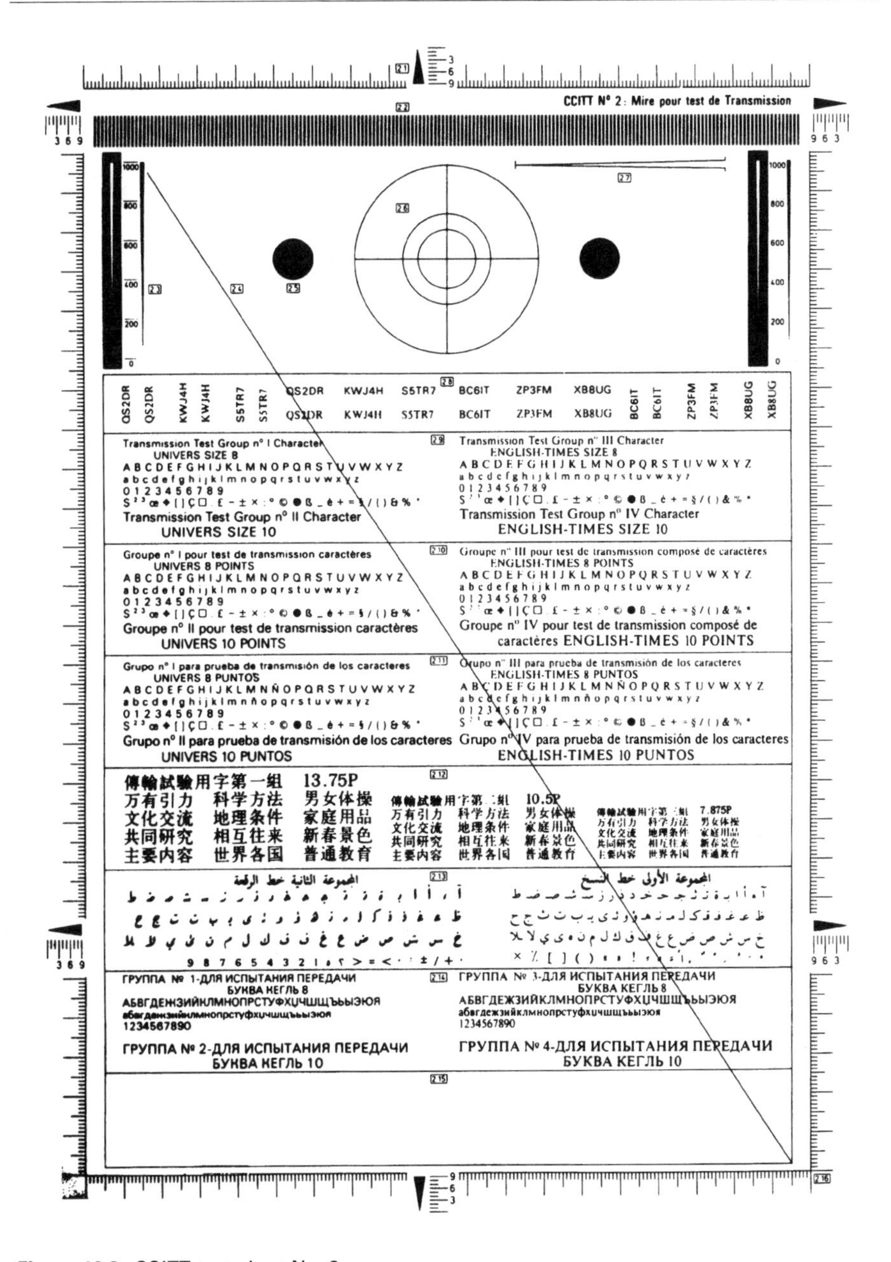

Figure 10.3 CCITT test chart No. 2.

10.1.4 CCITT Test Chart No. 3

Test chart No. 3 evaluated the technical quality parameters of fax units (Figure 10.4).

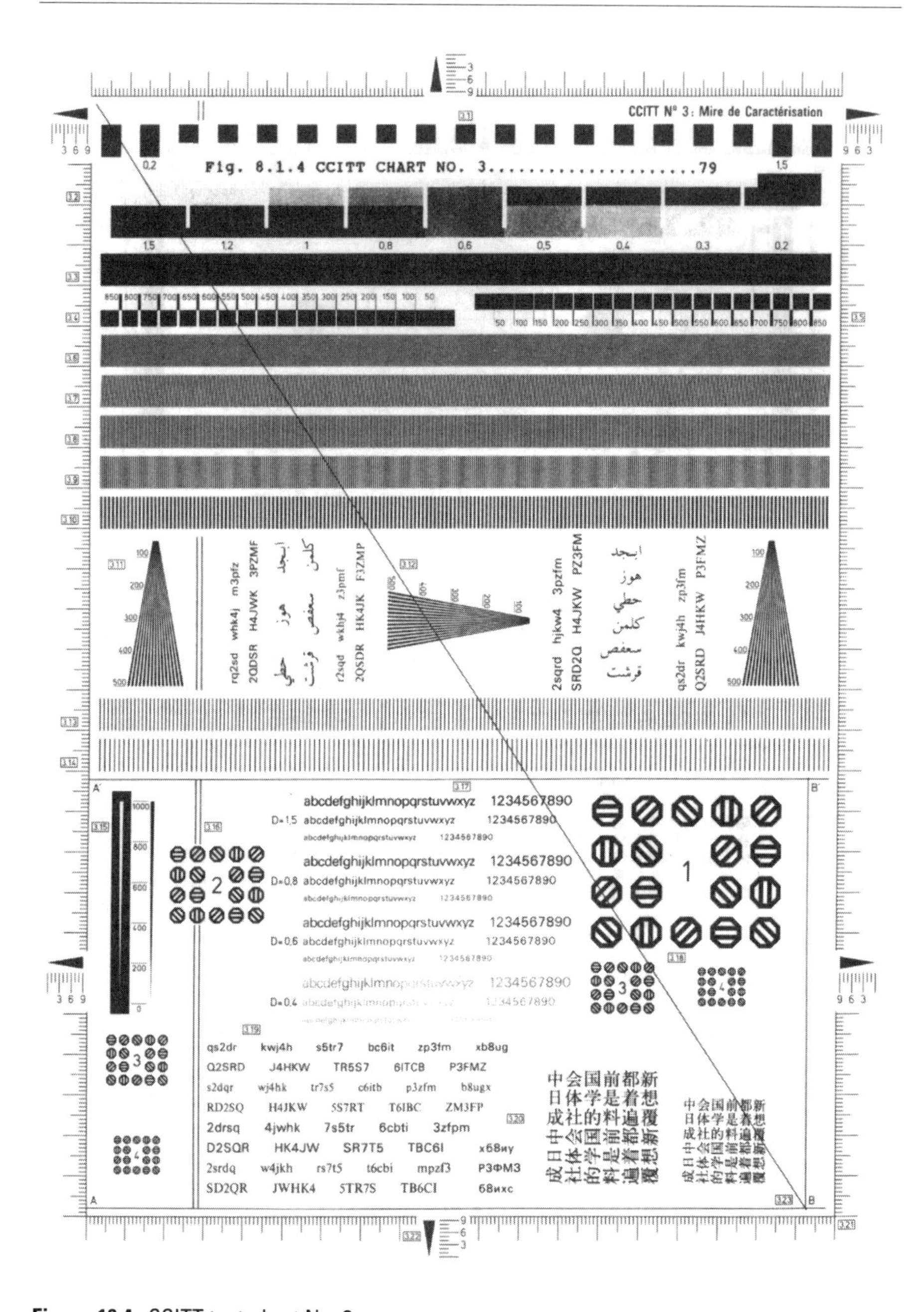

Figure 10.4 CCITT test chart No. 3.

This chart helps technicians maintain fax machines by detecting faults, making adjustments, and calibrating performance. Certain patterns show the system performance with thin, isolated black lines on white and white lines on black. A decreasing density character set allows for determination of the limits of reproducible density and for checking the effectiveness of the scanner adaptive threshold.

10.2 Development of Current Standardized Facsimile Test Charts

About 1990, the CCITT (now the ITU-T) determined that a new set of test charts with higher image sharpness would be required for testing Group 3 fax terminals as well as for the development and testing of Group 4 color fax terminals. No work on a color system for Group 4 had been started, and some thought that only a few markup colors would be needed since Group 4 was for sending business documents. The U.S. National Communications System (NCS) volunteered to undertake the task. All three authors of the third edition of *FAX* were instrumental in generating and producing these charts.

A single chart for both bilevel (black/white) and continuous-tone monochrome for photos was planned. It would be an extension of the IEEE-class of test chart, with more gray-scale steps and a larger-size page. The text and line patterns were fully saturated black with a sharpness that could measure performance of fax equipment operating at up to 600 lines/inch or more. At least 64 gray-scale steps were deemed necessary, preferably 128, using a continuous-tone printing process (no screening). After working on a number of alternatives, no method was found to print both bilevel and continuous images from the same negative. The bilevel imagery needed a high-gamma film, while the continuous-tone images needed a low-gamma film. Test charts with the desired characteristics possibly could be produced one at a time with direct printer control and no negatives, but that would be too expensive. It was decided that two charts would be needed instead of one, and the gray-scale steps were set at 48 after trying to get 64 or 128 steps to print properly from a negative.

10.3 ITU-T Chart No. 4: Bilevel (Black-White) Facsimile Test Chart (BW01)

BW01 of ITU-T Recommendation T.22 March 1993 is for use in image quality assessment of black and white document fax systems, especially the new higher dpi options of Group 4 and Group 3 fax (Figure 10.5).

This monochrome test chart with black-white patterns has only text and line work printed on high-gamma photographic paper for optimum sharpness

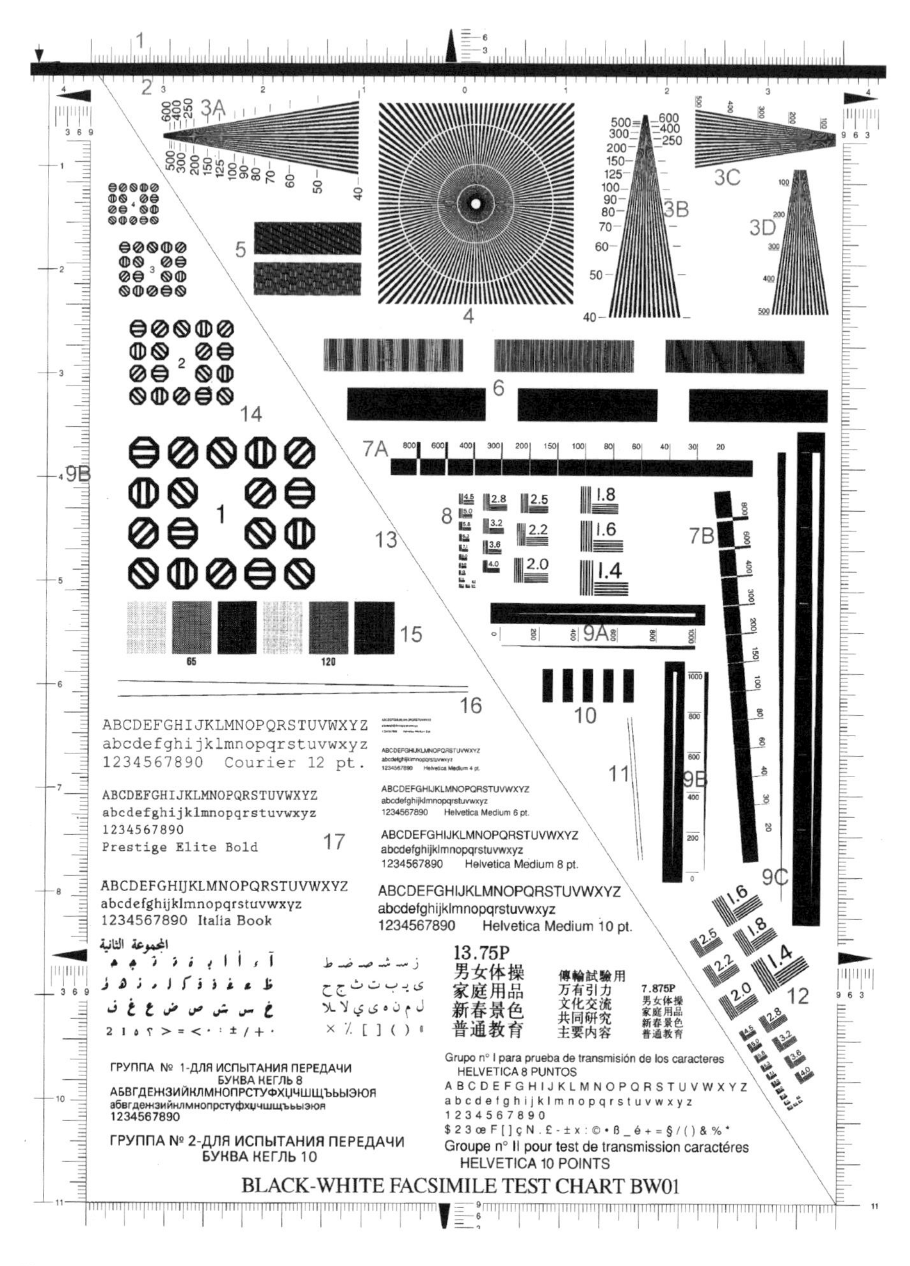

Figure 10.5 Black-white high-resolution facsimile test chart BW01.

and high contrast without fogging the white areas next to black markings. The finest line pattern (6F) on this chart is 600 line pairs/inch, adequate for testing 1,200-line/inch fax systems. The other patterns have the same image sharpness. The overall size of the printed chart is 308 mm long by 222 mm wide (11.875 by 8.875 inches) to allow for maximum paper size tolerances. The numbered patterns that make up the chart are defined next.

1. Border of four scales with millimeter markings. The arrows near the ends of the top border are 8.5 inches apart and centered on the page.
2. Black bar across full page width and a scale in inches across the top, starting from 0 in the middle of the page with 0.1-inch scale markings. The border at the left side of the chart is also marked in inches.
3. Four patterns of truncated fan-type multiple-line pattern with low taper rate. The larger two are calibrated in black plus white lines per inch, and the smaller ones are calibrated in microns.
4. Gurley-type Pestrecov star pattern with circles of 50, 100, and 200 lines/inch.
5. Alternating black and white lines. Upper pattern is 150 line pairs/inch, inclined at 3 degrees from vertical. The lower pattern is 200 line pairs/inch, inclined at 2 degrees from vertical. The angle is to allow the lines to drift through a match and a mismatch with the photosensor array elements.
6. Black-white bar patterns of 100, 150, 200, 300, 400, and 600 line pairs/inch.
7. Isolated black and white lines calibrated in microns. The vertical pattern is inclined at 5 degrees from vertical.
8. National Institute of Standards and Technology (NIST)–type image-sharpness pattern calibrated in line pairs (black plus white) per millimeter with the smallest patterns near the center of the chart.
9. Tapered isolated black and white line patterns with the line width calibrated in microns or inches.
10. Black-white bar pattern of 2 black plus white bars per inch.
11. Parallel lines inclined at 5 degrees from vertical.
12. NIST-type image-sharpness pattern calibrated in line pairs (black plus white) per millimeter with the smallest patterns near the edge of the chart.
13. Diagonal line for checking irregularities in vertical pitch. Printed fax line with errors will show breaks or steps of this line.
14. ISO-character hexagonal line patterns for readability testing.

15. Halftone dot screens of 10, 50, and 90% black. The 65 and 120 are the number of dots per inch measured at a 45-degree angle.
16. Line-crossing pattern. The center-to-center line separation is 0.15 inch on the left end and 0.05 inch on the right end. The number of scanning line crossings of both lines multiplied by 10 is the vertical line pitch in lines per inch.
17. Text in English, Arabic, Chinese, Russian, Spanish, and French. English text is in 12-, 10-, 8-, 6-, 4-, and 2-point sizes.

10.3.1 Notes on the Use of BW01

BW01 offers a quantitative means of assessing various technical quality parameters, detecting defects produced in received images, and evaluating the readability of text in various ITU official languages.

10.3.1.1 Image Sharpness, Resolution, and dpi

In a scanner, the number of pixels scanned per inch in either a horizontal or a vertical direction is known as dpi (dots per inch). In a printer, dpi is the number of dots printed in either the horizontal (h) or vertical (v) direction. In the era before the term *dpi* was used, *resolution* was the quantitative measurement of image sharpness. Now however, technical articles may equate resolution with dpi without regard to the sharpness of the image resolved by scanning or the image formed by the dots for viewing. In this chapter, we sometimes use the term *dpi* because of that ambiguity. Horizontal dpi is the number of pels per unit length along the scan line. In a fax transmitter or scanner, this number is usually the number of active cells in the photosensor divided by the active scan line length. In a fax receiver or printer, this number is the number of dots written per unit length. Vertical dpi is the number of lines written per unit length in the paper feed direction.

Image sharpness is a key factor in the received copy quality in a fax system. Image sharpness in the received display image or fax copy may be considerably less than the dpi, and it may vary between different fax transmitters, fax receivers, and displays. Design of the equipment and the adjustments made in the scanner, printer, or display device of the fax system affect the image sharpness achieved. The threshold setting of the A/D signal converter in the scanner also affects image sharpness achieved. Assuming maximum signal on white, threshold setting at a lower level will pick up thinner isolated black lines at the expense of losing the narrower white lines in a black background. Optimum threshold setting depends on the densities of image areas that should be sent compared with those that should not. Because most business documents contain mainly type fonts on a white background, the best compromise threshold may be a setting for isolated thin black lines.

Pages with gray or light-colored backgrounds usually need no change from the optimum setting for white background; ABC circuitry compensates for this. Group 3 or 4 machines often have limited or no operator control for this setting, but almost all PC-fax units have a software control for a precise adjustment. With the default setting, multiple-line image sharpness usually is lower than single-line image sharpness, and single white lines in a black background often have a lower image sharpness than single black lines in a white background.

Devices for displaying fax copy usually have a writing line length narrower or wider than the image sent, making the sending and receiving dpi different; in such cases, the sending dpi should be used. Even large monitors may have only 1,280 dots per line, compared with the 1,728 needed to display a standard Group 3 fax line. For the optional 400-dpi mode, 3,456 or more dots per line are needed. In either case, the image sharpness is considerably less than that stored in computer memory or recorded by a fax machine. Even worse, PC-fax software may not allow the monitor to use its highest resolution setting.

The transmission channel and modem design should have no effect on image sharpness in the received copy or display image, providing the digital signal produced in the fax receiver has no errors. However, common carriers that convert the fax signal sent by the internal digital modem (analog signal) to digital format for transmission might affect the signal quality, degrading the recording or image quality.

The setting of the black-white threshold control point in the fax transmitter influences the image sharpness of the received fax copy. A threshold setting that generates a black pel from a line thinner than the scanning pixel usually improves readability of very small type fonts and print-isolated thin black lines. It does, however, lose the ability to print isolated thin white lines and thin multiple-line patterns. Other factors, such as the line width printed at the fax receiver for a single line sent, have similar effects. The large number of patterns for testing image sharpness in the BW01 chart are intended to allow the user to select those best suited to the fax application.

Received fax copy image sharpness can be assessed by observation of patterns 1, 3, 4, 5, 6, 7, 8, 9, 12, 15, and 16.

Pattern 1

While this border scale is intended primarily for measurement of dimensional accuracy, it also provides an isolated line-image sharpness test. The large number of lines ensures that almost all degrees of alignment match between the lines in the pattern and the scanning pel samples. A fax system may miss some parts of these lines at 203h by 98v dpi, depending on where the transmitter black-white threshold is set.

Patterns 3A and 3C

The calibration markings of 3A are in black plus white lines per inch. The calibration markings of 3C are the black- or white-line width in microns. Vertical multiple-line image sharpness can be measured as the limiting position where a vertical line intercepts 15 black lines and all lower image sharpness positions intercept 15 black lines. The inch-based 3A and the metric-based 3C serve the same purpose but will align differently with the scanning pixels.

Patterns 3B and 3D

Patterns 3B and 3D are the same as patterns 3A and 2C, except that they are rotated 90 degrees to allow measurement of multiple-line image sharpness in the horizontal direction.

Pattern 4

Pattern 4 allows multiple-line image sharpness to be measured at all angles, reducing the measurement differences due to random alignment of the scanning pixel samples with the line edges. At 203h by 196v dpi, the good line formation limit is typically the third circle (150 lines per inch). Some fax receivers record multiple lines to increase the recording density, but that widens the line height and loses detail. 300h by 300v dpi should print with good line formation for the entire pattern, but a typical 203h-by-406v system shows little difference from a 203h-by-196v system. At 203h by 98v dpi, Moiré patterns in the received image between 50 and 100 lines/inch are common.

Pattern 5

These patterns are intended primarily for fax systems having at least 300-lines/inch capability. Reproduction of these patterns of 150 and 200 line pairs/inch (300 and 400 lines/inch) is beyond the capability of a standard Group 3 fax system.

Pattern 6

Only the top left pattern (200 lines/inch) can be expected to be printed by a Group 3 fax system, unless it has higher than 203-lines/inch capability.

Pattern 7

With many fax systems, as the line widths become narrower, isolated white lines disappear long before black lines of the same width do. That may relate to the black threshold setting of the transmitter being at a light gray, but it also can be affected by the optical system design. Because the lines of pattern 7A are perpendicular to the scanning line, changes in alignment between the line and the pel sample also may determine whether a thin line is printed. Pattern 7B is at an angle to eliminate the problem.

Note: Due to some error, the 40-micron black line is narrower than the 30-micron line. In fax system tests, received copies that barely show the 30 line do not show the 40 line.

Patterns 8 and 12

These patterns have groups of five parallel lines (NIST resolution test chart type) calibrated in black plus white line pairs per millimeter. Pattern 8 has the finest lines near the center of the test chart, while Pattern 12 has them near a corner of the chart. That allows image sharpness near the center of a page to be compared with sharpness near a corner for checking degradation due to lens distortion, illumination variation, and other factors. These patterns can test image sharpness for systems up to 900 dpi.

Multiple-line image sharpness for these patterns can be measured as the limiting line group where five black lines are recorded and all lower resolution groups have five black lines. In Pattern 8, the horizontal line image sharpness is often different from the vertical line image sharpness. Pattern 12 images are tilted 34 degrees to ensure that pel sampling positions vary along the test image.

Pattern 9

Pattern 9 shows three sets of a tapered isolated single black line and a matching white line. Pattern 9A and 9B line widths are calibrated in microns. The 9C upper mark line width is 0.06 inch, and the marks below are at 0.01 inch intervals. Black-white threshold is often set to print thinner black lines than white lines, favoring small font text lines at the expense of fill-in of thin white spaces.

Pattern 15

With higher dpi fax systems, the halftone dot screens print a better image sharpness. By reproducing smaller dots, the number of gray scale steps increases, and the Moiré pattern may disappear.

10.3.1.2 Scanning Line Pitch

Pattern 16

The gradually sloping lines of this pattern are almost parallel to the fax recording lines, producing staircase steps as they cross. Most Group 3 fax units have 200 line/inch capability and print two identical narrower lines when operating in 100 line/inch mode. This can be seen zooming in on the received fax. Pattern 16 will also show the difference between two varieties of 200-by-400 dpi fax machines. Originally, one was a proprietary (NSF) technique with 1/200-by-1/200 inch sensor elements across the page and produced 400 lines/inch in the vertical direction by stepping the paper 1/400 inch. The 200 dpi sensor overlapped the adjacent lines producing less image sharpness than expected. A true 200-by-400 dpi fax produces far better imnage sharpness, but must have a sensor that is 1/400 inch in the vertical direction, as in a 400 dpi fax machine. A pattern 16 fax recording will show the difference.

10.3.1.3 Limit of Readability of Characters

Pattern 14

Correct identification of ISO characters by character orientation may be easier to quantify than readability of printed text. All the sizes of these characters should be reproduced legibly even at 98-lines/inch resolution. Pattern 14 is considered readable when seven out of eight characters can be read correctly.

Pattern 17

Text is in English, Arabic, Chinese, Russian, Spanish, and French. By counting the number of characters that can be read correctly, a readability number can be generated. Readability of the same received fax copy by more than one observer usually produces different results. Even subsequent observations by the same observer may not produce consistent results.

10.3.1.4 Lines Printed With Errors

Pattern 13

Pattern 13 should record as a straight line. It shows a break or step when a fax line contains an error. When printing a line that has an error, some fax receivers print a white line. If the next line has an error, the previous good line may be printed. These error defects may be seen more easily in Pattern 13 than in text. This pattern can also be used to spot paper feed irregularities in the fax transmitter or receiver.

10.3.1.5 Distortion of Page Size

Patterns 1 and 2, Border Scales

The received page can be measured to assess size distortion. Photographic processing when printing the test chart originals produces small size distortion, increasing the scale length by about 0.4% but having almost no effect on the width. The precision of measurement may be improved by comparing the fax copy with a test chart original.

10.3.1.6 Print-Density Homogeneity

The pattern 2 black bar across the page width can be used to measure uniformity of black printing at the fax receiver. Nonuniformity of toner application is one possible cause.

10.3.1.7 Skew

Skew causes a mismatch in the length of diagonals across the rectangle formed by pattern 1, the border. Group 1 and 2 fax machines may have skew due to lack of absolute synchronism, but digital fax systems do not, unless there are mechanical or optical problems.

10.4 ITU-T Chart No. 5: Continuous-Tone Monochrome Facsimile Test Chart (CT01)

The continuous-tone monochrome chart (also contained in T.22) is designed for testing fax units with gray-scale capability and those with black-white only (Figure 10.6).

Low-gamma photographic material preserves the complete gray scale between paper white and black for the continuous-tone photographs. The black-white threshold setting of the fax scanner can be set quite accurately with this chart. The overall size of the chart is 279 mm long by 216 mm wide. The patterns that make up the chart are described next.

- Stepless continuous-tone strip from black on the left to white on the right;
- Stepless continuous-tone strip from white on the left to black on the right;
- Density-step tablet with 48 steps of 0.5-inch squares in three rows of 16 steps;
- Uniform-density strips across the page with 15 steps (plus white);
- Continuous-tone photograph—architectural photo showing fine detail;
- Continuous-tone photograph—portrait.

10.4.1 Notes on the Use of CT01

The CT01 chart offers a means of assessing technical quality parameters and detecting defects produced in received images when the original is continuous-

Figure 10.6 Continuous-tone monochrome test chart CT01.

tone gray scale. It is intended for digital fax systems that use multiple bits per pixel, for continuous-tone analog faxes, and systems that reproduce 1 bit/pixel, such as Groups 3 and 4 or fax equipment. Some fax systems may convert the image into a black-white-only rendition or send gray-scale information by an electronic screening process as clusters of black pels forming multiple-sized black dots. Color fax systems also can evaluate their accuracy of gray-scale reproduction. This chart is printed on photographic paper that is heavier than most documents sent by fax, making it necessary to manually assist feeding the chart through some fax machines.

10.4.1.1 Gray-Scale Reproduction

All the patterns in the CT01 chart can be used to evaluate the fidelity of gray-scale fax reproduction of continuous-tone images. For systems that simulate gray shading, this chart can be used to gauge the effectiveness of the electronic screening. When this test chart was designed, 64 gray-scale steps were used in the density-step wedges. That could be done if each test chart were printed directly from a computer, but the cost per chart would be too high. For generating reasonably priced charts, contact photographic printing from a negative is required. After many unsuccessful attempts to produce a negative that would reliably print 64 uniform gray-scale steps, the number of steps was reduced to 48, and even now some of the density steps are smaller than desired. For those users who may need 64, 128, or more density steps, this chart can be supplemented with special density step wedges available from other sources.

A density distortion may be caused with black-white transmission of the chart due to time constants of the ABC in the scanner. It may be noticed in pattern 2 as a shift to the left in the horizontal position where there is a change from white to light gray.

10.4.1.2 Vertical Streaks

With fax systems that record the gray-scale steps, vertical streaks may show at any gray density. With fax systems such as Group 3 and Group 4 operating in black-white mode, vertical black and white streaks may show in the recorded gray-scale steps that are near the threshold setting. These streaks are caused by sensitivity variation between individual photosensors.

10.4.1.3 Horizontal Streaks

Paper-feed irregularities cause underlap or overlap of adjacent recording lines. With some high-quality systems, even a very small irregularity is easily seen as a light or dark streak between recording lines. In halftone mode, photographs reproduce better than might be expected, but the gray-scale step patterns show the loss of information resulting from inadvertently making some steps larger than they should be, at the expense of eliminating other steps. The continuous-density strips at the top of the chart divide into the number of steps provided by the digitizing process or the electronic screening. If these steps are very small, some of them will not be visible, or there will be switching back and forth in density in an area that should be of constant density. Such switching can be caused by the slight density variations in each density step. Streaks caused by incompletely corrected photosensor sensitivity variations may appear as false gray-scale steps.

10.4.1.4 Irregular Printing Density

Test patterns 1, 2, 3, and 4 show irregular printing density. Variations in print density are much more easily seen in large areas of constant density than in ordinary photographs, where the distortion is masked by density changes in the image being sent. For certain classes of imagery, density-uniformity distortions may produce false data or mask the real data.

10.5 ITU Chart No. 6: Four-Color Printing Facsimile Test Chart (4CP01)

4CP01 of ITU-T Recommendation T.23 (April 1994) is a full-color chart for assessing the accuracy of receiving by fax an image with a full range of colors, including black text.[3] Figure 10.7 is a black and white copy of the color chart.

The printing process used for this chart represents the best commercial four-color offset printing available at the time. A full range of colors is achieved using small dots of cyan, magenta, yellow, and black (CMYK) printing inks. Test chart 4CP01 is used for testing color fax units that send four-color screen-printed material of brochures, business reports, and magazines. The pattern sharpness is adequate for testing fax equipment up to 600 lines/inch, but it is not as sharp as in the BW01 chart. The size of the chart is 302 by 222 mm (11.9 by 8.75 inches). The following numbered patterns make up the chart.

1. Border of four scales with millimeter markings. Red arrows near the ends are 8.5 inches apart and centered on the page. The top border is red (solid magenta and yellow printing). The right border is green (solid yellow and cyan printing). The left border, also marked in inches, is blue (solid cyan and magenta printing). The bottom border has three segments: cyan, magenta, and yellow.
2. A black bar at the top across the full page width and a red scale in inches under it, starting from 0 in the middle of the page with 0.1-inch scale markings.

3. IEEE Standard P167A.3 has the same images as the test chart contained in Recommendation T.23.

Figure 10.7 Four-color printing facsimile test chart 4CP01.

3. Gurley-type Pestrecov star of solid, nonscreened, tapered-width lines with circles of 50, 100, and 200 lines/inch (same as black-white fax

test chart BW01). The pattern is divided into four segments. Clockwise from the left are C, M, Y, and K.

4. Eighteen colors with eight intensities for each color, for a total of 144 different color patches. Two-color printing color primary patches are used for seven of the intensity columns. Column 8 has 25% black added to give warmer tones. Each row between the primaries has a fixed ratio of two printing primaries as the 175-line screen dot percentages decrease in steps.
5. Twelve light-color patches selected from the Macbeth color chart plus text and isolated line patterns in the same color. Line widths of the line patterns are 0.04, 0.01, and 0.005 inch (1.016, 0.254, and 0.127 mm). Half the lines are slanted 1 in 10 to provide a random match between the scanning line sample and the pattern. The edges of the typeface provide a similar function.
6. Black text simulating magazine text that might be on the same page as a screened color photo. The following PostScript™ fonts are used: headline, Bookman 0.40 in bold; heading, Helvetica 0.18 inch; first paragraph, New Century Schoolbook 0.13 inch; second paragraph, Kanji text; footnote, Helvetica Narrow 0.09 inch. The note is repeated in 4-point and 2-point type.
7. Solid unscreened blocks of printing ink colors. Starting at the left are cyan, magenta, yellow, red (magenta + yellow), green (yellow + cyan), blue (cyan + magenta). These are solid patches of the primary printing inks and two-color combinations. The last block is black.
8. Color patches for a combination of black and each of the printing color primaries to show darker colors not covered elsewhere in the chart. These patches were selected to give good steps to the eye. The left three patches are 100% primary overprinted by 60, 40, or 20% black dots. The next three are primary/black combinations of 60/20, 40/20, and 20/20% dots.
9. Three sets of screened gray scales with 90, 75, 50, 25, 10, and 5% dots. The first row is an 85-line screen. The next rows are 175- and 133-line screens.
10. Photograph of toys showing facial tones and wide color range.
11. A computer-generated simulation of spheres with shadings for three-dimensional effect.
12. Graphics image from a magazine cover showing a three-dimensional effect.

13. Five-step gray scale made only of CMY primary color dots, which can be used to check the color balance of the printing inks.

10.6 CCITT Group 3 Test Images (MH Coding Images)

During design of the MH code tables, a set of eight test images were used to determine which code words should be shortest (Figure 10.8). The eight images were scanned and a histogram made showing the number of times that each white-run length and each black-run length occurred. The MH code assigned short code words to those runs that occurred most frequently. Test image No. 1, a short business letter, is used by vendors to measure the time taken to send a page. The time listed does not include the initial handshake, which takes about 15 seconds, or the few seconds between pages. Initially, the CCITT (now the ITU-T) did not recognize these test images as official standards, although they were considered by many to be de facto standards. Table 10.1 lists the pel data over an area of 8.64-by-11.70 inches for each of the eight reference images as scanned at 200, 300, 400, and 600 pels/inch for the T.24 CD-ROM. The "Width" and "Height" dimensions in pels are the same for all images. The "Bytes" column lists the size of the uncompressed image. "Black Pels" and "White Pels" columns give the total counts per image. The last columns are the percentage of the image that is colored black and white.

10.7 ITU T.24 CD-ROM Test Charts and Images

When first published in November 1994, T.24 was a single CD-ROM that contained scanned versions of the active ITU test charts listed here plus many other test images. It was revised in June 1998 to add 28 more images, and the new CD-ROM should be available by the time this book is published. The purpose of T.24 is to provide a test image suite to evaluate existing fax and other image systems and to provide a consistent baseline for future work. Results of compression algorithm experiments and image-quality tests can be compared by a broad range of users, knowing that the input image data are identical. Images included are the eight ITU-T "test images" (referred to for years as the "CCITT test documents"), two bilevel test charts, a gray-scale test chart, various screened halftone images, electronically dithered images, computer generated images, gray-scale images, and color images. The T.24 CD-ROM is available from the ITU.

THE SLEREXE COMPANY LIMITED

SAPORS LANE . BOOLE . DORSET . BH 25 8 ER

TELEPHONE BOOLE (945 13) 51617 . TELEX 123456

Our Ref. 350/PJC/EAC

18th January, 1972.

Dr. P.N. Cundall,
Mining Surveys Ltd.,
Holroyd Road,
Reading,
Berks.

Dear Pete,

Permit me to introduce you to the facility of facsimile transmission.

In facsimile a photocell is caused to perform a raster scan over the subject copy. The variations of print density on the document cause the photocell to generate an analogous electrical video signal. This signal is used to modulate a carrier, which is transmitted to a remote destination over a radio or cable communications link.

At the remote terminal, demodulation reconstructs the video signal, which is used to modulate the density of print produced by a printing device. This device is scanning in a raster scan synchronised with that at the transmitting terminal. As a result, a facsimile copy of the subject document is produced.

Probably you have uses for this facility in your organisation.

Yours sincerely,

Phil.

P.J. CROSS
Group Leader - Facsimile Research

Figure 10.8(a) CCITT Group 3 test image: English letter.

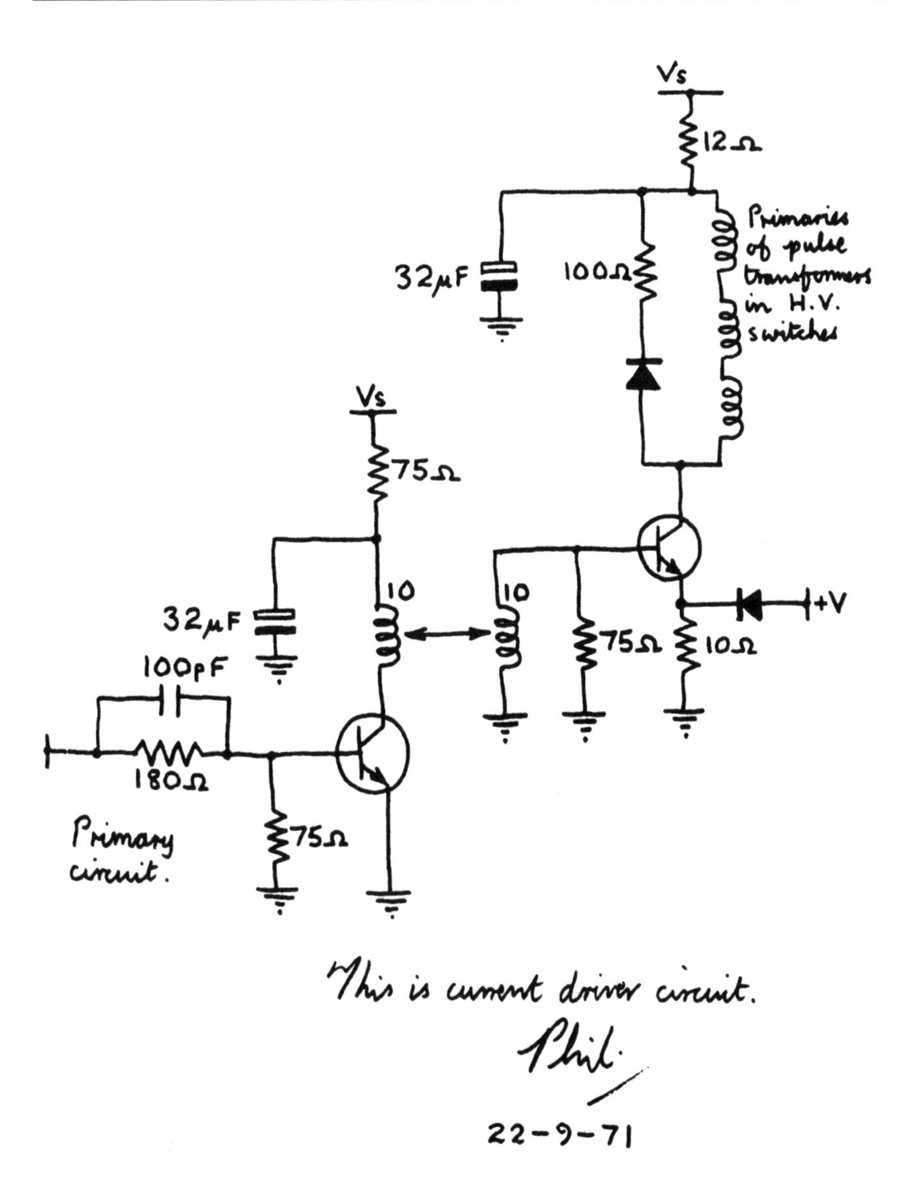

Figure 10.8(b) CCITT Group 3 test image: Circuit drawing.

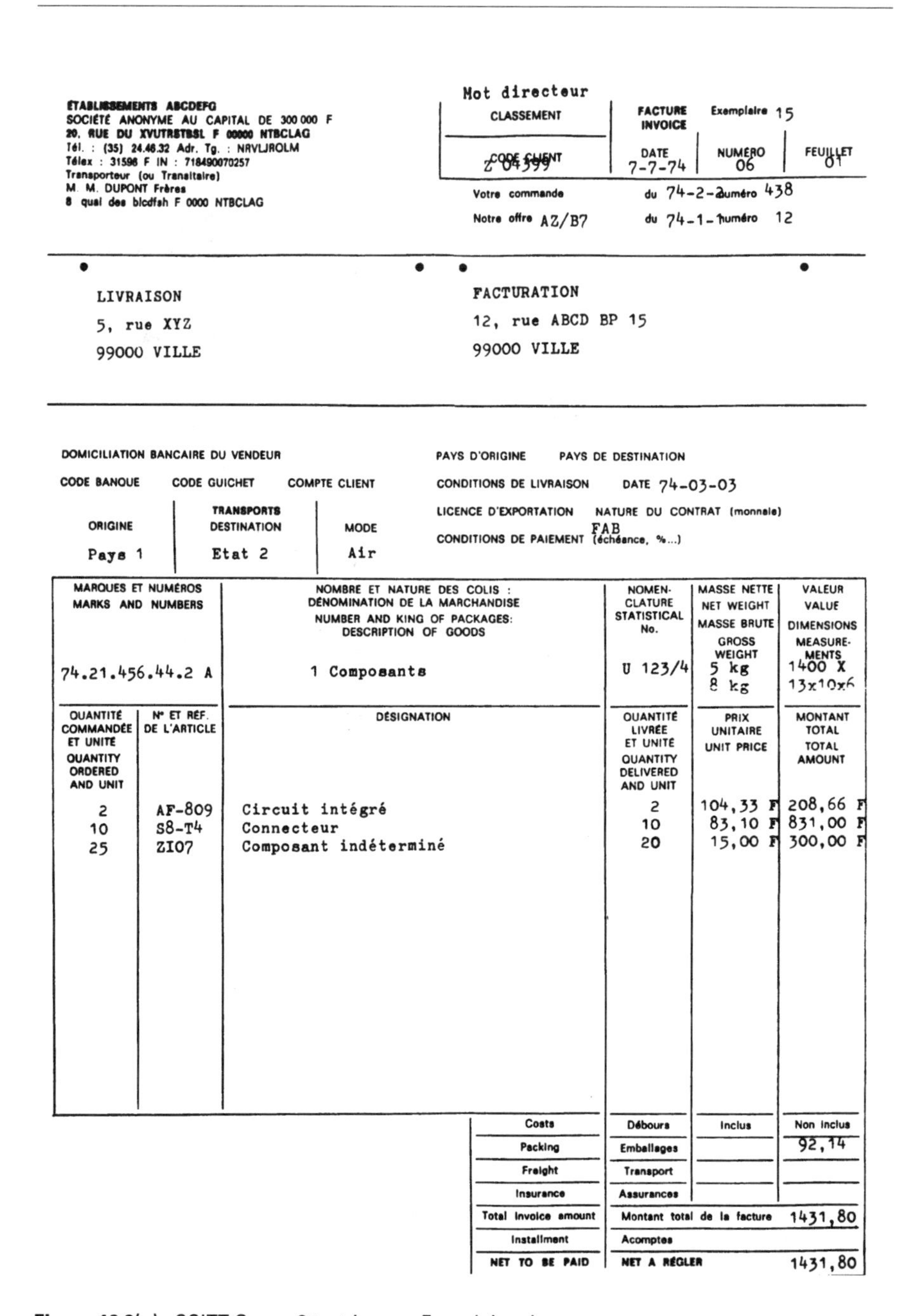

ÉTABLISSEMENTS ABCDEFG
SOCIÉTÉ ANONYME AU CAPITAL DE 300 000 F
20, RUE DU XVUTRSTBSL F 00000 NTBCLAG
Tél. : (35) 24.46.32 Adr. Tg. : NRVLJROLM
Télex : 31596 F IN : 718490070257
Transporteur (ou Transitaire)
M. M. DUPONT Frères
8 quai des blcdfsh F 0000 NTBCLAG

Mot directeur

CLASSEMENT	FACTURE INVOICE	Exemplaire 15	
CODE CLIENT Z 04399	DATE 7-7-74	NUMERO 06	FEUILLET 01

Votre commande du 74-2-2 numéro 438
Notre offre AZ/B7 du 74-1-1 numéro 12

LIVRAISON
5, rue XYZ
99000 VILLE

FACTURATION
12, rue ABCD BP 15
99000 VILLE

DOMICILIATION BANCAIRE DU VENDEUR

CODE BANQUE CODE GUICHET COMPTE CLIENT

ORIGINE	TRANSPORTS DESTINATION	MODE
Pays 1	Etat 2	Air

PAYS D'ORIGINE PAYS DE DESTINATION
CONDITIONS DE LIVRAISON DATE 74-03-03
LICENCE D'EXPORTATION NATURE DU CONTRAT (monnaie)
CONDITIONS DE PAIEMENT (échéance, %...) FAB

MARQUES ET NUMÉROS MARKS AND NUMBERS	NOMBRE ET NATURE DES COLIS : DÉNOMINATION DE LA MARCHANDISE NUMBER AND KING OF PACKAGES: DESCRIPTION OF GOODS	NOMENCLATURE STATISTICAL No.	MASSE NETTE NET WEIGHT MASSE BRUTE GROSS WEIGHT	VALEUR VALUE DIMENSIONS MEASUREMENTS
74.21.456.44.2 A	1 Composants	U 123/4	5 kg 8 kg	1400 X 13x10x6

QUANTITÉ COMMANDÉE ET UNITÉ QUANTITY ORDERED AND UNIT	N° ET RÉF. DE L'ARTICLE	DÉSIGNATION	QUANTITÉ LIVRÉE ET UNITÉ QUANTITY DELIVERED AND UNIT	PRIX UNITAIRE UNIT PRICE	MONTANT TOTAL TOTAL AMOUNT
2	AF-809	Circuit intégré	2	104,33 F	208,66 F
10	S8-T4	Connecteur	10	83,10 F	831,00 F
25	ZI07	Composant indéterminé	20	15,00 F	300,00 F

Costs	Débours	Inclus	Non inclus
Packing	Emballages		92,14
Freight	Transport		
Insurance	Assurances		
Total invoice amount	Montant total de la facture		1431,80
Installment	Acomptes		
NET TO BE PAID	NET A RÉGLER		1431,80

Figure 10.8(c) CCITT Group 3 test image: French invoice.

L'ordre de lancement et de réalisation des applications fait l'objet de décisions au plus haut niveau de la Direction Générale des Télécommunications. Il n'est certes pas question de construire ce système intégré "en bloc" mais bien au contraire de procéder par étapes, par paliers successifs. Certaines applications, dont la rentabilité ne pourra être assurée, ne seront pas entreprises. Actuellement, sur trente applications qui ont pu être globalement définies, six en sont au stade de l'exploitation, six autres se sont vu donner la priorité pour leur réalisation.
Chaque application est confiée à un "chef de projet", responsable successivement de sa conception, de son analyse-programmation et de sa mise en oeuvre dans une région-pilote. La généralisation ultérieure de l'application réalisée dans cette région-pilote dépend des résultats obtenus et fait l'objet d'une décision de la Direction Générale. Néanmoins, le chef de projet doit dès le départ considérer que son activité a une vocation nationale donc refuser tout particularisme régional. Il est aidé d'une équipe d'analystes-programmeurs et entouré d'un "groupe de conception" chargé de rédiger le document de "définition des objectifs globaux" puis le "cahier des charges" de l'application, qui sont adressés pour avis à tous les services utilisateurs potentiels et aux chefs de projet des autres applications. Le groupe de conception comprend 6 à 10 personnes représentant les services les plus divers concernés par le projet,et comporte obligatoirement un bon analyste attaché à l'application.

II - L'IMPLANTATION GEOGRAPHIQUE D'UN RESEAU INFORMATIQUE PERFORMANT

L'organisation de l'entreprise française des télécommunications repose sur l'existence de 20 régions. Des calculateurs ont été implantés dans le passé au moins dans toutes les plus importantes. On trouve ainsi des machines Bull Gamma 30 à Lyon et Marseille, des GE 425 à Lille, Bordeaux, Toulouse et Montpellier, un GE 437 à Massy, enfin quelques machines Bull 300 TI à programmes câblés étaient récemment ou sont encore en service dans les régions de Nancy, Nantes, Limoges, Poitiers et Rouen ; ce parc est essentiellement utilisé pour la comptabilité téléphonique.
A l'avenir, si la plupart des fichiers nécessaires aux applications décrites plus haut peuvent être gérés en temps différé, un certain nombre d'entre eux devront nécessairement être accessibles, voire mis à jour en temps réel : parmi ces derniers le fichier commercial des abonnés, le fichier des renseignements, le fichier des circuits, le fichier technique des abonnés contiendront des quantités considérables d'informations.
Le volume total de caractères à gérer en phase finale sur un ordinateur ayant en charge quelques 500 000 abonnés a été estimé à un milliard de caractères au moins. Au moins le tiers des données seront concernées par des traitements en temps réel.
Aucun des calculateurs énumérés plus haut ne permettait d'envisager de tels traitements.
L'intégration progressive de toutes les applications suppose la création d'un support commun pour toutes les informations, une véritable "Banque de données", répartie sur des moyens de traitement nationaux et régionaux, et qui devra rester alimentée, mise à jour en permanence, à partir de la base de l'entreprise, c'est-à-dire les chantiers, les magasins, les guichets des services d'abonnement, les services de personnel etc.
L'étude des différents fichiers à constituer a donc permis de définir les principales caractéristiques du réseau d'ordinateurs nouveaux à mettre en place pour aborder la réalisation du système informatif. L'obligation de faire appel à des ordinateurs de troisième génération, très puissants et dotés de volumineuses mémoires de masse, a conduit à en réduire substantiellement le nombre.
L'implantation de sept centres de calcul interrégionaux constituera un compromis entre : d'une part le désir de réduire le coût économique de l'ensemble, de faciliter la coordination des équipes d'informaticiens; et d'autre part le refus de créer des centres trop importants difficiles à gérer et à diriger,et posant des problèmes délicats de sécurité. Le regroupement des traitements relatifs à plusieurs régions sur chacun de ces sept centres permettra de leur donner une taille relativement homogène. Chaque centre "gèrera" environ un million d'abonnés à la fin du VIème Plan.
La mise en place de ces centres a débuté au début de l'année 1971 : un ordinateur IRIS 50 de la Compagnie Internationale pour l'Informatique a été installé à Toulouse en février ; la même machine vient d'être mise en service au centre de calcul interrégional de Bordeaux.

Figure 10.8(d) CCITT Group 3 test image: French text.

Cela est d'autant plus valable que $T\Delta f$ est plus grand. A cet égard la figure 2 représente la vraie courbe donnant $|\phi(f)|$ en fonction de f pour les valeurs numériques indiquées page précédente.

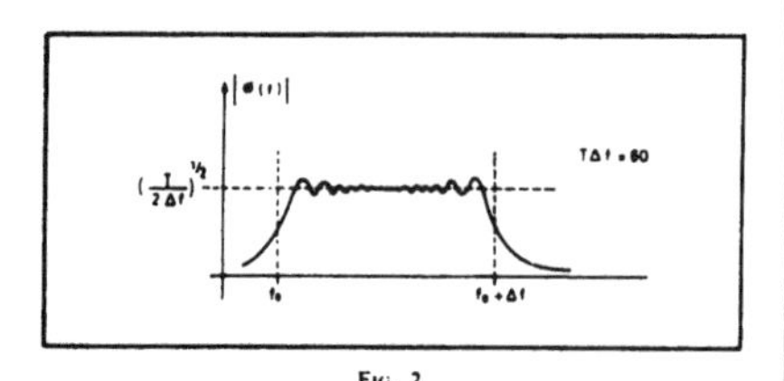

FIG. 2

Dans ce cas, le filtre adapté pourra être constitué, conformément à la figure 3, par la cascade :

— d'un filtre passe-bande de transfert unité pour $f_0 \leqslant f \leqslant f_0+\Delta f$ et de transfert quasi nul pour $f < f_0$ et $f > f_0+\Delta f$, filtre ne modifiant pas la phase des composants le traversant ;

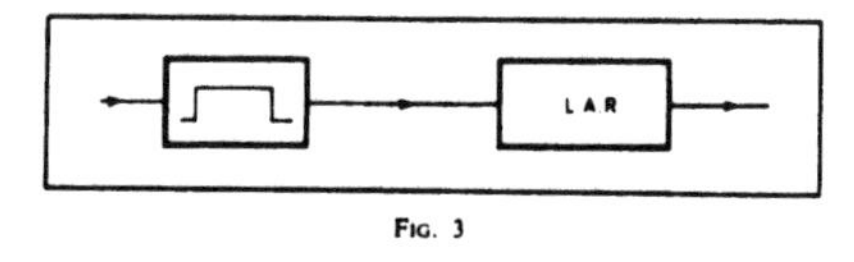

FIG. 3

— filtre suivi d'une ligne à retard (LAR) dispersive ayant un temps de propagation de groupe T_R décroissant linéairement avec la fréquence f suivant l'expression :

$$T_R = T_0+(f_0-f)\frac{T}{\Delta f} \quad (\text{avec } T_0 > T)$$

(voir fig. 4).

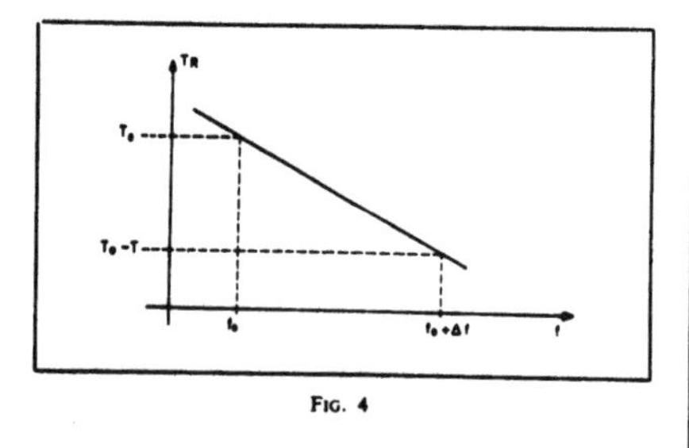

FIG. 4

telle ligne à retard est donnée par :

$$\varphi = -2\pi \int_0^f T_R \, df$$

$$\varphi = -2\pi\left[T_0+\frac{f_0 T}{\Delta f}\right]f + \pi\frac{T}{\Delta f}f^2$$

Et cette phase est bien l'opposé de $/\phi(f)$, à un déphasage constant près (sans importance) et à un retard T_0 près (inévitable).

Un signal utile $S(t)$ traversant un tel filtre adapté donne à la sortie (à un retard T_0 près et à un déphasage près de la porteuse) un signal dont la transformée de Fourier est réelle, constante entre f_0 et $f_0+\Delta f$, et nulle de part et d'autre de f_0 et de $f_0+\Delta f$, c'est-à-dire un signal de fréquence porteuse $f_0+\Delta f/2$ et dont l'enveloppe a la forme indiquée à la figure 5, où l'on a représenté simultanément le signal $S(t)$ et le signal $S_1(t)$ correspondant obtenu à la sortie du filtre adapté. On comprend le nom de récepteur à compression d'impulsion donné à ce genre de filtre adapté : la « largeur » (à 3 dB) du signal comprimé étant égale à $1/\Delta f$, le rapport de compression est de $\frac{T}{1/\Delta f} = T\Delta f$

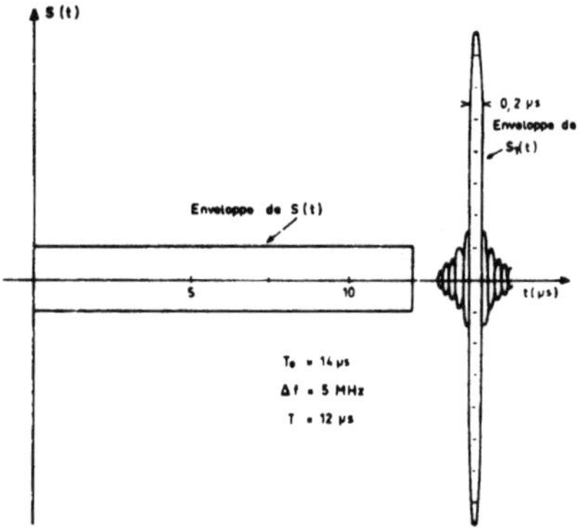

FIG. 5

On saisit physiquement le phénomène de compression en réalisant que lorsque le signal $S(t)$ entre dans la ligne à retard (LAR) la fréquence qui entre la première à l'instant 0 est la fréquence basse f_0, qui met un temps T_0 pour traverser. La fréquence f entre à l'instant $t=(f-f_0)\frac{T}{\Delta f}$ et elle met un temps $T_0-(f-f_0)\frac{T}{\Delta f}$ pour traverser, ce qui la fait ressortir à l'instant T_0 également. Ainsi donc, le signal $S(t)$

Figure 10.8(e) CCITT Group 3 test image: French text with figures.

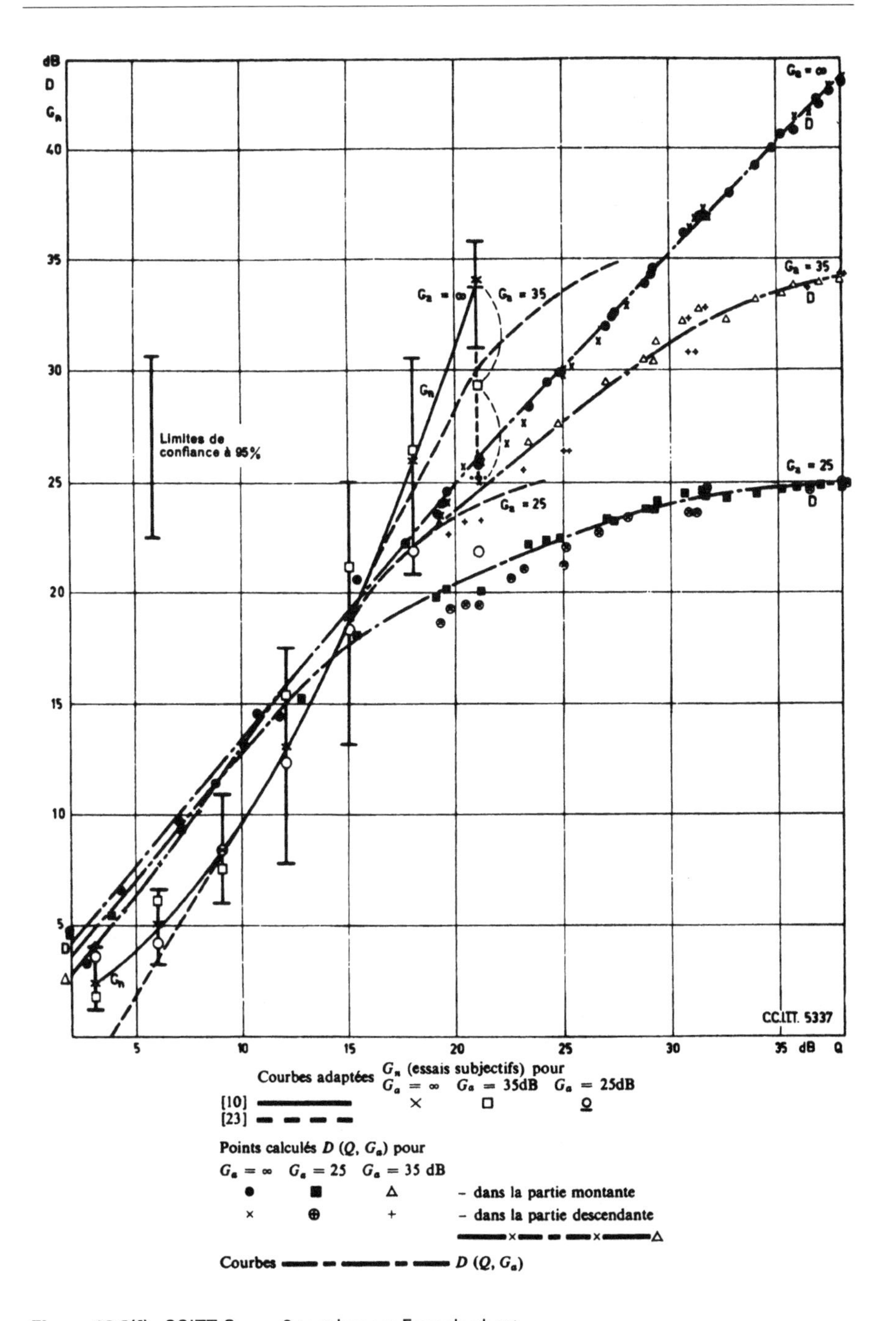

Figure 10.8(f) CCITT Group 3 test image: French chart.

ＣＣＩＴＴの概要

沿革

ＣＣＩＴＴは、国際電気通信連合（ＩＴＵ）の四つの常設機関（事務総局、国際周波数登録委員会、ＣＣＩＲ、ＣＣＩＴＴ）の一つとして、ＩＴＵの中でも、世界の国際通信上の諸問題を真先に取上げ、その解決方法を見出して行く重要な機関である。日本名は、国際電信電話諮問委員会と称する。

ＣＣＩＴＴの前身は、ＣＣＩＦ（国際電話諮問委員会）とＣＣＩＴ（国際電信諮問委員会）である。ＣＣＩＦは、１９２４年にヨーロッパに「国際長距離電話通信諮問委員会」が設置され、これが１９２５年のパリ電信電話会議のとき、正式に、「国際電話諮問委員会」として万国電信連合の公式機関となったものである。ＣＣＩＴは、同じく１９２５年の会議のとき、ＣＣＩＦと併立するものとして設置された。

そして、ＣＣＩＦは、１９５６年の12月に第18回総会が開催されたのち、ＣＣＩＴは、同年同月に第8回総会が開催されたのち、併合されて現在のＣＣＩＴＴとなった。このＣＣＩＴＴは、ＣＣＩＦとＣＣＩＴが解散した直後、第1回総会を開催し、第2回総会は、１９６０年にニューデリーで、第3回総会は、１９６４年、ジュネーブで、第4回総会は、１９６８年、アルゼンチンで開催された。

ＣＣＩＦとＣＣＩＴが合併したのは、有線電気通信の分野、とくに伝送路について電信回線と電話回線とを技術的に分ける意味がなくなってきたこと、各国とも大体において、電信部門と電話部門は同一組織内にあること、ＣＣＩＦの事務局とＣＣＩＴの事務局の合併による能率増進等がおもな理由であった。

ＣＣＩＴＴは、上述のように、ヨーロッパ内の国ぐにによって、ヨーロッパ内の電信・電話の技術・運用・料金の基準を定め、あるいは統一をはかってきたので、現在でも、その影響を受け、会合参加国は、ヨーロッパの国が多く、ヨーロッパで生起する問題の研究が多い。たとえば、１９６０年のＣＣＩＴＴ勧告の中で、技術上配慮する距離は約２，５００kmであったが、これはヨーロッパ内領域を想定したものである。

しかしながら、１９５６年9月に敷設された大西洋横断電話ケーブルは、大陸間電話通信の自動化および半自動化への技術的可能性を与え、ＣＣＩＴＴがこの問題を取り上げるに及び、ＣＣＩＴＴの性格は漸次、汎世界的色彩を実質的に帯びるに至った。この汎世界的性格は第2次世界大戦後目ざましくなったアジア・アフリカ植民地の独立に伴ってＩＴＵの構成員の中にこれらの国が加わり、ＩＴＵの中に新しい意見が導入されたことにも起因して、技術面、政治面の双方から導入されてきた。ＣＣＩＴＴの汎世界化は、１９６０年の第2回総会がニューデリーで開催されたことにもあらわれている。この総会までは、ＣＣＩＴ、ＣＣＩＦのいずれにしろ、アメリカやアジアで総会が開催されたことがなく、ＣＣＩＴＴ委員長も、ニューデリー総会の準備文書で、この点には注目すべきであるとのべている。

任務

ＩＴＵは、全権委員会議、主管庁会議を始めとして、七つの機関をもち、それぞれの機関の権限と任務は国際電気通信条約に明記されている。そこで条約を参照してみるならば、ＣＣＩＴＴの任務は、つぎのとおりとなっている。

「国際電信電話諮問委員会（ＣＣＩＴＴ）は、電信および電話に関する技術、運用および料金の問題について研究し、および意見を表明することを任務とする。」（１９６５年モントルー条約第１８７号）

「各国際諮問委員会は、その任務の遂行に当たって、新しい国または発展の途上にある国における地域的および国際的分野にわたる電気通信の創設、発達および改善に直接関連のある問題について研究し、および意見を作成するように妥当な注意を払わなければならない。」（同第１８８号）

「各国際諮問委員会は、また、関係国の要請に基づき、その国内電気通信の問題について研究し、かつ、勧告を行なうことができる。」（同第１８９号）

上記第１８７号と第１８８号にいわれる「意見」とは、フランス語の Avis から訳したもので、英語では、「勧告（Recommendation）」となっている。ＣＣＩＴＴの表明する意見は、国際法的には強制力をもたないものであって、この点が、条約、電信規則、電話規則等各国を拘束する力をもっているものと異なる。もっとも意見とは称しても、技術的分野では、電信規則のごとき、各国政府が承認してその内容を実施する強制規則をもたないので、実際にある機器の仕様を定める場合には、多くの国の意見が統一されたこの「意見」に従わなければ、円滑な国際通信を行なうことができない場合が多い。この意見（または勧告）は、国際通信を行なう場合各国が直面する問題について、具体的意見を表明するもので、たとえば、大陸間ケーブルで大陸間通話を半自動化しようとする場合、その信号方式や取り扱う通話の種類および料金は、どのようにするかを研究して意見を表明する。したがって、ＣＣＩＴＴの活動は、つねに時代の最先端を行くもので、ＣＣＩＴＴの活動方向は、そのまま世界の国際通信の活動方向であるともいえる。

この意見は、また、電信規則以下のその他の規則のごとく、数年以上の間隔をもって開催される主管庁会議というような大会議の決定をまたなくても表明することができ、また、その改正も容易であるので、現在のように進歩の早い国際通信界では、関係国の意見を統一した国際的見解としては非常に便利である。

Figure 10.8(g) CCITT Group 3 test image: Kanji.

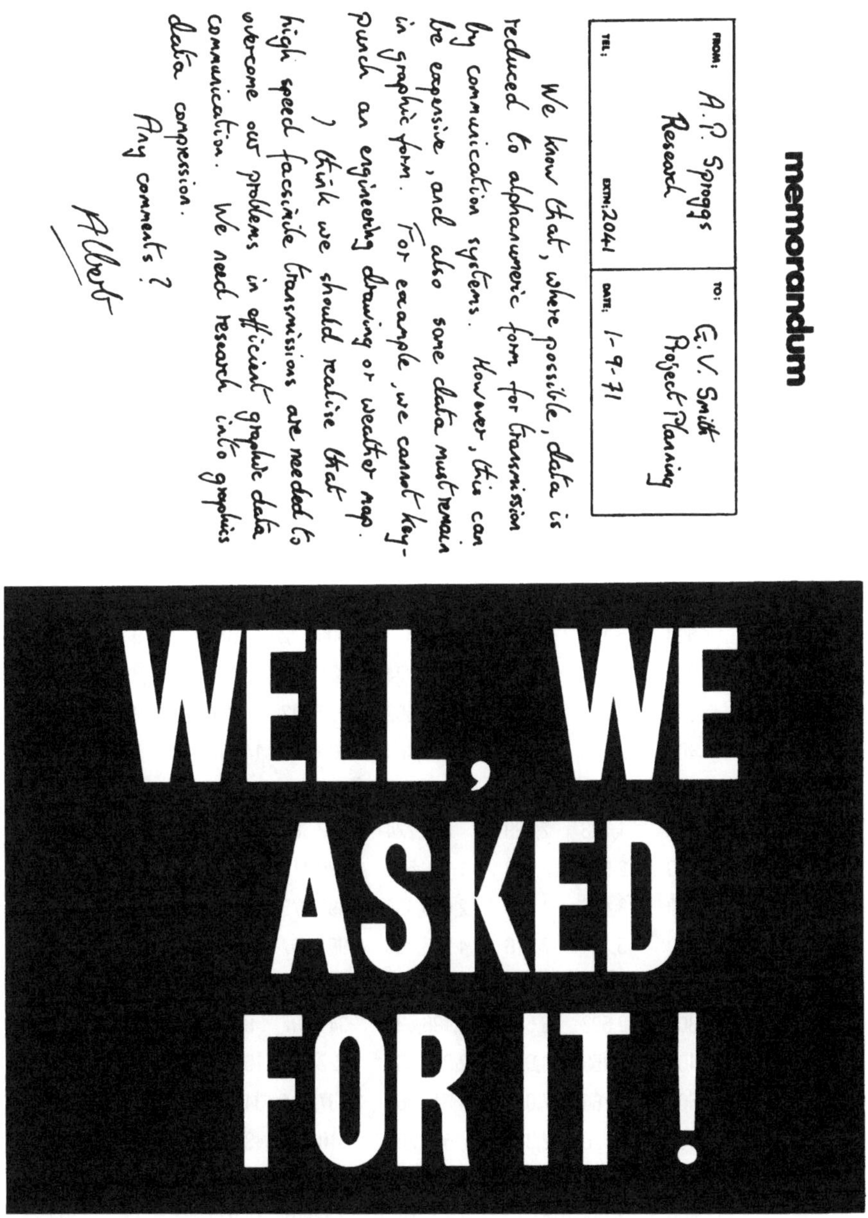

memorandum

FROM: A.P. Spriggs Research	TO: G.V. Smith Project Planning
TEL: EXTN: 2041	DATE: 1-9-71

We know that, where possible, data is reduced to alphanumeric form for transmission by communication systems. However, this can be expensive, and also some data must remain in graphic form. For example, we cannot key-punch an engineering drawing or weather map.

I think we should realise that high speed facsimile transmissions are needed to overcome our problems in efficient graphic data communication. We need research into graphics data compression.

Any comments?

Albert

WELL, WE ASKED FOR IT!

Figure 10.8(h) CCITT Group 3 test image: Handwritten memorandum.

Table 10.1
ITU Reference Images

Figure	Image	Pels/ in	Width	Height	Bytes	Black pels	White pels	% Black	% White
10.8a	No. 1	200	1,728	2,339	505,224	104,990	3,936,802	2.60	97.40
	English letter	300	2,592	3,508	1,136,592	236,943	8,855,793	2.61	97.39
		400	3,456	4,677	2,020,464	426,637	15,737,075	2.64	97.36
		600	5,184	7,016	4,546,368	980,552	35,390,392	2.70	97.30
10.8b	No. 2	200	1,728	2,339	505,224	158,713	3,883,079	3.93	96.07
	Circuit drawing	300	2,592	3,508	1,136,592	350,335	8,742,401	3.85	96.15
		400	3,456	4,677	2,020,464	643,328	15,520,384	3.98	96.02
		600	5,184	7,016	4,546,368	1,446,165	34,924,779	3.98	96.02
10.8c	No. 3	200	1,728	2,339	505,224	222,697	3,819,095	5.51	94.49
	French invoice	300	2,592	3,508	1,136,592	498,319	8,594,417	5.48	94.52
		400	3,456	4,677	2,020,464	905,106	15,258,606	5.60	94.40
		600	5,184	7,016	4,546,368	2,004,822	34,366,122	5.51	94.49
10.8d	No. 4	200	1,728	2,339	505,224	371,671	3,670,121	9.20	90.80
	French text	300	2,592	3,508	1,136,592	837,842	8,254,894	9.21	90.79
		400	3,456	4,677	2,020,464	1,539,573	14,624,139	9.52	90.48
		600	5,184	7,016	4,546,368	3,587,602	32,783,342	9.86	90.14
10.8e	No. 5	200	1,728	2,339	505,224	222,306	3,819,486	5.50	94.50
	French figures	300	2,592	3,508	1,136,592	490,419	8,602,317	5.39	94.61
		400	3,456	4,677	2,020,464	892,675	15,271,037	5.52	94.48
		600	5,184	7,016	4,546,368	1,987,057	34,383,887	5.46	94.54
10.8f	No. 6	200	1,728	2,339	505,224	154,711	3,887,081	3.83	96.17
	French chart	300	2,592	3,508	1,136,592	341,837	8,750,899	3.76	96.24
		400	3,456	4,677	2,020,464	622,225	15,541,487	3.85	96.15
		600	5,184	7,016	4,546,368	1,387,214	34,983,730	3.81	96.19
10.8g	No. 7	200	1,728	2,339	505,224	310,743	3,731,049	7.69	92.31
	Kanji	300	2,592	3,508	1,136,592	690,828	8,401,908	7.60	92.40
		400	3,456	4,677	2,020,464	1,239,891	14,923,821	7.67	92.33
		600	5,184	3,035	1,966,680	1,184,951	14,548,489	7.53	92.47

Table 10.1 (continued)

Figure	Image	Pels/ in	Width	Height	Bytes	Black pels	White pels	% Black	% White
8h	No. 8	200	1,728	2,339	505,224	1,603,283	2,438,509	39.67	60.33
	Memo-	300	2,592	3,508	1,136,592	3,613,143	5,479,593	39.74	60.26
	randum	400	3,456	4,677	2,020,464	6,337,111	9,826,601	39.21	60.79
		600	5,184	7,016	4,546,368	14,259,312	22,111,632	39.21	60.79

11

HF Radio Fax Systems

This chapter describes fax systems designed to operate over the often noisy HF radio channels of 3–30 MHz instead of telephone lines. These systems provide an alternative method of faxing, one that avoids use of the PSTN where it is not available or when for some reason it is not desirable to use. It also avoids use of satellite radio systems, whose services can be very expensive. On an HF radio channel, signal transmission conditions are quite different from those on telephone channels. The received fax signal is often distorted due to atmospheric and manmade noise, interference from other HF radio stations, fading, and multipath transmission. The fax systems covered are for point-to-point communication (as with Group 3) rather than broadcast radio facsimile, as discussed in Chapter 2. Some of these systems can be operated over other types of radio channels, such as VHF, UHF, or even noisy telephone lines.

11.1 Group 3 HF Radio Fax

Group 3 fax was not designed for operation over HF radio channels. To do so, its communication protocol and modem type must be changed to use HF radio protocol with an HF modem. To avoid changing the fax into a nonstandard unit, the HAL FAX-4100[1] system has been developed with an interface that looks like another Group 3 fax on one side and an HF data modem on the other side. Normal fax functions control operation of an HF radio transceiver

1. HAL Communications Corporation, 1201 West Kenyon Street, Urbana, IL 61801, (217) 367-7373.

as used by some U.S. federal government agencies and by amateur (ham) radio operators. It is a proprietary scheme with two boxes connected in tandem between a Group 3 fax machine and an HF radio transceiver (Figure 11.1).

The FAX-4100 controller includes a built-in V.29 Group 3 fax modem and emulates a fax machine. The fax signals from the modem connect to the HF data modem, which interfaces with the HF radio transceiver. The modem

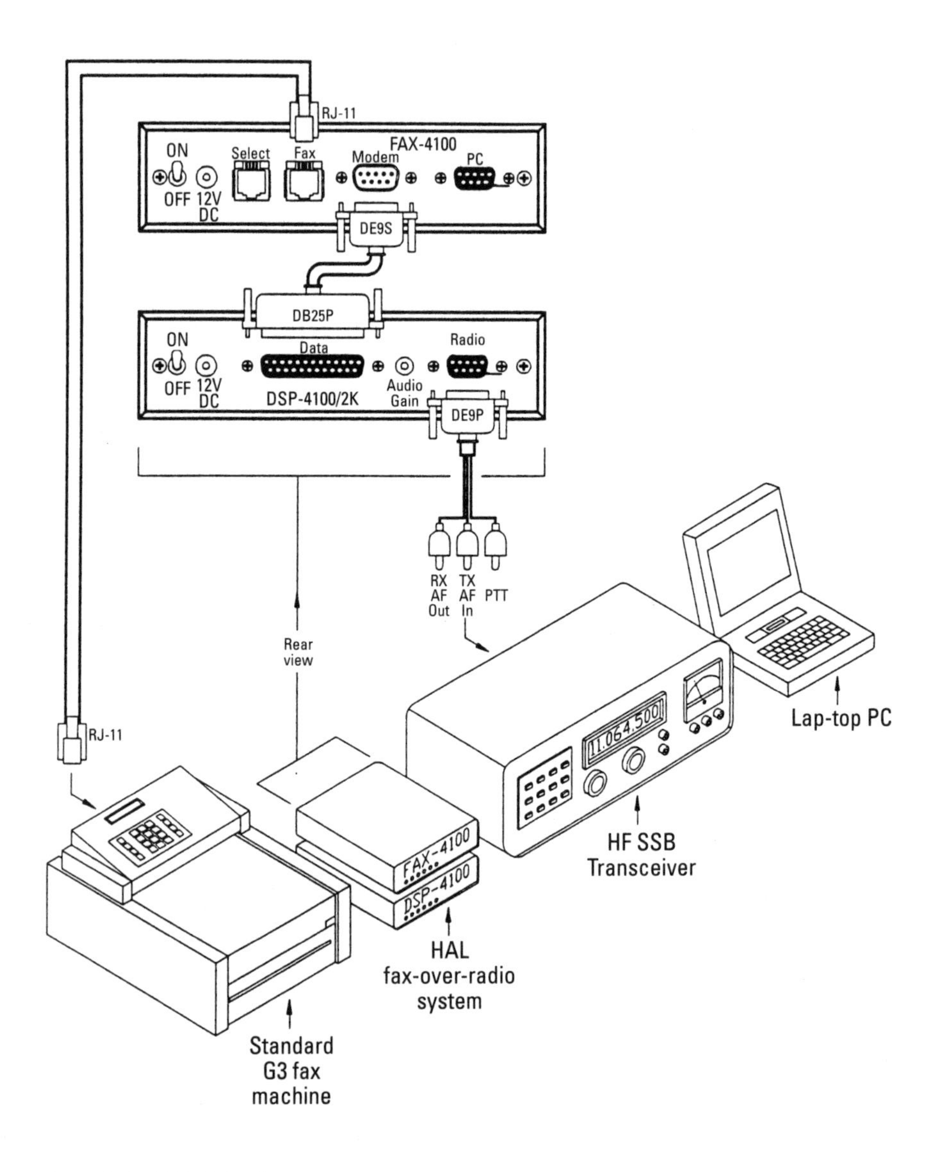

Figure 11.1 HAL FAX-4100 system.

uses a far different transmission technique to successfully send fax data over an HF radio channel.

A call, based on the radio phone number, is initiated from the fax unit keypad just as if it were connected to a phone line. The controller at the initiating end answers the ring from the originating fax and establishes an HF radio link between the radio data modems at both ends. A ring signal command passes through to generate a ring at the receiving fax unit to start it, and the standard handshake between the fax units follows. Fax image data then pass from the sending fax unit into the controller's memory at the originating end. The controller next passes the fax data through to memory in the other controller at the receiving end, then to its V.29 fax modem, and finally into the receiving Group 3 fax unit.

Sending a fax is a simple three-step process: (1) Place the page(s) in the fax machine document feeder, (2) enter the radio phone number, (3) push "GO" on the fax machine. A full page can be sent in 2 to 6 minutes, depending on ionospheric conditions and the amount of information on the page to be transmitted. The entire link setup and maintenance procedure is transparent to the fax operator, who need not know nor care that an HF radio system is part of the fax link. It all works just like a standard fax telephone transmission, except that it takes longer. Housekeeping-control functions and indications (e.g., feeding messages back to the fax machine when the radio link fails or the other station is not available) also are automated.

11.2 Fax 480/TIA 668 (PC-VGA Fax)

Since the Group 3 standard was adopted in 1980, fax has become more and more closely integrated with computer technology. One example of that trend was development of a fax system by amateur (ham) radio operators who used PCs with standard VGA screen images. The audio frequency shift (AFS) modem uses analog tones with black and white pixels transmitted as 1,500 Hz and 2,300 Hz, respectively. Gray-scale values are represented by frequencies between those black and white limits.

FAX480 is an unofficial standard for sending PC screen images in 640-by-480 VGA mode by fax between PCs over the amateur (ham) radio transceivers instead of the PSTN. FAX480 was developed by Dr. Ralph E. Taggert, WB8DQT, and published in his 1993 articles in *QST* magazine [1]. The image size is 480 pixels vertical by 512 pixels horizontal. Each pixel can have 64 or more shades of gray, producing far better quality for photographs than with Group 3, where more than 64 electronically screened black and white pels are needed to simulate the proper gray value of one pixel. FAX480 was based on slow-scan TV (SSTV) and HF radio weather facsimile (WEFAX) standards.

TIA/EIA-668-A, *High Frequency Radio Facsimile*, December 1998 is an official standard version of FAX480. It was issued by the TIA, TR-29, Facsimile Systems and Engineering Equipment Committee. This standard came about after the NCS recognized that the transmission of fax over HF radio would be useful for federal government departments and agencies but was limited by the lack of an official standard.[2] Because TR-29 was an accredited committee of ANSI, NCS requested that it take on the task of obtaining approval of a new U.S. national standard that could be used by federal agencies [2]. TIA/EIA-668-A has the same specifications as FAX480 except for adding a frequency tolerance for the clock crystal. The pixel clock frequency is 1,953.125 MHz (2,048 divider from 4-MHz crystal). Neither standard is patented and no licensing agreement is required.

11.2.1 The System

One version of FAX480 uses a simple electronics box with parts costing about $15. It plugs into the PC serial port, and a cable connects to the HF radio transceiver. A 741 operational amplifier in the box runs at very high gain, making it an audio frequency limiter (clipper) similar to the radio frequency limiter of an FM radio receiver (Figure 11.2).

In receiving mode, the 1,500- to 2,300-Hz fax signals from the radio transceiver pass through it, and the square-wave output goes to the PC. The limiter maintains a constant output signal level even under severe fading of the received HF radio signal. The computer software (available as freeware) measures the time between the zero crossings to determine the frequency of the received signal, converting it to the proper brightness for each pixel for the fax image display or print. Some later versions substitute for the box a card that plugs into a spare PC bus slot. Others use the PC sound card. A major advantage of FAX480 is that gray-scale images of 16, 32, 64, 128, or more gray-scale steps are sent at the same 2 minutes and 18 seconds per image. Even color images can be sent between compatible units using a nonstandard Colorfax

2. Among the responsibilities assigned to the NCS is the management of the Federal Telecommunication Standards Program, which identifies, develops, and coordinates proposed federal standards that either contribute to interoperability of functionally similar systems or achieve a compatible and efficient interface between computer and telecommunication systems. That includes initiating and pursuing joint standards development efforts with appropriate technical committees of the ISO and the ITU. For more information, see the NCS Web site at http://www.ncs.gov.

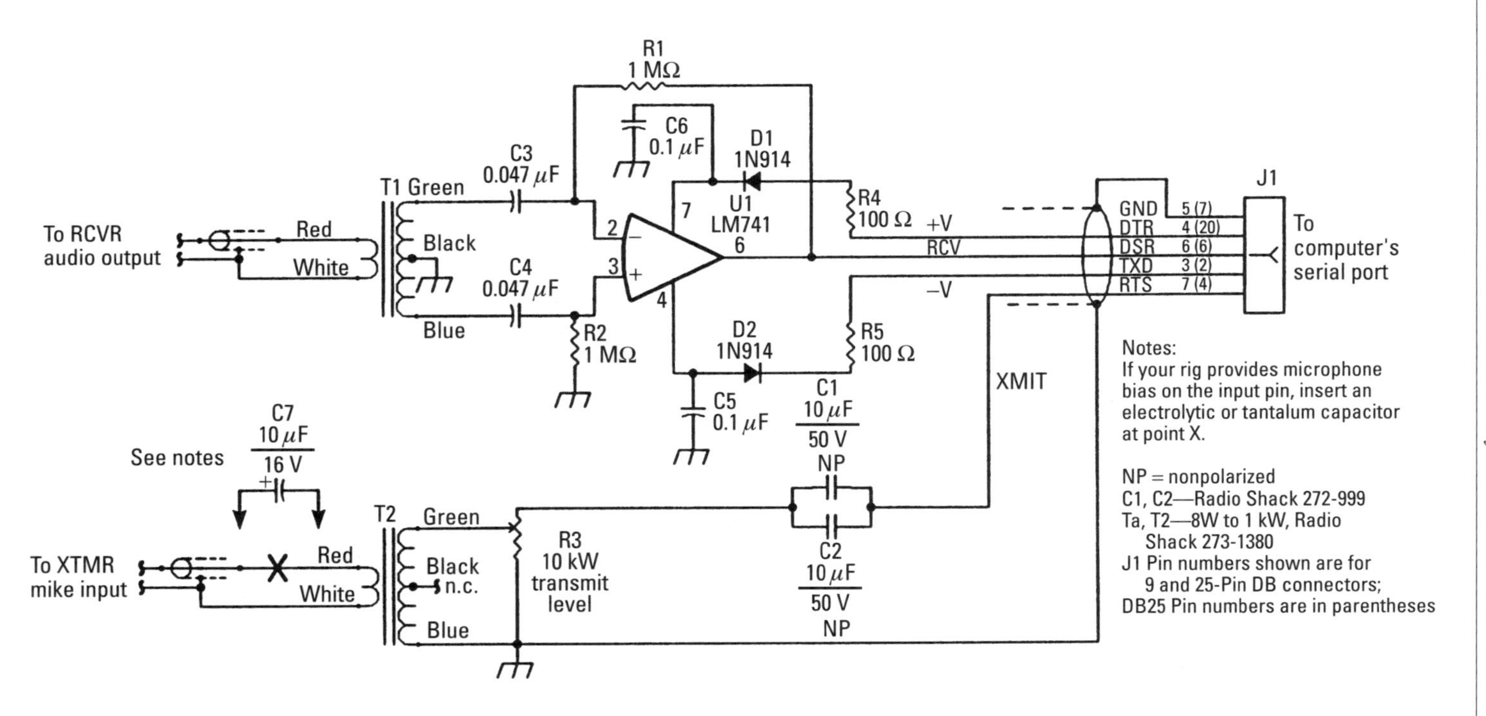

Figure 11.2 FAX480 circuit diagram.

mode of FAX480 developed by Vester. It has the same line timing as FAX480 and uses SSTV line sequential color format.

11.2.2 Transmitting and Receiving Images

The standardized transmission sequence is as follows:

1. Start signal: 5 seconds (four clock cycles of 2,300 Hz, then four clock cycles of 1,500 Hz at a 244 Hz rate).
2. Phasing pulses: 20 phasing lines (5.12 ms of sync frequency at the beginning of each line followed by 262.144 ms of white).
3. Picture signal: 480 image lines. Each line has a 5.12-ms sync pulse followed by 262.144 ms (512 pixels) of video.

Because the fax signal is sent by radio without handshake, it can also be copied by others. The fax receiving equipment is not standardized and may use the transmitted nonvideo signals in different manners to lock in the image sent. A good method of receiving is to use the system clock, because that provides noise immunity for starting each received line at the proper time and placing each pixel in the proper spot. An alternative method locks each received line to its sync pulse. The pulse makes it possible to start reception after the fax picture has started and still get the remainder of the image. A receiving system, which starts only after the start signal and phasing signals, misses an entire frame if it misses the beginning. Without a high-stability clock, timing inaccuracies can result in a noticeable skew or slant to the received picture, even causing it to wrap back around if the relative timing difference between transmitter and receiver is too great. Taggart's original implementation [1] avoids that by using a crystal oscillator for timing accuracy. TIA/EIA-668-A adds a clock tolerance of 25 ppm. Vester has a system without an external crystal oscillator that locks on the first received horizontal sync pulse. He also has a system that replaces the start signal and phasing signals with a vertical interval signal, allowing unattended reception and picture storage (like SSTV). The Internet and *QST* magazine have many sources for detailed information on building or buying the systems described here.

11.2.3 Background

FAX480 and EIA/TIA-668-A use the same 1,500- to 2,300-Hz audio frequencies and methods that were successful for sending radiophotos, messages, and weather maps by HF radio in the 1940s (see Chapter 2). Author McConnell was instrumental in setting those fax standards, designing and building the first equipment, and in later adapting it to broadcast weather maps by HF radio using frequency shift (FS) transmission. An earlier U.S. Army Signal Corps system used with their FX-1 fax, operated at 1,800 Hz black to 3,000 Hz white frequencies.

Many ham radio operators contributed to the use of PCs, first to display HF radio fax signals and later to communicate, within the assigned amateur radio bands. In 1986, Keith Sueker, W3VF, used a home computer for copying WEFAX signals from a weather satellite [3]. His clipper and one-bit A/D converter provided a black and white image. Ben Vester, K3BC, later added gray scale by counting the number of time-clock pulses between the zero crossings of the signal and used the computer's internal time clock count as a reference for timing the fax lines [4]. Sueker's and Vester's contributions were significant because they used only the computer, instead of full PC boards, to demodulate the AFS signals, provide synchronization timing control, and process the signal for display.

11.3 U.S. Government and Military Secure Fax

11.3.1 Encryption Basics

Fax encryption devices available for business or U.S. government digital fax use can provide a very high degree of security. It would take a highest level workstation computer hundreds of years to break the code of a good encryption unit. The basic concept of encrypting a digital fax message is simple. Instead of sending the stream of bits normally fed to the modem input (clear text), each 0 or 1 in the bit stream is either passed through unchanged or inverted so a 0 is sent as a 1 or a 1 as a 0 (see Figure 11.3). The decision whether to invert a particular bit is controlled in a predetermined pseudo-random manner.

If truly random inversions were used, it would not be possible to decode the received fax signal. The pseudo-random pattern used is so long that it is essentially the same as a "never" repeating one. The receiving encryption unit must have the same pattern to decode the received fax signals by again inverting the same bits. Identical patterns may be generated from pseudo-random number generators in both the sending and the receiving encryption units by using matching codes. Changing the code periodically further decreases the chances of reading an intercepted fax message. The new code must be inserted at the same time at all the linked encryption devices. Group 3 machines have an analog signal and cannot be encrypted for U.S. government secure use. At one

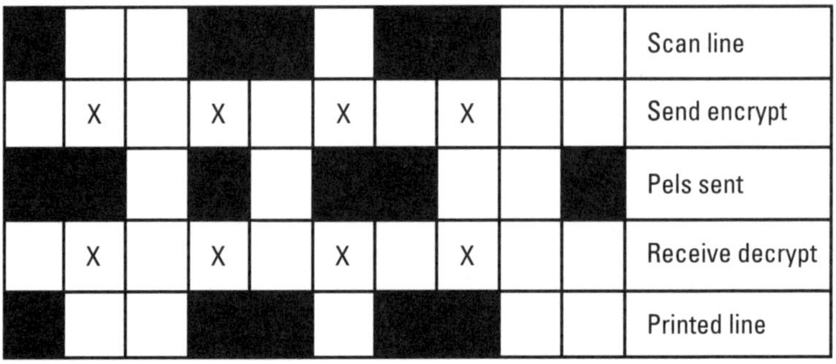

										Scan line
	X		X		X		X			Send encrypt
										Pels sent
	X		X		X		X			Receive decrypt
										Printed line

Figure 11.3 Digital fax encryption.

time, some manufacturers bypassed the fax modem and used their own non-standard encryption protocol. That resulted in interoperability problems, and the U.S. Department of Defense (DOD) does not allow them to be used.

11.3.2 MIL-STD-188-161D Fax

Military Standard MIL-STD-188-161D, Interoperability and Performance Standards for Digital Facsimile Equipment, is mandatory for the U.S. Department of Defense (DOD) fax equipment to ensure interoperability among all services and agencies within DOD.[3] This standard covers both military tactical fax equipment and long-haul fax. Fax machines built per MIL-STD-188-161 or North Atlantic Treaty Organization (NATO) Standardization Agreement (STANAG) 5000 are interoperable. Classified documents are sent through a digital interface to an encryption unit such as a KG-90 or a STU III telephone. The interface is bit-by-bit asynchronous with an external clock, at data rates of 2.4, 4.8, 9.6, and 16.0 Kbps. Dual-mode protocols permit secure fax equipment that has this feature to communicate with regular Group 3 fax units in nonsecure mode. Group 3 protocols cannot be used for secure fax communication. The interface to encryption unit prevents classified traffic from passing to the outside world through an unsecured path. For sending secure black and white information, including screened photographs, Type I mode (not to

3. Copies of federal and military specifications, standards, and handbooks are available from the Standardization Document Order Desk, 700 Robbins Avenue, Building 4, Section D, Philadelphia, PA 19111. Copies of Federal Information Processing Standards (FIPS) are available to DOD activities from the Commanding Officer, Naval Publications and Forms Center, 5801 Tabor Avenue, Philadelphia, PA 19120-5099. Others must request copies of FIPS from the National Technical Information Service, 5285 Port Royal Road, Springfield, VA 22161-2171.

be confused with Type 1 secure phones, discussed in Section 11.3.3) is used. For sending gray-scale documents with continuous tone such as photographs, Type II mode is used.

Type I and Type II modes operate with a robust digital protocol, enabling operation over noisy channels. One of three modes is used, depending on the error rate of the transmission channel:

- MH coding, compressed mode, provides the best performance for digital networks or for external modems operating on the PSTN. Group 3 coding tables are used.
- Bose Chandhuri Hocquenghem (BCH), an FEC mode, together with MH coding is used for operating over communication channels with high error rates (1 in 100 or worse). A bit-interleaving buffer improves the error correcting performance, especially for transmission-bit errors clustered in bursts.
- Uncompressed mode is best for sending fax through channels with even higher error rates.

Type II–mode fax equipment provides for the transmission and reception of gray-scale information, printing one of 16 shades of gray-scale (including black and white) for each pixel rather than simulating gray areas with a cluster of black and white pels. The image is represented by four image patterns (planes) of 0s and 1s. Each pattern is then sent as a black and white image using the same MH compression mode as for Type I, except that the pels from two adjacent scanned lines in memory are sampled in wobble mode. This mode is particularly effective in photographs, where adjacent pixels often have the same shade of gray. The coding sends these areas with 1 bit rather than 4, taking advantage of both horizontal and vertical correlations of adjacent pels to give higher compression of the image data. The sampling sequence is: pel 1/line 1, pels 1&2/line 2, pels 2&3/line 1, pels 3&4/line 2, continuing until 1,728 pels are sampled. Compression is increased further by adaptively reducing the resolution in areas that have little detail. At the receiver, each pixel is set to the same gray shade as the original, giving much higher quality recording than Group 3 fax recording. Alternatively, eight or four gray-scale shades can be used to shorten the transmission time.

11.3.3 TEMPEST Fax

A significant portion of the U.S. federal budget is spent to ensure that all government classified material sent, including fax messages, cannot be decoded and read by anyone other than the intended recipient. Some fax machines that send U.S. government-classified material employ special construction techniques to control compromising emissions. They must have government approval for TEMPEST, meeting applicable requirements of National Telecommunications and Information Systems Security Instruction (NTISSI) 7000 (U) (SECRET). Fax machines specially manufactured or modified to meet TEMPEST security requirements are furnished by at least three U.S. companies.

Probably the easiest way to encrypt fax is to use a secure digital telephone that has a built-in data jack. The telephone plugs into a regular telephone jack on the PSTN, and the fax machine plugs into the data jack. The telephone called STU III is made in four different classes, depending on the degree of security authorized. For new phones, the STU III design is being replaced by STE phones. The phones are controlled by the National Security Agency (NSA). All the classes make it extremely difficult and expensive to decode and display intercepted fax signals even though an interloper might have a STU III or STE phone.

Type 4, the lowest security level, has the fewest restrictions on its use. It can be used by businesses worldwide to protect sensitive data from falling into unauthorized hands, but export control approval is required for overseas use. Proprietary encryption algorithms are used. It is not approved for sending U.S. government-classified data.

Type 3 uses the NIST Data Encryption Standard (DES). This class of phone is generally limited to use within the United States, but some banks are authorized to use this class of phone for international funds transfer. Involvement in government programs is not required.

Type 2 can be used by federal government contractors who need to send certain classes of government-classified documents. The fax machine must be a special type that has been approved by the cognizant government agency for this use. The user of the phone is registered and issued a special key. Use of the secure phone with the fax machine connected to it generally is limited to that person. The phone must be operated only in a secure environment that prevents interception of clear text signals before encoding. The secure phones will also operate with another secure phone at one level lower (Type 1 can work with Type 2 phones and so forth).

Type 1, the highest level of security, is used for sending government-classified documents by authorized government personnel only. Operating security restrictions are similar to those for Type 2 phones.

References

[1] Taggart, R., "A New Standard for Amateur Radio Analog Facsimile," *QST*, Feb. 1993, pp. 31–36; http://taggart.glg.msu.edu/fax480.htm.

[2] National Communications System Technical Information Bulletin 96-3, *High Frequency Radio Facsimile*, Jan. 1996.

[3] Sueker, K., "Real-Time HF WEFAX Maps on a Dot-Matrix Printer," *QST*, Mar. 1986, pp. 15–20.

[4] Vester, B., "Improved HF Weather Facsimile Programs," Technical Correspondence, *QST*, Sept. 1991, pp. 40–41.

12

The Future of Fax

What next? Let us review the forecasts made in the previous edition of *FAX* seven years ago. Then let us state the realities of 1999 and make new predictions.

Forecast: In 1944, when the 100th anniversary of the telegraph was being celebrated, the *New York Times* heralded it as an "outstanding invention" that along with radio has brought about the "shrinking of the planet" and is "part of the technological unification of mankind."

In 1992, one year short of the 150th anniversary of facsimile, "fax" dominates the business world's means of communication and could well have been substituted for "telegraph" if the *Times* were to publish any such tribute again. The telegraph has long been obsolete; fax, the sleeping giant, is shrinking the planet in ways not even conceived in 1944.

Reality: Advancing to 1999, the year of the third edition of *FAX*, that trend is continuing. There has been a further large reduction in machine costs and the costs of fax communication. The fax machine has become more of a specialized computer, and fax functions are being added to many multifunctional communication units. The Internet is starting to be used for fax communication and is further lowering the communication costs. Along with the integration of fax and PCs, the PCs are developing other communication means that compete with fax in certain areas. E-mail offers a choice of data or fax messaging, but fax still dominates for immediate delivery and legal records.

12.1 The Future of Group 3 Fax

12.1.1 Group 3 Standalone Fax

Forecast: Group 3 with a configuration similar to the current one will be in great demand for many years.

Reality: That prediction is still valid, not only for fax machines but also for PC-fax. Fortunately, Group 3 fax protocol was arranged to add options as technology evolves, without making earlier designs obsolete.

Forecast: The laser printer fax devices currently available as add-ons will be built into standard laser printers.

Reality: Some PC printer designs are now available with built-in fax receiving capability, but there are many more multifunction designs that provide fax receiving and sending capability plus functions for PC printing, scanning, and copying. Most of these have plain-paper printing by thermal transfer, laser, or ink jet. A popular multifunction unit has scanning, printing and Group 3 fax capabilities. Some print on plain paper, others can print with color (see Chapter 5).

12.1.2 High-Performance Group 3 Fax

Forecast: Many enhancements have been added to Group 3 fax since the initial recommendations were adopted in 1980. New enhancements being tested by manufacturers and circuit providers will allow Group 3 fax to do many of the tasks on the PSTN that Group 4 now does only on digital networks. Already, ECM with MMR compression is a standardized option, providing faster error-free fax copies on PSTN calls.

Reality: Almost all the functions that Group 4 was designed for are now available on Group 3, operating over digital networks and the PSTN.

Forecast: Anticipated additional features are higher speeds and higher resolutions of 300 and 400 lines/inch. The 300- and 400-lines/inch resolutions match those of Group 4.

Reality: Higher speeds, up to 33.6 Kbps, plus higher resolutions of 300 and 400 lines/inch, are now available for Group 3. In addition, similar millimeter-based resolutions allow matching to either metric or inch-based fax units. A fax standard for 600 lines/inch, as well as related nonsquare resolutions, is anticipated for 1999.

Forecast: Even 600 lines/inch may become important after the 64-Kbps ISDN interface is available.

Reality: Although 600-lines/inch resolution is now common in both scanners and printers for PCs, only a small number of fax units are on ISDN lines. A fax standard for 600 lines/inch is expected in 1999.

Forecast: Work is under way on a simplified Group 3 mixed mode for use with computers that should reach standardization soon.

Reality: The mixed mode standard has been adopted, but it is not likely to be used.

Forecast: Higher modem speeds for Group 3, up to 28.8 Kbps will soon be available, doubling the 14.4-Kbps transmission rate of the V.17 modem now being used. In the United States, the EIA/TIA modem group, TR-30.1, tested suitable high-speed modem schemes. CCITT Study Group XVII [now ITU SG16] calls the modem V.$_{fast}$, [now V.34]. A probe technique may be used to check the communication channel quality quickly and select the modem speed. A channel that fails to meet the minimum quality requirements is dropped automatically and the call redialed. Channel equalization may require sending a training signal in both directions at the same time. Fax data rates higher than 14.4 Kbps are unlikely to be approved by the CCITT before 1994.

Reality: Even faster modem speeds of up to 33.6 Kbps are now available for Group 3, using modems V.34-V8/V8.$_{8bis}$ or V.34Q. The 56-Kbps ITU V.90 modem is not likely to be used for fax, but other higher speed transmission systems for Group 3 fax are probable.

Forecast: Also under consideration are other factors that limit the number of pages that can be sent per minute, including optimization of the protocol by shortening the initial handshake time from 15 sec and other delays. Other improvements are appending the postmessage commands to the fax message transmission and using the 2.4-Kbps signaling rate for binary coded commands and responses. The 3- to 5-sec time between pages becomes significant when page transmission times can be much less than 10 sec per page. On channels where the error rate is very low, the protocol could provide switching to a mode of transmitting multiple pages before acknowledging receipt.

Reality: Those features have been available for some time as private NSF features that work between pairs of Group 3 fax with the same special protocol. V.34 half-duplex fax uses a full-duplex control channel at 1,200 or 2,400 bps, substantially reducing handshake and end-of-page times.

12.1.3 Group 3 on ISDN at 64 Kbps

Forecast: A 64-Kbps interface for operation of Group 3 fax on the new ISDN is in the process of becoming a standard (using one of the B channels). Error correction is based on the selective repeat option, now a part of T.30. Full-duplex operation is proposed to speed up the error correction process. The fax call is set up in the ISDN circuit-switched mode for DTE-to-DTE communication per CCITT [now ITU] Recommendation T.90 using the session elements of its procedure rules of paragraphs 2.2.1, 2.2.2, and 2.2.4 with the HDLC IE code of Group 4 facsimile class 1. The error correction method is the one used for Group 3 fax Recommendation T.4, Annex A, transmitting documents as pages and partial pages with a frame size of 256 octets. With full-duplex operation in the 64-Kbps mode, page transmission continues without waiting for a response to the previous page. For calling a normal Group 3 fax connected to the PSTN, the ISDN 3.1-kHz audio bearer capability interface fallback is operated in the same manner as voice is handled. Half-duplex error correction is then used.

Reality: This method became a standard option for Group 3 fax in March 1993, but fax machines and PC-fax units offering the feature are not available. The lack of demand is caused by the small number of fax units that have ISDN connections.

Forecast: Another optional capability of accessing the ISDN, known as Group 3 unrestricted digital information (UDI), was also approved. This option allows communication between Group 3 fax machines and Group 4 Class 1 fax machines by utilizing Group 4 protocols.

Reality: Because there are so few Group 4 fax units, the UDI feature is seldom, if ever, used.

12.1.4 Group 3 Fax Integrated Into Other Products

Forecast: Fax capability will find its way into communication products such as PCs in the same way in which digital clocks found their way into all kinds of appliances. In 1980, a fax modem required 10 LSI chips and 150 discrete components. Today, the same function can be found on one chip. In a few years, fax design should improve to further reduce the chip count per machine and thus the manufacturing costs. The same chip may contain T.30 protocol, T.4 image compression, scanner interface, and printer interface. This is especially good news for PC users. Addition of fax capability to a new computer should be quite inexpensive and does not even require a plug-in slot. Some manufacturers are already offering PCs with fax built-in. The fax engine will also make it possible to lower the cost of fax machines significantly. It might be the boost needed for the home fax market to take off.

Reality: As expected, fax capability has found its way into PCs and many other new products, from personal digital assistants (PDAs) to cellular telephones.

12.1.5 Small Group 3 Fax

Forecast: When manufacturers started selling the smaller A5 and A6 fax units, they expected it might open up a new home fax market. They were looking for something to pick up the slack when the demand for standard-page-size business fax units tapered off with a maturing market. Since their introduction in Japan in about 1987, interest in Europe or North America has been slight. Although these small fax machines might find some niche in the market, none is evident yet.

Reality: A market never did develop for the small A5 and A6 formats, and they were removed from the Group 3 fax options. It is possible, however, that something similar to them will see future use for screen-to-screen communication between PCs. These formats are better suited to PC screens than are the larger Group 3 formats.

12.2 Innovations on the Horizon

Forecast: Now that Group 3 fax finally has been recognized as essential to business communication, what is next? Will Group 3 fax keep pace with newer technology or fall behind and be replaced?

Reality: There is no indication that Group 3 is yet falling behind. Its open architecture allows new features to be added as the fax art evolves while it maintains compatibility with earlier Group 3 fax.

Forecast: Visionaries foresee computers that will use radio waves and become miniaturized to the size of a pen. Such an ultra-portable device would send and receive faxes, store and retrieve computer files, recognize handwriting, and respond to voice commands.

Reality: Small PDAs are now available that can provide these features, except for the extremely small size (it is not certain that very small size units would find many applications).

12.3 Immediate Future

In the more immediate future, what can we expect? Changes are occurring so rapidly that any predicted advances are likely to be in effect by the time you read this statement. With that in mind, the following items are in the works or are proposed for the coming years.

12.3.1 Integrated Fax Equipment

Forecast: The layout of the business office may change when most PCs are equipped to handle faxing, either on LANs or as standalone devices. Several functions, including fax, will be integrated into one multitasking unit. As these become standard home equipment, more workers will have the option of conducting their affairs away from the office, what the media refers to as telecommuting.

Reality: Most new PCs are now capable of faxing on LANs or the PSTN. Multitasking units with fax capability are proliferating, and their prices have dropped considerably. The number of telecommuting workers has increased rapidly and should continue to do so. Fax is essential for many of them, in addition to PCs. This trend should continue.

12.3.2 The Home Market for Facsimile

Forecast: In England in mid-1991, British Telecom (BT) tested a new service designed for the home fax user. Using a 900 number, a customer can call for an updated fax report from the BT database on the "Around the World" yacht race. A two- or three-page report is then faxed. This could be a start for fax access to many databases.

Reality: Currently, many specialty information sources provide text and graphics to interested customers by fax or Internet-provided data from Web sites. Internet access via PCs finds more users than fax for the retrieval of documents.

12.3.3 Facsimile Newspaper to the Home

Forecast: At one time, more than a half-dozen morning papers in the United States were experimenting with afternoon summaries of the news, which they faxed to subscribers wanting a sneak preview of the next day's news. Most of these ventures were discontinued when the publishers found that readers were unwilling to pay premium prices for condensed stories and were content to wait for the complete morning edition. Among the papers that stopped their fax editions were the Knoxville (Tennessee) *News-Sentinel*, the Minneapolis *Star Tribune*, and the Chicago *Tribune*. A few newspapers press on, sending to areas where its daily product is not available. Japanese and Caribbean cruise ships regularly receive the *New York Times* facsimile edition of six to eight pages. The military personnel engaged in the 1991 war in the Persian Gulf were kept informed of events back home by fax newspapers. The *Hartford Courant* finds a limited but loyal audience for its fax sheet, especially at local businesses. "We use it as an early-warning notification about news of the company or of the opposition," says a spokesman for Travelers Corporation, Hartford, Connecticut.

Reality: This application is expected to continue but not to become more than a niche service for the foreseeable future. Some cruise ships have switched from fax to PC reception (see Chapter 2).

12.3.4 Telepublishing

Forecast: Some innovative publishing companies are using fax board systems to distribute newsletters and other time-critical information to subscribers with fax machines or boards. Such emerging applications are fueling rapid growth in the PC-fax board market.

Reality: This is another niche service, an extension from the fax newspaper. The PC-fax market continues to grow rapidly for use on LANs, the PSTN, and other networks.

12.3.5 Much Higher Fax Speeds

Forecast: Optical fiber to the home has been an economic reality for three years, with or without cable TV in the pipe. T-3 networking with business customer access at 45 Mbps (equivalent to 672 voice telephone circuits) is starting. Optical broadband ISDN draft standards have been recommended by the CCITT [now ITU] and completed standards are expected in 1992. Fujitsu has demonstrated a 1.8-Gbps (equivalent of 24,000 voice telephone circuits) optical fiber subscriber loop conforming to the CCITT draft standard. Bell Labs is experimenting with a device to transmit 350 billion light pulses a second in glass fiber, in contrast to 2.5 billion pulses for the fastest commercial systems.

Reality: PC-fax should be able to utilize almost any type of high-speed channel available. Internet fax is just one example where much higher transmission speeds can be used today. Many other fast channels being made available include cable TV and new systems that use the customer's existing copper telephone wires at much higher speeds than ISDN.

12.4 Summaries for Group 3 Options

Forecast: Group 3 fax has proved its ability to keep up with the rapid advances in technology and will continue its dominance for many years.

Reality: Group 3 open-system protocols will still allow extensions to be added without making older fax units obsolete.

Forecast: The technology for color fax is now here, but production will be very slow until there is a proven business need that will pay for the higher cost fax machines. The situation is somewhat like the use of office copiers compared to color copiers.

Reality: Low-cost color scanners and printers have made it possible to have a low-cost color fax system now. The Internet has seen a rapid growth of color-image transmission, and with recent Internet standards for fax and MRC, color fax will also grow rapidly.

Forecast: Fax will continue on the upswing far into the coming century.

Reality: Fax will still be a leading communication tool well past the millennium. The flexibility built into the open-ended Group 3 fax standards has proved that compatibility with earlier designs can be maintained while remaining on the cutting edge of technology.

Also, new standards are available for fax operation over the Internet. Fax images can be sent via e-mail using ITU-T T.37 and IETF RFC 2305 standards. They are also being extended to permit color-document transfer over the Internet using TIFF image files as defined in RFC 2301 and color space conventions per T.42, T.43, and T.44. In addition, T.38 provides a means to send Group 3 fax over IP packet networks. This is expected to be a popular complement to voice over IP services recently started.

12.5 The Future of Group 4 Fax

Forecast: Although plans called for Group 4 fax to be operating by 1984 on PSTN, it is not yet available. There is a good possibility that Group 4 will be limited to digital data channels and that PSTN operation will never be offered. The future for extensive use of this standard is unclear. Digital fax for networking, store-and-forward, and image databases is in the offing, and Group 4 might fill that slot. Development of a viable mixed mode with teletex under the OSI protocol is unlikely to be of much importance in the foreseeable future, but computer mixed mode is probable. Group 4 will have very little impact on Group 3.

Reality: Little if anything has been done to extend the Group 4 capabilities or applications. There is little chance that Group 4 fax use will increase significantly in the foreseeable future, because Group 3 fax with its high-quantity production provides the same features at lower costs.

Forecast: Group 4 fax will continue to be used mainly by large corporations and will still account for much less than 1% of the fax machines used for some years to come.

Reality: Because the techniques developed for Group 4 fax are available in lower cost, mass-produced Group 3 fax, it is expected that very few new Group 4 fax machines will be manufactured.

13

Fax Standards Development

Standards development plays a key role with respect to the use of fax for business and personal applications throughout the world. According to Webster's dictionary, standards constitute "carefully drawn specifications" that have been "established by authority, custom, or general consent as proper and adequate." Without standards, chaos would likely result not only with respect to fax but also with respect to other types of telecommunications [1]. Clearly, standards development commands considerable attention in modern telecommunications.

This chapter provides an overview of the principal international and national organizations engaged in fax standardization development. The overview highlights relevant fax technical groups and their respective fax technical focus areas to guide readers into the world of fax standardization. Finally, this chapter provides useful reference information such as Web sites for the benefit of readers who wish to obtain additional or more recent information on fax standards development.

13.1 International Telecommunication Union

Today's International Telecommunication Union (ITU) was created in 1865 under the name International Telegraph Union [2]. In 1934, the International Telegraph Union members agreed to a name change for their organization. The new name, International Telecommunication Union, better reflected the union's responsibility in all forms of communications. In 1947, the ITU became a specialized agency of the United Nations, with its headquarters in Geneva.

In 1993, the first World Telecommunications Standardization Conference (WTSC) of the ITU formally approved changes to restructure the ITU. Foremost among the changes was a consolidation of activities formerly performed by the ITU's International Telegraph and Telephone Consultative Committee (CCITT) and the International Radio Consultative Committee (CCIR). This consolidation represents a timely response to the integration of wireline and wireless communications. The new structure positions the ITU as a lead telecommunications player in the global economy.

The new structure divides the ITU into three sectors: Radiocommunication, Telecommunication Standardization, and Development (Figure 13.1). Each sector contains study groups that focus on development of international "recommendations," sometimes unofficially referred to as international voluntary standards (Tables 13.1, 13.2, and 13.3). ITU recommendations stem from study questions assigned to a study group in one of the three ITU sectors.

Study questions encompass group efforts for resolution of problems in key telecommunications subject areas during a four-year ITU study period. The ITU's WTSC in October 1996 formally assigned study questions to sector study groups for the 1997–2000 study period. With respect to fax, three Telecommunications Standardization Sector study groups (see Table 13.2) have been assigned a number of study questions that likely will result in new ITU recommendations for fax. In future study periods, study groups of the other two ITU sectors also may address the development of fax recommendations. Tables 13.4, 13.5, 13.6, and 13.7 list the ITU-T study groups that have been assigned study questions related to fax. Those tables include the titles of the study questions that affect the development of fax recommendations. (Appendix A.1 lists ITU-T T-series recommendations for terminal equipment and protocols for telematic services.) The ITU Web home page address, listed in the References section at the end of this chapter, is an excellent source for information about ITU fax developments. Links on the home page include detailed ITU information on study groups, study questions, and standardization development organization. The reader is encouraged to check the ITU home page links for the latest ITU fax information.

The October 1996 ITU WTSC designated Study Group 8 of ITU-T responsible for the studies of telematic terminal characteristics and related service aspects as well as the lead study group on facsimile. Consequently, Study Group 8 will remain the prime focus of attention by world fax standards experts during the 1997–2000 ITU study period. Certain questions before Study Groups 2, 7, and 16 also relate to fax and likely will have some influence

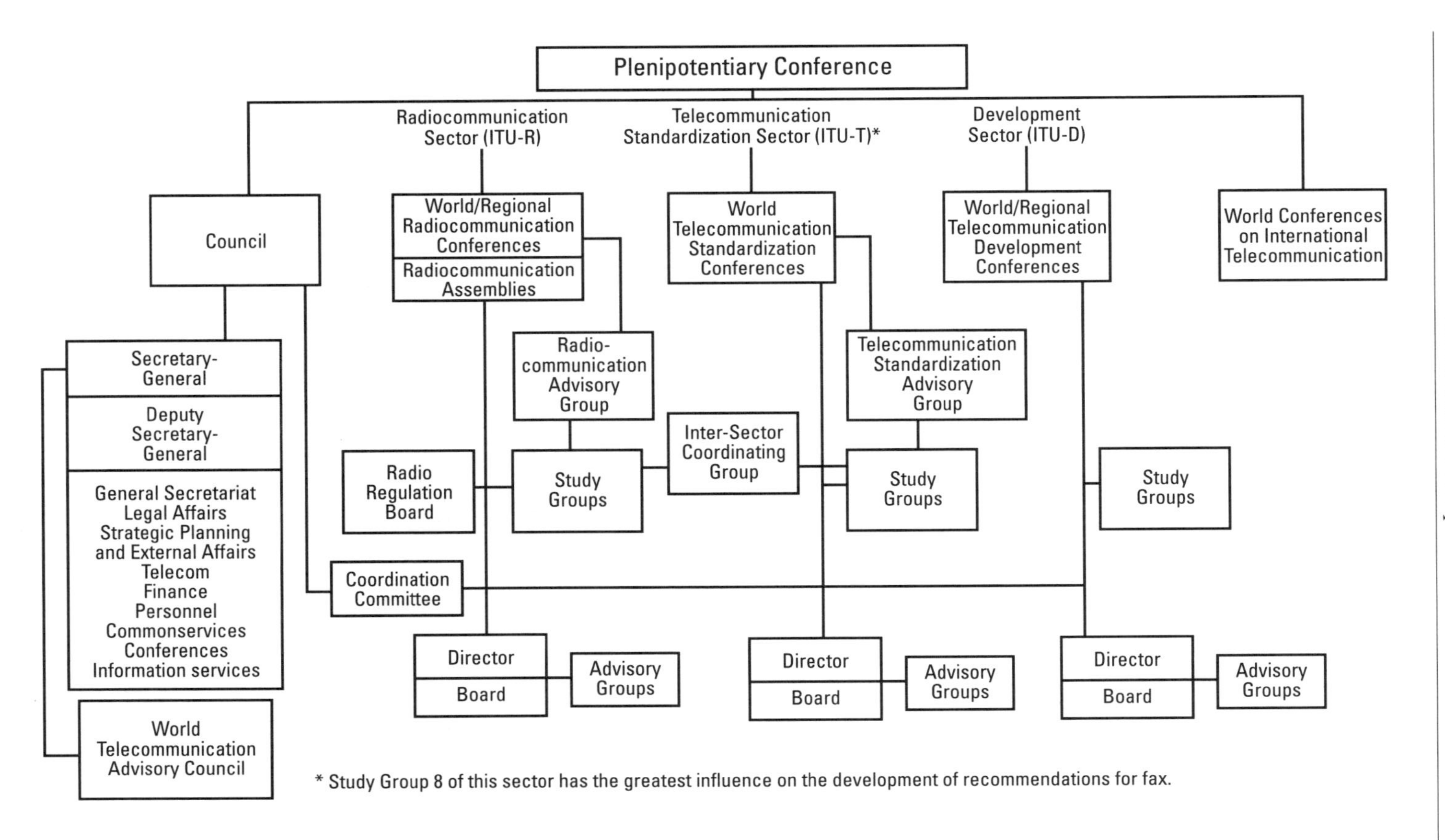

* Study Group 8 of this sector has the greatest influence on the development of recommendations for fax.

Figure 13.1 Structure of the ITU.

Table 13.1
ITU Radiocommunication Sector Study Groups

Study Group 1	Spectrum Management
Study Group 3	Radiowave Propagation
Study Group 4	Fixed-Satellite Service
Study Group 7	Science Services
Study Group 8	Mobile, Radiodetermination, Amateur, and Related Satellite Services
Study Group 10	Broadcasting Service (Sound)
Study Group 11	Broadcasting Service (Television)

Table 13.2
ITU Telecommunication Standardization Sector Study Groups

Study Group 2	Network and Service Operation*
Study Group 3	Tariff and Accounting Principles
Study Group 4	Telecommunications Management Network (TMN) and Network Maintenance
Study Group 5	Protection Against Electromagnetic Environment Effects
Study Group 6	Outside Plant
Study Group 7	Data Networks and Open System Communications*
Study Group 8	Characteristics of Telematic Systems*
Study Group 9	Television and Sound Transmission
Study Group 10	Languages and General Software Aspects for Telecommunications Systems
Study Group 11	Signaling Requirements and Protocols
Study Group 12	End-to-End Transmission Performance of Networks and Terminals
Study Group 13	General Network Aspects
Study Group 15	Transport Networks, Systems, and Equipment
Study Group 16	Multimedia Services and Systems*

*Study groups relevant to fax recommendation development

Table 13.3
ITU Telecommunication Development Sector Study Groups

Study Group 1	Telecommunication Development Strategies and Policies
Study Group 2	Development, Harmonization, Management, and Maintenance of Telecommunication Networks and Services Including Spectrum Management

Table 13.4
Fax Study Questions of ITU-T Study Group 8*

Question Number	Study Question
1/8	Facsimile terminals
2/8	Facsimile test chart and test images
3/8	Cooperative document handling
4/8	Document communications services
5/8	Color for telematic applications
6/8	Common components for image communication
7/8	Coded character sets and control functions for telematic and other ITU-T services

*Study questions for this study group have the greatest influence on development of ITU recommendations for fax

Table 13.5
Study Questions of ITU-T Study Group 2 With Fax Relationships

Question Number	Study Question
3/2	Service quality of networks
13/2	Mobile/personal telephone, telegraph, telematic, data, audiovisual, and multimedia services

Table 13.6
Study Questions of ITU-T Study Group 7 With Fax Relationship

Question Number	Question Title
2/7	Network performance and quality of service
8/7	Nonnative-mode terminal access DTE/DCE interface procedures

on the development of fax recommendations. Tables 13.5, 13.6, and 13.7 list Study Questions before Study Groups 2, 7, and 16 that relate to fax. However, these questions have less influence on the development of fax recommendations than the Study Questions of Study Group 8.

Table 13.7
Study Questions of ITU-T Study Group 16 With Fax Relationships

Question Number	Question Title
4/16	Modems for switched telephone network and telephone-type leased circuits
5/16	ISDN terminal adapters and interworking of DTEs on ISDNs with DTEs on other networks
7/16	DTE-DCE protocols
11/16	Circuit-switched network (CSN) multimedia systems and terminals
12/16	B-ISDN multimedia systems and terminals
13/16	Packet-switched multimedia systems and terminals
14/16	Common protocols, MCUs, and protocols for interworking with H.300 series terminals
16/16	Harmonization of multimedia systems, applications, and services
17/16	AVMMS coordination
18/16	Interaction of high-speed voice-band data systems with signal processing equipment in the PSTN
22/16	Software and hardware tools for signal processing standardization activities

13.2 Internet Society and Internet Engineering Task Force

The Internet Society (ISOC) [3] is a nonprofit, nongovernmental, international professional membership society. It provides leadership in addressing issues that confront the future of the Internet and is the organization home for groups responsible for Internet infrastructure standards, including the Internet Engineering Task Force (IETF). ISOC is a sector member of the ITU. The IETF [4] is a large, open international community of network designers, operators, vendors, and researchers concerned with the evolution of the Internet architecture and the smooth operation of the Internet. IETF working groups are grouped into areas and managed by area directors. There is an active working group (WG) on Internet facsimile, which is currently chaired by James Rafferty. Facsimile serves as a reliable, inexpensive global communications service. As the Internet becomes pervasive, integrating fax and Internet services is appealing in terms of cost savings and opportunities for functional enhancements. The Internet fax WG will pursue a review and specification for enabling standardized messaging-based fax over the Internet. In that regard, the Internet fax WG has developed a series of requests for comments (RFCs) that pertain to facsimile via the Internet; they are listed in Table 13.8. These RFCs have been adopted by the ITU-T Study Group 8 in Recommendations T.37 and T.38. For further information, the reader can consult the ISOC and IETF home pages, listed in the References section at the end of this chapter.

13.3 Telecommunications Industry Association

Besides the ITU, another organization attracts considerable attention with respect to fax standardization development. This organization, a full-service trade group called the Telecommunications Industry Association (TIA), was formed in April 1988 after a merger of the United States Telephone Association (USTA) and the Information and Telecommunications Technologies Group of the Electronic Industries Association (EIA/ITG) [5]. TIA develops voluntary industry standards for a wide variety of telecommunications products under American National Standards Institute (ANSI) accreditation [6].

TIA provides numerous services, including the following: government relations, market support activities, standards-setting activities, and educational programs. TIA's standards and technology department comprises five divisions, as listed in Table 13.9. The User Premises Equipment Division is divided into four committees. One of those committees, TR-29, Facsimile Systems and Equipment, develops standards that cover facsimile areas of the following: terminal equipment, systems, interfaces, and transmission media. TR-29 divides into four subcommittees, as listed in Table 13.10, to develop ANSI-accredited facsimile standards. (Appendix A.2 provides a list of the TIA fax standards in use when this book was submitted for publication.) The reader is encouraged to access the appropriate Web sites listed in the References section at the end of this chapter for the most recent TIA fax standards.

Table 13.8
ISOC/IETF RFCs Relating to Fax

RFC 2301	File Format for Internet Fax
RFC 2302	Tag Image File Format (TIFF)—Image/TIFF MIME Subtype Registration
RFC 2303	Minimal PSTN Address Format in Internet Mail
RFC 2304	Minimal Fax Address Format in Internet Mail
RFC 2305	A Simple Mode of Facsimile Using Internet Mail

Table 13.9
TIA Standards and Technology Divisions

Fiber Optics	
User Premises Equipment*	
TR-29	Facsimile Systems and Equipment
TR-30	Data Transmission Systems and Equipment
TR-32	Personal Radio Equipment
TR-41	User Premises Telephone Equipment Requirements
Network Equipment	
Mobile and Personal Communications	
Satellite Communications	

*All divisions are subdivided into committees, but this table displays only those committees related to fax

Table 13.10
Subcommittees and Task Groups of TIA Committee TR-29

TR-29.1	Facsimile and File Transfer Protocols
TR-29.2	Facsimile Digital Interfaces
TR-29.3	Audiographics Teleconferencing
TR-29.4	Secure Facsimile

In addition to American fax standards, the subcommittees often develop important standards contributions for submission to ITU-T Study Group 8. However, the relevant subcommittee first gains approval from Study Group D of the Department of State's International Telecommunications Advisory Committee—Telecommunications (ITAC-T) before actually submitting contributions to the ITU as official U.S. ITU-T contributions [7]. Table 13.11 shows the ITAC-T structure for the 1997–2000 ITU study period.

13.4 European Telecommunications Standards Institute

The European Telecommunications Standards Institute (ETSI), recognized by the European Council of Ministers, was created in 1988 to set standards

Table 13.11
U.S. Department of State ITAC-T Study Group Structure

Study Group	Relevant ITU-T Study Group	Focus
A	2, 3, 12	Telephony services
B	4, 6, 10, 11, 13, 15	Network infrastructure
D	5, 7, 8, 9, 16	New services applications (includes fax)

for Europe in telecommunications [8]. ETSI has 12 technical committees (Table 13.12). With respect to fax, the Terminal Equipment (TE) technical committee develops fax standards in one of two subcommittees dedicated to terminal equipment. The TE technical committee coordinates with the European Committee for Standardization (CEN) [9]. Table 13.12 also lists the second subcommittee of the TE committee for reference. (Appendix A.3 lists the ETSI fax standards available at publication time of this book.) The ETSI Web site, listed in the References section at the end of this chapter, provides the latest information on ETSI committees and on fax standards.

Table 13.12
ETSI Technical Committees and the TE Subcommittees

Business Telecommunications (BTC)
Methods for Testing and Specifications (MTS)
Transmission and Multiplexing (TM)
Signaling, Protocols, and Switching (SPS)
Terminal Equipment (TE)
Subcommittee TE2 of Telematics: Facsimile, Document Handling, and Transfer
Audiovisual and Voice Terminals
Network Aspects (NA)
Radio Equipment and Systems (RES)
Satellite Earth Stations and Systems (SES)
Special Mobile Group (SMG)
Equipment Engineering (EE)
Human Factors (HF)
Joint Technical Committee of the European Broadcasting Union of ETSI

13.5 Japanese Telecommunications Standards

The Japanese Telecommunication Technology Committee (TTC) was established in 1985 as a private standardization organization [10]. In Japan, the TTC plays the principal role with respect to standardization of fax in Japan. The TTC's organizational structure comprises five technical committees (TCs). With respect to fax, Working Group 2 of TC4 develops Japanese fax standards. Table 13.13 lists the five TCs and the working groups of TC4, including the relevant fax working group (WG4-2). (Appendix A.4 lists fax standards developed and listed by the TTC at time of publication of this book.) The reader can check the TTC's Web home page, listed in the References section at the end of this chapter, to find the most current information on TTC fax standards. Internationally, Japan influences the development of fax recommendations in the ITU via a Japan Telecommunications Standardization Committee within a Telecommunications Technology Council of the Ministry of Posts and Telecommunication [11].

Figure 13.2 shows the key committees, groups, and others with relevance to fax in the Japanese Organization of Telecommunications Standardization. The figure illustrates that the Telecommunications Standardization Committee makes international contributions to the ITU-T. Domestically speaking, however, the Japan TTC receives technical information, via ITU recommendations, that it uses to develop standards, including fax standards, in Japan. The

Table 13.13
TTC Technical Committees

Technical Committee 1 (TC1): Network-network interfaces, mobile communications
Technical Committee 2 (TC2): User-network interfaces
Technical Committee 3 (TC3): PBX, LAN
Technical Committee 4 (TC4): Higher layer protocols
Working Group 1 (WG4-1): Interface protocol standards for OSI network management, MHS, directory systems, etc.
Working Group 2 (WG4-2): Interface protocol and standards for telematics, document interchange, etc.
Working Group 3 (WG4-3): Interface protocol for interconnection of messaging systems, etc.
Working Group 4 (WG-4): Infrared communication interface
Working Group on Object Identifier (WG-Obj): Technical matter on object identifier related to TTC standards, etc.
Technical Committee 5 (TC5): Voice and video signal coding scheme and systems

figure also illustrates that the TTC, along with user and manufacturer groups, contributes to the promotion conference of Harmonization of Advanced Telecommunication Systems (HATS) [12]. (In Japan, communications tests of the HATS conference and the Communications Industry Association of Japan (CIAJ) maintain the interoperability among different manufacturers' products [13]). Finally, Figure 13.2 illustrates the role of user and manufacturer groups in standardization.

13.6 International Organization for Standardization and International Electrotechnical Commission

The International Organization for Standardization (ISO) was created in 1947 to promote the development of standardization and related activities [14]. ISO facilitates the international exchange of goods and services while developing cooperation in the spheres of intellectual, scientific, technological, and economic activity. In 1978, ISO developed the open systems interconnection (OSI) model for the logical structure of networked DCE (which includes fax). The OSI model provides a basis for open systems interconnection in which

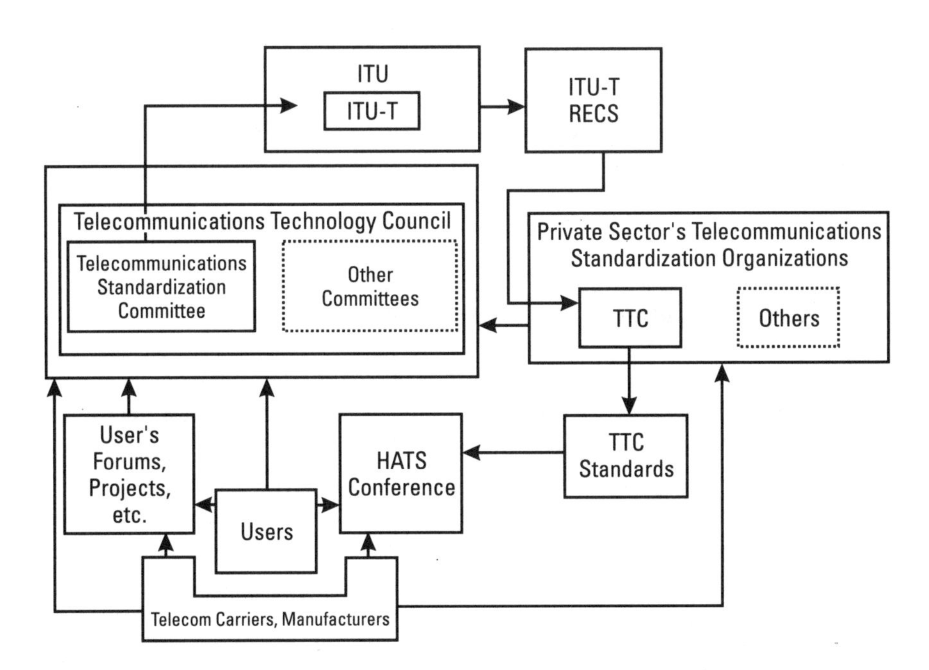

Figure 13.2 Japanese Organization of Telecommunications Standardization with relevance to fax.

control of a network resides in various nodes. However, the network still operates according to common standards.

The International Electrotechnical Commission (IEC) was created in 1906 to promote international cooperation on all questions of standardization and related matters in the fields of electricity, electronics, and related technologies [15]. The IEC cooperates with numerous other international organizations, particularly the ISO but also with the ITU. In 1987, ISO and IEC established the Joint Technical Committee 1 (JTC-1) to cover the vast and expanding field of information technology. ANSI serves as the U.S. member body representative to JTC-1.

Subcommittee (SC) 28 (Office Equipment) of JTC 1 develops standards relating to fax. (These standards are listed in Appendix A.5.) The reader is encouraged to check the ISO Web home page, listed in the References section at the end of this chapter, for the latest list of JTC-1 fax standards. The reader may also want to check the standards on related office equipment, such as printers and copy machines.

13.7 Defense Information Systems Agency

The Defense Information Systems Agency (DISA) identifies standards and participates in the development of standards suitable for use by the U.S. Department of Defense (DOD). The agency also updates existing military standards (MIL-STDS) used by the DOD. The DISA Center for Standards (CFS) lists the MIL-STDS and identifies MIL-STD-188-161D as the military standard relevant to fax [16].

13.8 National Communications System

The Cuban missile crisis of 1962 exposed some difficulties with U.S. government communications. After the crisis, President Kennedy made a decision to strengthen the communications support of all major functions of government. He created the National Communication System (NCS) via a presidential memorandum on August 21, 1963. On April 3,1984, President Reagan signed Presidential Executive Order 12472, which reaffirms the NCS and expands the NCS mission to include National Security and Emergency Preparedness (NSEP) telecommunications functions [17]. Part of reaffirming the NCS mission included management of the Federal Telecommunications Standards Program (FTSP) and "ensuring whenever feasible that existing or evolving industry, national, and international standards are used as the basis for Federal telecommunication standards" [18]. Therefore, the NCS, through the FTSP, develops federal telecommunication standards, including fax standards to improve NSEP for the United States. (Appendix A.6 lists the Federal Telecommunication fax standards.) The reader is encouraged to check the NCS Web home page, listed in the References section at the end of this chapter, to find updates of federal telecommunication fax standards.

13.9 Summary

This chapter presented an overview of fax standards development by principal standards organizations that include the following: ITU, TIA, ETSI, TTC, ISO/IEC, DISA, and NCS. Although ITU-T Study Group 8 remains the prime focus of attention by world fax standards experts, other groups and organizations play notable roles with respect to the development of fax standards. The utility of World Wide Web information on fax standards facilitates understanding of the most recent fax developments for worldwide applications.

References

[1] Cohen, E. J., and W. B. Wilkens, "The IEEE Role in Telecommunication Standards," *IEEE Communication Magazine*, Vol. 23, No. 1, Jan. 1985, p. 31. Institute of Electrical and Electronics Engineers (IEEE) World Wide Web home page, http://www.ieee.org/.

[2] International Telecommunication Union (ITU), World Wide Web home page, http://www.itu.ch/.

[3] Internet Society (ISOC) World Wide Web home page, http://www.isoc.org/.

[4] Internet Engineering Task Force (IETF), World Wide Web home page, http://www.ietf.org/.

[5] Telecommunications Industry Association (TIA), World Wide Web home page, http://www.tiaonline.org/.

[6] American National Standards Institute (ANSI), World Wide Web home page, http://www.ansi.org/.

[7] Department of State (DOS), World Wide Web home page, http://www.state.gov/.

[8] European Telecommunications Standards Institute (ETSI), World Wide Web home page, http://www.etsi.fr/.

[9] The European Committee for Standardization (CEN), World Wide Web home page, http://www.cenorm.beldefault.htm.

[10] Telecommunication Technology Committee (TTC), World Wide Web home page, http://www.ttc.or.jp/.

[11] Ministry of Posts and Telecommunication (MPT), World Wide Web home page, http://www.mpt.jp/.

[12] Harmonization of Advanced Telecommunication Systems (HATS), World Wide Web home page, http://www.ciaj.or.jp/hats/.

[13] Communications Industry Association of Japan (CIAJ), World Wide Web home page, http://www.ciaj.or.jp/.

[14] International Organization for Standardization (ISO), World Wide Web home page, http://www.iso.ch/.

[15] International Electrotechnical Commission (IEC), World Wide Web home page, http://www.iec.ch/.

[16] Center for Standards (CFS), World Wide Web home page, http://www.itsi.disa.mil/.

[17] National Communications System (NCS), World Wide Web home page, http://www.ncs.gov/.

[18] Bain, G. P., "The Federal Telecommunication Standards Program of the National Communications System," *Proc. Pacific Telecommunications Council, PTC'95 Convergence: Closing the Gap*, Jan. 22–25, 1995, Honolulu, pp. 160–167. Pacific Telecommunications Council (PTC), World Wide Web home page, http://www.ptc.org/.

Appendix: Fax Recommendations and Standards Lists

A.1 ITU-T T-Series Fax-Relevant Recommendations

Series T Terminal Equipment and Protocols for Telematic Services (Study Group 8)

Recommendation	Title
T.0 (7/96)	Classification of Facsimile Terminals for Document Transmission Over the Public Networks
T.1 (11/88)	Standardization of Phototelegraph Apparatus (Blue Book Fascicle VII.3)
T.4 (7/96)	Standardization of Group 3 Facsimile Apparatus for Document Transmission
T.4 (7/97)	Standardization of Group 3 Facsimile Terminals for Document Transmission —Amendment 1
T.4 (10/97)	Standardization of Group 3 Facsimile Terminals for Document Transmission —Amendment 2
T.4 (6/98)	Standardization of Group 3 Facsimile Terminals for Document Transmission —Corrigendum 1
T.6 (11/88)	Facsimile Coding Schemes and Coding Control Functions for Group 4 Facsimile Apparatus (Blue Book Fascicle VII.3)
T.10 (11/88)	Document Facsimile Transmissions on Leased Telephone-Type Circuits (Blue Book Fascicle VII.3)

Recommendation	Title
T.10bis (11/88)	Document Facsimile Transmissions in the General Switched Telephone Network (Blue Book Fascicle VII.3)
T.22 (3/93)	Standardized Test Charts for Document Facsimile Transmissions
T.23 (4/94)	Standardized Color Test Chart for Document Facsimile Transmissions
T.24 (11/94)	Standardized Digitized Image Set
T.30 (7/96)	Procedures for Document Facsimile Transmission in the General Switched Telephone Network
T.30 (7/97)	Procedures for Document Facsimile Transmission in the General Switched Telephone Network—Amendment 1
T.T.30 (10/97)	Procedures for Document Facsimile Transmission in the General Switched Telephone Network—Amendment 2
T.30 (6/98)	Procedures for Document Facsimile Transmission in the General Switched Telephone Network—Amendment 3
T.31 (8/95)	Asynchronous Facsimile DCE Control—Service Class 1
T.31 (7/96)	Asynchronous Facsimile DCE Control—Service Class 1 Annex B—Procedure for Service Class 1 Support of V.34 Modems—Amendment 1
T.32 (8/95)	Asynchronous Facsimile DCE Control—Service Class 2
T.32 (7/96)	Asynchronous Facsimile DCE Control—Service Class 2—Amendment 1
T.32 (10/97)	Asynchronous Facsimile DCE Control—Service Class 2—Corrigendum 1
T.33 (7/96)	Facsimile Routing Utilizing the Subaddress
T.35 (1/91)	Procedure for the Allocation of CCITT Defined Codes for Non-Standard Facilities
T.37 (6/98)	Procedures for the Transfer of Facsimile Data via Store and Forward on the Internet
T.38 (6/98)	Procedures for Real Time Group 3 Facsimile Communication Over IP Networks
T.42 (10/96)	Continuous Tone Color Representation Method for Facsimile
T.50 (9/92)	International Reference Alphabet (IRA) (Formerly International Alphabet No. 5 or IA5)—Information Technology—7-Bit Coded Character Set for Information Interchange
T.51 (9/92)	Latin Based Coded Character Sets for Telematic Services
T.51 (8/95)	Latin Based Coded Character Sets for Telematic Services—Amendment 1
T.52 (3/93)	Non-Latin Coded Character Sets for Telematic Services
T.52 (10/96)	Non-Latin Coded Character Sets for Telematic Services—Amendment 1
T.53 (4/94)	Character Coded Control Functions for Telematic Services
T.60 (3/93)	Terminal Equipment for Use in the Teletex Service

Recommendation	Title
T.62 (3/93)	Control Procedures for Teletex and Group 4 Facsimile Services
T.62bis (3/93)	Control Procedures for Teletex and Group 4 Facsimile Services
T.63 (3/93)	Provisions for Verification of Teletex Terminal Compliance
T.64 (3/93)	Conformance Testing Procedures for the Teletex Recommendations
T.65 (11/88)	Applicability of Telematic Protocols and Terminal Characteristics to Computerized Communication Terminals (CCTs) (Blue Book Fascicle VII.5)
T.70 (3/93)	Network-Independent Basic Transport Service for the Telematic Services
T.71 (11/88)	Link Access Protocol Balanced (LAPB) Extended for Half-Duplex Physical Level Facility (Blue Book Fascicle VII.5)
T.80 (9/92)	Common Components for Image Compression and Communication—Basic Principles
T.81 (9/92)	Information Technology—Digital Compression and Coding of Continuous-Tone Still Images—Requirements and Guidelines
T.82 (3/93)	Information Technology—Coded Representation of Picture and Audio Information—Progressive Bi-level Image Compression
T.82 (3/95)	Technical Corrigendum 1 to Recommendation T.82—Information Technology—Coded Representation of Picture and Audio Information—Progressive Bi-level Image Compression
T.83 (11/94)	Information Technology—Digital Compression and Coding of Continuous-Tone Still Images: Compliance Testing
T.83 Encl (11/94)	Compliance Test Data Processable Format on 3 diskettes 3-1/2" MS-DOS format.
T.84 (7/96)	Information Technology—Digital Compression and Coding of Continuous-Tone Still Images—Extensions
T.85 (8/95)	Application Profile for Recommendation T.82—Progressive Bi-level Image Compression (JBIG Coding Scheme) for Facsimile Apparatus
T.85 (10/96)	Amendment 1 to Recommendation T.85—Application Profile for Recommendation T.82—Progressive Bi-level Image Compression (JBIG Coding Scheme) for Facsimile Apparatus
T.85 (10/97)	Amendment 2 to Recommendation T.85—Application Profile for Recommendation T.82—Progressive Bi-level Image Compression (JBIG Coding Scheme) for Facsimile Apparatus
T.85 (2/97)	Corrigendum 1 to Recommendation T.85—Application Profile for Recommendation T.82—Progressive Bi-level Image Compression (JBIG Coding Scheme) for Facsimile Apparatus
T.90 (2/92)	Characteristics and Protocols for Terminals for Telematic Services in ISDN
T.90 (11/94)	Amendment 1 to Recommendation T.90—Characteristics and Protocols for Terminals for Telematic Services in ISDN

Recommendation	Title
T.90 (7/96)	Amendment 2 to Recommendation T.90—Characteristics and Protocols for Terminals for Telematic Services in ISDN
T.90 (6/98)	Amendment 3 to Recommendation T.90—Characteristics and Protocols for Terminals for Telematic Services in ISDN: Cause Value for a G4 Fax Fall-back
T.100 (11/88)	International Information Exchange for Interactive Videotex (Blue Book Fascicle VII.5)
T.101 (11/94)	International Interworking for Videotex Services
T.102 (3/93)	Syntax-based Videotex End-to-End Protocols for the Circuit Mode ISDN
T.103 (3/93)	Syntax-based Videotex End-to-End Protocols for the Packet Mode ISDN
T.104 (3/93)	Packet Mode Access for Syntax-Based Videotex via PSTN
T.105 (11/94)	Syntax-Based Videotex Application Layer Protocol
T.106 (3/93)	Framework of Videotex Terminal Protocols
T.107 (8/95)	Enhanced Man Machine Interface for Videotex and Other Retrieval Services (VEMMI)
T.120 (7/96)	Data Protocols for Multimedia Conferencing
T.121 (7/96)	Generic Application Template
T.122 (3/93)	Multipoint Communication Service for Audiographics and Audiovisual Conferencing Service Definition
T.123 (10/96)	Network Specific Data Protocol Stacks for Multimedia Conferencing
T.124 (8/95)	Generic Conference Control
T.125 (4/94)	Multipoint Communication Service Protocol Specification
T.126 (8/95)	Multipoint Still Image and Annotation Protocol
T.127 (8/95)	Multipoint Binary File Transfer Protocol
T.150 (11/88)	Telewriting Terminal Equipment Blue Book Fascicle VII.5
T.171 (10/96)	Protocols for Interactive Audiovisual Services: Coded Representation of Multimedia and Hypermedia Objects
T.174 (10/96)	Application Programming Interface (API) for MHEG-1
T.190 (8/95)	Cooperative Document Handling (CDH)—Framework and Basic Services
T.191 (7/96)	Cooperative Document Handling—Joint Synchronous Editing (Point-to-Point)
T.200 (10/96)	Programming Communication Interface for Terminal Equipment Connected to ISDN: General Architecture
T.300 (11/88)	General Principles of Telematic Interworking (Blue Book Fascicle VII.5)
T.330 (11/88)	Telematic Access to Interpersonal Message System (Blue Book Fascicle VII.5)
T.351 (11/88)	Imaging Process of Character Information on Facsimile Apparatus (Blue Book Fascicle VII.5)

Recommendation	Title
T.390 (11/88)	Teletex Requirements for Interworking With the Telex Service (Blue Book Fascicle VII.5)
T.411 (3/93)	Information Technology—Open Document Architecture: Introduction and General Principles
T.411 (10/97)	Corrigendum 1 to Information Technology—Open Document Architecture: Introduction and General Principles Technical Corrigendum 1
T.412 (3/93)	Information Technology—Open Document Architecture (ODA) and Interchange Format: Document Structures
T.412 (10/97)	Corrigendum 1 to Information Technology—Open Document Architecture (ODA) and Interchange Format: Document Structures Technical Corrigendum 1
T.412 (10/97)	Corrigendum 2 to Information Technology—Open Document Architecture (ODA) and Interchange Format: Document Structures Technical Corrigendum 2
T.413 (11/94)	Information Technology—Open Document Architecture (ODA) and Interchange Format: Abstract Interface for the Manipulation of ODA Documents (Common text with ISO/IEC.)
T.414 (3/93)	Information Technology—Open Document Architecture (ODA) and Interchange Format: Document Profile (Common Text With ISO/IEC.)
T.414 (10/97)	Corrigendum 1 to Information Technology—Open Document Architecture (ODA) and Interchange Format: Document Profile Technical Corrigendum 1
T.414 (10/97)	Corrigendum 2 to Information Technology—Open Document Architecture (ODA) and Interchange Format: Document Profile Technical Corrigendum 2
T.415 (3/93)	Information Technology—Open Document Architecture (ODA) and Interchange Format: Open Document Interchange Format (ODIF)
T.415 (10/97)	Corrigendum 1 to Information Technology—Open Document Architecture (ODA) and Interchange Format: Open Document Interchange Format Technical Corrigendum 1
T.415 (10/97)	Corrigendum 2 to Information Technology—Open Document Architecture (ODA) and Interchange Format: Open Document Interchange Format Technical Corrigendum 2
T.416 (3/93)	Information Technology—Open Document Architecture (ODA) and Interchange Format: Character Content Architectures (Common Text With ISO/IEC.)
T.416 (10/97)	Corrigendum 1 to Information Technology—Open Document Architecture (ODA) and Interchange Format: Character Content Architectures Technical Corrigendum 1
T.417 (3/93)	Information Technology—Open Document Architecture (ODA) and Interchange Formats: Raster Graphics Content Architectures

Recommendation	Title
T.417 (10/97)	Amendment 1 to Information Technology—Open Document Architecture (ODA) and Interchange Formats: Raster Graphics Content Architectures Amendment 1
T.417 (10/97)	Corrigendum 1 to Information Technology—Open Document Architecture (ODA) and Interchange Formats: Raster Graphics Content Architectures Technical Corrigendum 1
T.418 (3/93)	Information Technology—Open Document Architecture (ODA) and Interchange Format: Geometric Graphics Content Architecture (Common Text With ISO/IEC.)
T.419 (8/95)	Information Technology—Open Document Architecture (ODA) and Interchange Format: Audio Content Architectures (Common Text With ISO/IEC.)
T.421 (11/94)	Information Technology—Open Document Architecture (ODA) and Interchange Format: Tabular Structures and Tabular Layout (Common Text With ISO/IEC.)
T.422 (8/95)	Information Technology—Open Document Architecture (ODA) and Interchange Format—Identification of Document Fragments (Common Text With ISO/IEC.)
T.424 (7/96)	Information Technology—Open Document Architecture (ODA) and Interchange Format: Temporal Relationships and Non-Linear Structures (Common Text With ISO/IEC)
T.431 (9/92)	Document Transfer and Manipulation (DTAM)—Services and Protocols—Introduction and General Principles
T.432 (9/92)	Document Transfer and Manipulation (DTAM)—Services and Protocols—Service Definition
T.432 (8/95)	Amendment 1 to Document Transfer and Manipulation (DTAM)—Services and Protocols Amendment 1: Revisions of T.432 to Support G4 Color and File Transfer
T.433 (9/92)	Document Transfer and Manipulation (DTAM)—Services and Protocols—Protocol Specification
T.433 (8/95)	Amendment 1 to Document Transfer and Manipulation (DTAM)—Services and Protocols—Protocol Specification—Revisions of T.433 to Support G4 Color and File Transfer Amendment 1
T.434 (7/96)	Binary File Transfer Format for the Telematic Services
T.435 (8/95)	Document Transfer and Manipulation (DTAM)—Services and Protocols—Abstract Service Definition and Procedures for Confirmed Document Manipulation
T.436 (8/95)	Document Transfer and Manipulation (DTAM)—Services and Protocols—Protocol Specifications for Confirmed Document Manipulation

Recommendation	Title
T.441 (11/88)	Document Transfer and Manipulation (DTAM)—Operational Structure (Blue Book Fascicle VII.7)
T.501 (3/93)	Document Application Profile MM for the Interchange of Formatted Mixed Mode Documents
T.502 (11/94)	Document Application Profile PM-11 for the Interchange of Simple Structure, Character Content Documents in Processable and Formatted Forms
T.503 (1/91)	A Document Application Profile for the Interchange of Group 4 Facsimile Documents
T.503 (11/94)	Annex B: Extension for Continuous-Tone Color and Gray-Scale Image Documents—Amendment 1
T.503 (8/95)	Amendment 2 to Recommendation T.503—A Document Application Profile for the Interchange of Group 4 Facsimile Documents
T.503 (7/97)	Amendment 2 to Recommendation T.503—A Document Application Profile for the Interchange of Group 4 Facsimile Documents
T.503 (6/98)	Corrigendum 1 to Recommendation T.503—A Document Application Profile for the Interchange of Group 4 Facsimile Documents Corrigendum 1
T.504 (3/93)	Document Application Profile for Videotex Interworking
T.505 (11/94)	Document Application Profile PM-26 for the Interchange of Enhanced Structure, Mixed Content Documents in Processable and Formatted Forms
T.506 (8/93)	Document Application Profile PM-36 for the Interchange of Extended Document Structures and Mixed Content Documents in Processable and Formatted Forms
T.510 (3/93)	General Overview of the T.510-Series Recommendations
T.521 (11/94)	Communication Application Profile BT0 for Document Bulk Transfer Based on the Session Service

A.2 TIA Fax Standards

800 MHZ Cellular Systems TDMA Services Async Data and Fax (April, 1995)

TIA/EIA/465A and TIA/EIA/466A Conformity Test Standard (TSB-85)

Asynchronous Facsimile DCE Control Standard (ANSI/TIA/EIA-592-93) (May 1993)

Routing of Group 3 Facsimile Messages Utilizing the Subaddress (Oct. 1994)

Facsimile Coding Schemes and Coding Control Functions for Group 4 Facsimile Equipment (ANSI/EIA-538-88) (Aug. 1988)

Facsimile Digital Interfaces—Voice Control Interim Standard for Asynchronous DCE (TIA/EIA/IS-101)

Facsimile DCE-DTE Packet Protocol Standard (ANSI/TIA/EIA-605-92) (Dec. 1992)

Facsimile Digital Interfaces Asynchronous Facsimile DCE Control Standard, Service Class 1 (ANSI/TIA/EIA-578-A-95) (May 1995)

Group 3 Facsimile Apparatus for Document Transmission (ANSI/TIA/EIA 465A 1995) (June, 1995)

Procedures for Document Facsimile Transmission (ANSI/TIA/EIA 466A)(May 97)

A.3 ETSI Fax Standards

DE/TE-02015-1, ETS, 300 243-1:1995-11, Programmable Communication Interface (PCI) APPLI/COM for Facsimile Group 3, Facsimile Group 4, Teletex and Telex Services

DE/TE-02015-2, ETS, 300 243-2: 1995-11, Programmable Communication Interface (PCI) APPLI/COM for Facsimile Group 3, Facsimile Group 4, Teletex and Telex Services

DTR/TE-02017, ETR, 195: 1995-08, Analysis of Mechanisms for Selection Between Voice and G3 Facsimile on One Line

RE/TE-02035, ETS, 300 280/A1: 1996-02, Facsimile Group 4 Class 1 Equipment on the Integrated Services Digital Network (ISDN)

RE/TE-02036, ETS, 300 242: 1997-02 Edition, Group 3 Facsimile Equipment

T/TE 05-07, ETS, 300 112: 1994-05, Facsimile Group 4 Class 1 Equipment on the ISDN End-to-End Protocols

T/TE 05-08, ETS, 300 155: 1995-02, Facsimile Group 4 Class 1 Equipment on the ISDN End-to-End Protocol Tests

T/TE 05-09, ETS, 300 087: 1994-05, Facsimile Group 4 Class 1 on the ISDN Functional Specification of the Equipment

T/TE 05-10, ETS 300 280: 1994-02, Facsimile Group 4 Class 1 Equipment on the Integrated Services Digital Network (ISDN)

A.4 TTC Fax Standards

Standard Number/ Version	Title
JT-T4/6	Standardization of Group 3 Facsimile Apparatus for Document Transmission
JT-T30/8	Protocols for Document Facsimile Transmission in the General Switched Telephone Network
JT-T42/1	Continuous Tone Color Representation Method for Facsimile
JT-T85/1	Application Profile for Recommendation T.82—Progressive Bi-level Image Compression (JBIG Coding Scheme) for Facsimile Apparatus
JT-T503/4	A Document Application Profile for the Interchange of Group 4 Facsimile Document
JT-T563/6	Terminal Characteristics for Group 4 Facsimile Apparatus

A.5 ISO/IEC JTC 1 Fax Standards

ISO/IEC DIS 15404-1, Office Machines—Facsimile Equipment—Part 1: Concepts and Classification

ISO/IEC DIS 15404-2, Office Machines—Facsimile Equipment—Part 2: Minimum Requirements for Documents to be Transmitted

ISO/IEC DIS 15775, Office Machines—Facsimile Equipment—Part 3: Minimum Requirements for Received Copies

A.6 Federal Telecommunication Fax Standards

Federal Telecommunication Recommendation (FTR) Number	Title
FTR-1062-1997	Group 3 Facsimile Apparatus for Document Transmission (CCITT T.4 & EIA RS-465A)
FTR-1063-1997	Procedures for Document Facsimile Transmission (CCITT T.4 & EIA RS- 466A)

Federal Information Processing Standard (FIPS) Publication Number	Title
FIPS 150	Facsimile Coding Schemes and Coding Control Functions for Group 4 Facsimile Apparatus (CCITT T.6 & EIA-538-1988)

List of Acronyms and Abbreviations

A/D Analog to digital

ABC Automatic background control

ACELP Algebraic codebook excited linear prediction

AFS Audio frequency shift

AIEE American Institute of Electrical Engineers

AL Adaptation layer

AM Amplitude modulation

ANSI American National Standards Institute

AP Associated Press

APDU Application protocol data unit

API Application programming interface

ARQ Automatic repeat request

ASCII American Standard Code for Information Interchange

ASN.1 Abstract Syntax Notation

ASVD Analog simultaneous voice plus data

ASVF Analog simultaneous voice and facsimile

AT Adaptive context template

AT&T American Telephone and Telegraph

ATM Asynchronous transfer mode

BAS Bit-rate allocation signal

bpp Bits per pixel

BCH Bose Chandhuri Hocquenghem

BFICC British Facsimile International Consultative Committee

BFO Beat frequency oscillator

BFT Binary file transfer

BIC Bump in cord

B-ISDN Broadband ISDN

BT British Telecom; also bulk transfer

C&I Control and indication

CAPI Communication application programming interface

CBF Computer-based fax

CCD Charge-coupled device

CCIR International Radio Consultative Committee

CCITT International Telegraph and Telephone Consultative Committee

CD-ROM Compact disk read only memory

CED Called station identification

CFR Confirmation to receive

CFVD Constant frequency variable dot

CIAJ Communication Industry Association of Japan

CIE Commission Internationale de l'Éclairage

CIS Contact image sensor

CMY(K) Cyan, magenta, yellow, and black

CNG Calling tone

CRT Cathode ray tube

CSA Canadian Standards Association

CSDN Circuit switched digital network

CSI Called station identification

CT Computer telephony

CW Continuous wave

DATAM Document transfer and manipulation

DCE Data circuit-terminating equipment; also known as data communication equipment

DCN Disconnect

DCS Digital communication signal

DCT Discrete cosine transform

DDD Direct distance dialing

DES Data Encryption Standard

DIS Digital identification signal

DISA Defense Information Systems Agency

DOD Department of Defense

DP Deterministic prediction

dpi dots per inch

DSL Digital subscriber line

DSN Delivery service notification

DSVD Digital simultaneous voice and data

DSVF Digital simultaneous voice and facsimile

DTAM Document transfer and manipulation

DTE Data terminal equipment

DTMF Dual-tone multiple frequency

DWA Defense Weather Andrews

ECM Error correction mode

ECTF Enterprise Computer Telephony Forum

EIA Electronic Industries Association

EOL End of line

EOP End of procedure

ETSI European Telecommunication Standards Institute

FAS Frame alignment signal

fax facsimile

FCC Federal Communications Commission

FEC Forward error correction

FIF Facsimile information field

FPAD Facsimile packet assembly/disassembly

FS Frequency shift

FTAM File transfer, access, and management

FTP File transfer protocol

FTSP Federal Telecommunications Standards Program

G1 Group 1 facsimile

G2 Group 2 facsimile

G3 Group 3 facsimile

G4 Group 4 facsimile

GCC Generic conference control

GSTN General switched telephone network

HATS Harmonization of Advanced Telecommunication Systems

HDLC High-level data link control

HF High frequency

HFX40 Hawthorne Facsimile Cipher

HKM Hawthorne Key Management

H-MLP High-speed multilayer protocol

HSD High-speed data

HTTP Hypertext transport protocol

Hz hertz

IAF Internet-aware facsimile

IEC International Electrotechnical Commission

IEEE Institute of Electrical and Electronics Engineers

IETF Internet Engineering Task Force

IFax Internet fax

IFP Internet facsimile packet

IMAP Internet mail access protocol

IMC Information Management Consultants

INP International News Photos

I/O Input/output

IP Internet protocol

IRE Institute of Radio Engineers

ISDN Integrated services digital network

ISO International Organization for Standardization

ISOC Internet Society

ISP Internet service provider

ITAC-T International Telecommunications Advisory Committee—Telecommunications

ITU International Telecommunication Union

ITU-T International Telecommunication Union—Telecommunications Standardization Sector

JBIG Joint Bi-Level Image Group

JPEG Joint Photographic Experts Group

Kbps kilobits per second

kHz kilohertz

LAN Local area network

LCD Liquid crystal display

LCU Line connection unit

LED Light-emitting diode

LSD Low-speed data

LSI Large-scale integrated circuit

MBFT Multipoint binary file transfer

MCF Message confirmation

MCU Minimum coding unit

MFP Multifunction peripheral

MFPI Multifunction peripheral interface

MGCS Matsushita Graphic Communication Systems

MH Modified Huffman

MHS Message handling services

MIL-SPEC Military specification

MIL-STDS Military standards

MIME Multipurpose Internet mail extension

MLP Multilayer protocol

MM Mixed mode

MMR Modified modified READ

MP-MLQ Multipulse maximum likelihood quantizer

MR Modified READ

MRC Mixed raster content

MSLT Minimum scan line time

MUX Multiplex

NCS National Communications System

N-ISDN Narrowband ISDN

NIST National Institute of Standards and Technology

NSA National Security Agency

NSC Nonstandard command

NSF Nonstandard facilities

NSS Nonstandard setup

NTISSI National Telecommunications and Information Systems Security Instruction

OCR Optical character recognition

ODIF Office document interchange format

OSI Open System Interconnection

OWI Office of War Information

PAM Pulse amplitude modulation

PC Personal computer

PC-fax Personal computer facsimile

PCI Programming communication interface

PDA Personal digital assistant

PDU Protocol data unit

pel Picture element

PMT Photomultiplier tube

POP Post office protocol

POTS Plain old telephone service

PPR Partial page request

PSDN Packet-switched data network

PSTN Public switched telephone network

PTT Post Telephone and Telegraph

QoS Quality of service

RCA Radio Corporation of America

READ Relative address

RF Radio frequency

RFC Request for comments

RGB Red, green, blue

RNR Receive not ready

RSA Rivest Shamir Adleman

RTC Return to control

SAC Standalone controller

SBS Satellite Business Systems

SCFM Subcarrier frequency modulation

SCSI Small computer serial interface

SFU Store-and-forward unit

SMTP Simple mail transfer protocol

SPIFF Still picture interchange format

SSTV Slow-scan TV

SVD Simultaneous voice and data

SVF Simultaneous voice and facsimile

sync Synchronizing, synchronous, or synchronization

TCP/IP Transport control protocol/Internet protocol

TFC Times Facsimile Corporation

TIA Telecommunications Industry Association

TIFF Tagged image file format

TP Typical prediction

TSI Transmit subscriber identification

TTC Telecommunication Technology Committee

UDI Unrestricted digital information

UDP User datagram protocol

UDPTL UDP transport layer

UPI United Press International

USTA United States Telephone Association

WAN Wide area network

WBAN Weather Bureau–Air Force–Navy

WEFAX Weather facsimile

WG Working group

WTSC World Telecommunication Standardization Conference

WU Western Union

WWW World Wide Web

XID Exchange identification

Glossary

Analog facsimile In analog facsimile equipment, the signals that represent marking densities of the original are stepless, and each pixel value is transmitted and recorded without digital fax processing, although the communication network over which the signal travels may employ digital processing. Densities between black and white are shown as gray pixels.

American National Standards Institute (**ANSI**) The U.S. member organization of the International Organization for Standards (ISO). ANSI is responsible for adopting U.S. national standards. In the facsimile area, the ANSI-accredited Telecommunications Industry Association (TIA) develops national standards.

Baud The number of changes per second in signal state (symbols) sent by a digital modem. Many modems send 4 or more bits per baud. Unfortunately, some people equate baud with the number of bits per second (bps).

Binary file transfer (**BFT**) The Telecommunications Standardization sector of the International Telecommunication Union (ITU-T) T.434 recommendation for transferring binary files via facsimile protocols or other telecommunications services. The U.S. version, TIA/EIA-614, was its predecessor.

Bit The contraction for *binary digit*, the smallest amount of information in a binary system; a "0" or a "1" condition.

bps Bits per second.

Broadcasting The procedure for sending a fax document to multiple receiving locations.

Byte A string of 8 bits used as a basic unit in a digital computer for memory storage and data processing. Known as an octet in ITU recommendations.

CCITT The International Telegraph and Telephone Consultative Committee; now the International Telecommunication Union (ITU).

CIELAB A color space defined by the Commission Internationale de l'Éclairage (CIE), used by the Joint Photographic Experts Group (JPEG) to represent image pixel values. This color space has approximately equal visually perceptible differences between equispaced points throughout the space.

Codec A two-way device or software that converts analog signals such as video, voice, or fax into digital code (coder or A/D) and converts digital code into analog signals (decoder or D/A). It may include compression and decompression functions.

Color facsimile A facsimile system that produces the recorded copy in more than one color. Group 3 fax has more than one mode of producing full-color fax copies.

Compatibility In facsimile, matching facsimile transmitter and facsimile receiver characteristics to a degree that produces an acceptable but not necessarily duplicate fax copy.

Compression ratio In digital facsimile, the ratio of the total bits used to represent the original to the total number of encoded bits.

Continuous-tone image (analog gray-scale image) An image in which each resolvable element is represented by one of a continuous range of gray densities between black and white.

Contouring Density-step border lines in recorded copy resulting from quantization of an original image that has observable gray shadings between adjacent quantization intervals.

Contrast The range of density between the lightest and darkest portions of an image.

Cover sheet An optional page with addressing information that is faxed before the first document page. It usually contains the names of the sender and the addressee, the subject, comments, the call-back telephone number, and the number of pages sent.

Data circuit-terminating (aka communication) equipment (DCE) A device that connects a computer, a fax machine, or other data terminal equipment (DTE) to the public switched telephone network (PSTN) or data communication network. For PC-fax, the DCE could be an internal modem on its bus, an external modem, or other device connected to a digital port.

Data terminal equipment (DTE) Any terminal or computer that can provide commands to operate data circuit-terminating equipment (DCE). For fax, the DTE can be a computer of any size.

Default A predetermined selection of settings for a fax device or program. A default may be the original manufacturer's setting or one customized by the user. For a PC-fax scanner, default settings may include dots per inch (dpi), contrast, brightness, sharpening, and screening algorithms.

Density (D) An optical darkness measurement of an area on a document, where $D = \log_{10}(100/\text{light reflectance})$. *Note:* For 10% reflected light, $D = 1$; for 1% reflected light, $D = 2$.

Digital facsimile The form of facsimile in which densities of the original are sampled and quantized as a digital signal for processing, transmission, and storage.

Direct inward dialing (DID) A method for a caller to dial directly a specific private branch exchange (PBX) internal number or extension within a company, bypassing the PBX operator.

Direct recording The type of recording in which a visible recorded copy is produced without subsequent processing.

Dither coding A means of converting a gray-level image to a bilevel image by varying the threshold from pixel to pixel, with the result that the human eye perceives the input gray scale.

Document A set of one or more pages that can be transmitted as a unit.

Dots per inch (dpi) The number of dots (pels or pixels) per inch used by a fax device to scan or form an image.

Electrolytic recording An outmoded recording method in which signal-controlled current passes through an electrolyte in wet recording paper and deposits metallic ions to produce a mark.

Electronic shading An electronic method of compensating for variations in sensitivity of the individual sensors of an array and for spatial variation in illumination of copy being scanned. The signal from each sensor is compensated under the control of information stored previously while a white line is scanned.

Electrosensitive recording An outmoded recording method in which signal-controlled current produces a mark by burning off a light-colored surface coating, exposing a black coating underneath at the point where the current passes into the recording paper.

Electrostatic recording Recording by means of a signal-controlled electrostatic field. A toner is required to make the image visible.

End-of-line (EOL) code In Group 3 digital facsimile systems, a sequence of digital symbols introduced at the beginning of a scanning line to establish synchronization of decoding and for error detection.

Encryption The process of pseudorandomly changing 0s and 1s, thereby scrambling a digital signal such as fax, by a coding pattern known only by a decoding unit at the fax receiver. Encryption prevents unauthorized interception of fax information.

Error correction mode (ECM) An optional error-correcting method of Group 3 with fax data sent in high-level data link control (HDLC) frames. The receiver detects frames that contain a check error and requests retransmission of those frames after a block of frames has been received.

European Telecommunications Standards Institute (ETSI) A group of manufacturers that provides telecommunications standards for use in Europe.

Facsimile The process by which a document is scanned or a rasterized image file is read and the resulting electrical signals are transmitted to another location, where the image is recorded or displayed as a copy of the original.

Facsimile copy A fax-recorded copy of an original produced by a facsimile recorder or a PC-fax printer.

Facsimile receiver The apparatus used to translate picture signals from the communications channel into a facsimile copy of the original.

Facsimile recorder The part of the facsimile receiver that performs the final conversion of electrical picture signals to an image of the original on the record medium.

Facsimile signal See *picture signal.*

Facsimile transmitter The apparatus employed to translate the original into picture signals suitable for delivery to the communication system.

Fax The abbreviation for *facsimile*; to send a document by facsimile; a document that has been received by facsimile.

Fingerprint facsimile Facsimile equipment used to transmit fingerprint cards. *Note:* Some systems send 8-by-8-inch cards at 192 lines/inch.

Ghost In analog facsimile, a spurious image resulting from echo, envelope delay distortion, or multipath reception.

Group 1 (obsolete) Analog facsimile equipment per CCITT Recommendation T.2 for sending an A4 page in 6 minutes over a voice-grade telephone line using frequency modulation with 1,300 Hz corresponding to white and 2,100 Hz to black of the original.

Group 2 (obsolete) Analog facsimile equipment per CCITT Recommendation T.3 for sending an A4 or an 8-1/2-by-11-inch page in 3 minutes over a voice-grade telephone line using 2,100-Hz amplitude, phase, vestigial sideband modulation.

Group 3 The digital facsimile standard per ITU Recommendations T.4 and T.30 that is used today by virtually all fax units worldwide.

Group 4 A little-used 1984 standard for digital facsimile equipment using digital transmission signals per ITU-T Recommendations T.5, T.6, and others.

Halftone image An image that has been converted from a continuous-tone image into a black-and-white image of small dots that retains the appearance of a continuous-tone image when viewed at a distance from which the individual dots are not visible.

Handshake An exchange of signals between two facsimile units to determine which of a set of specifications will be used for faxing and to verify that the fax receiving unit is ready. Here is a simplified example:

beep, beep, beep ("I am a fax calling.")

whistle (fax receiver answering), *bzzpbzzpbzzp* ("My telephone number is 123-456-7890. Here is a menu of features you can select.")

bzzpbzzpbzzp ("Here are my selections and the signals to set up your modem for receiving. Are you ready?")

bzzp ("Yes.")

bzzp ("Here comes the fax.") *hiss, hiss, hiss, hiss, ..., bzzp* ("Here is the end of page. No more pages to send.")

bzzp ("Received it OK; good-bye.")

bzzp ("Disconnect; good-bye.")

Horizontal resolution The number of dpi or dots per millimeter in the direction of scanning or recording.

Integrated services digital network (ISDN) A digital telephone network. Basic user service includes two 64-Kbps channels for voice or data plus one 16-Kbps channel.

International Computer Facsimile Association (ICFA) Formed in 1991, a group whose members include the leading companies of the computer and communications industries.

International Organization for Standardization (ISO) The Paris-based organization that develops standards for international and national data communications. The U.S. representation to the ISO is the American National Standards Institute (ANSI).

International Telecommunication Union (ITU) A United Nations intergovernmental advisory organization that sets worldwide recommendations as standards for data communications systems and equipment, including facsimile. As of 1993, all the former CCITT telecommunications standards activities are handled by the ITU Telecommunications Standardization sector (ITU-T).

Internet facsimile The use of the Internet to carry facsimile transmission in addition to or instead of the PSTN. Two basic modes have been defined, a store-and-forward mode and a real-time mode.

ITU-T recommendations Standards generated and published by various ITU Study Groups. Study Group 8, Characteristics of Telematic Services, is responsible for the T-series of recommendations that cover fax.

Jitter Error in the position of the recorded spot along the recorded line. Jitter, which usually is caused by mechanical imperfections in single-spot scanners and printers, is noticeable on the recording of a vertical line.

Joint Bi-level Imagery Group (JBIG) The ITU-T and ISO experts who developed the bilevel image compression standard ITU-T Recommendation T.82 (ISO/IEC IS 11544).

Joint Photographic Experts Group (JPEG) The ITU-T and ISO experts who developed ITU-T Recommendation T.81 (ISO/IEC IS 10918-1), the standard for compressing continuous-tone images (color or black and white). Used by fax and still-frame video.

K factor In modified READ (MR) coding for Group 3 facsimile, the number of facsimile scanning lines in a set used for coding. At the most, $K - 1$ lines are coded two-dimensionally to limit the disturbed area in the event of transmission errors. In ITU-T Group 3, $K = 2$ for 3.85 lines/mm, $K = 4$ for 7.7 lines/mm. In Group 4 facsimile, $K = \infty$.

Lines per inch (or millimeter) The number of scanning or recording lines per unit length measured perpendicular to the direction of scanning.

Line-to-line correlation The correlation of image information from scanning line to scanning line. Useful for two-dimensional coding, for example, modified READ (MR).

Maximum keying frequency The frequency equal to one-half the number of picture elements scanned per second.

Modem A device that converts serial digital data from a fax or data transmitter terminal into analog signals suitable for sending over a telephone channel; at the receiver terminal, it converts the signals back into digital format. (The term *modem* is shorthand for modulator/demodulator.)

Modified Huffman (MH) coding A one-dimensional image coding scheme for Group 3 fax that improves transmission speed. MH coding reduces the number of bits used to represent the image by using digital words to represent the numbers of black pels and white pels in each run along the scanning line.

Modified modified READ (MMR) coding A two-dimensional coding scheme that improves transmission speed over MR for fax systems that have no transmission errors.

Modified READ (MR) coding A two-dimensional image coding scheme. MR provides an improved transmission speed over MH by coding correlation of like-color pels in adjacent scanning lines.

Multifunctional Peripheral Interface (MFPI) MFPI is an interface for a new class of hybrid devices that may have fax, print, scan, and copy capabilities, as well as local storage. It has been approved by TIA TR-29.2 as interim standard IS-650.

Newsphoto facsimile Facsimile equipment used to transmit photographs for newspaper or magazine publishing.

Nonstandard facilities (NSF) Facsimile receiver that has the ability to use proprietary features not covered by the T-Series recommendations.

Octet Terminology used by the ITU to define a string of 8 bits. Known in the U.S. as a *byte.*

Original In facsimile usage, the page or stored computer image that is transmitted.

PC-fax A computer or multifunction communication device capable of sending and/or receiving Group 3 fax signals.

Pel A picture element that contains only black and white information (no gray shading). See *pixel.*

Photographic recording Recording by the exposure of a photosensitive surface to a signal-controlled light beam or spot.

Picture element The smallest area of the original that is sampled and represented by an electrical signal. See also *pel* and *pixel.*

Picture signal The signal that results from the scanning process or an electronically generated equivalent.

Pixel A picture element that has more than two levels of gray-scale information or color. See also *pel.*

Pixel interpolation Generation of additional unscanned pixels by logical comparison of nearby scanned pixels to simulate increased resolution. Interpolation may be one- or two-dimensional.

Polling A feature that allows a called fax machine to automatically transmit documents intended for the calling fax machine. To prevent unauthorized access, a polling password is required.

Printer In facsimile usage, a device that converts electrical signals representing images or characters into dots replicated on paper.

Protocol A set of format and timing conventions for exchange of handshaking signals and messages over a communications network.

Public switched telephone network (PSTN) The telephone system commonly used for making local and long-distance calls. Sometimes called the general switched telephone network (GSTN) or the "plain old telephone system" (POTS).

Quantizing levels In a digital facsimile system, the number of different gray steps representing a continuous-tone image. See also *contouring.*

Recommendation A term used by the ITU to describe an international standard approved by its members (e.g., Recommendation ITU-T T.4).

Record medium The physical medium on which the facsimile recorder forms a fax image of the original.

Record sheet The medium used to produce a fax hard copy. The record medium and the record sheet may be identical.

Recording The process of converting the picture signal in a facsimile receiver to an image on the record medium.

Recording spot The image area corresponding to a picture element formed on the record medium.

Redundancy reduction Coding for elimination of redundant information in the picture signal, reducing the transmission time and memory needed for storage.

Resolution A recent usage is merely the number of dpi used to produce the image, but that ignores the substantial differences observed in image sharpness from different scanners with the same dpi and from different threshold settings with the same scanner. The older usage relates to measurement of sharpness observed with a standardized test chart, such as ITU No. 4.

Resolution No. 1 The method for accelerated approval of a proposed ITU-T recommendation.

Scanner The part of the facsimile transmitter that systematically translates the densities of the original into a signal-wave form.

Scanning The process of analyzing successively the densities of the original according to a predetermined pattern.

Scanning direction Normal direction is from left to right and top to bottom of the original, as when one reads a page of print.

Scanning spot The area on the original viewed instantaneously by one photosensor of the scanner.

Service Class 1 (ITU-T.31 and TIA/EIA/ANSI-578) A PC-fax standard in which the PC provides most of the Group 3 fax functions via software and CPU processing.

Service Class 2 (ITU-T.32 and TIA/EIA/ANSI-592) A PC-fax standard in which the PC manages the session and provides image data in T.4 format. The PC-fax board provides the T.30 protocol.

Simultaneous voice and facsimile (SVF) An in-channel signaling procedure that provides for simultaneous transmission of voice and facsimile on the same telephone line. Calls may start as voice calls, add a data/facsimile channel during the call, start as a data call and add voice later, or start as an SVF call and switch to voice only.

Skew In analog fax systems, deviation of the recorded copy from rectangularity due to asynchronism between scanner and recorder. Also, the angular misalignment of the original from the paper feed direction.

Soft copy Image of the received fax on a cathode ray tube (CRT) or similar display.

Synchronizing In analog facsimile, maintenance of the proper position of the recording spot while it is writing to produce an undistorted recorded copy of the original.

Telecommunications Industry Association (TIA) A trade association that sponsors accredited engineering standards committees in the United States. The committees set U.S. standards for data communications technology and formulate the U.S. positions taken at ITU study group meetings.

Telefax The European term for *fax*; also referred to as *telecopier*.

Thermal recording Recording produced when signal-controlled current selectively heats a line of very small resistors arranged in a row across the recording page. Direct thermal recording marks directly on a thermosensitive record medium that turns black when heated. In transfer thermal recording, a hot spot on the thermal print head melts solid ink on an overlay sheet or web and transfers a dot, marking a white sheet of paper to form the recorded copy.

TR-29 The TIA-sponsored committee responsible for developing U.S. industry standards for fax technologies and for generating U.S. positions and contributions to the ITU.

TR-29.1 The TR-29 subcommittee that develops facsimile and file transfer protocols.

TR-29.2 The TR-29 subcommittee that develops standards for programmable facsimile digital interfaces (notably for controlling interaction between computers and fax devices).

TR-29.3 The TR-29 subcommittee that develops standards for audiographics teleconferencing.

TR-29.4 The TR-29 subcommittee that develops standards for secure facsimile.

Transmission time The time for sending a single page (elapsed time between the start of picture signals and the detection of the EOP signal by the facsimile receiver). This definition gives a shorter time per page than can be achieved for sending, because the times for needed handshaking signals are ignored.

Units/25.4 mm Expression used in ITU for 10-characters/inch printers with inch-based fax standards, becoming 10 characters/25.4 mm, thus avoiding the word *inch.*

V.8 An ITU presession negotiations method used between two modems for selecting and negotiating the session.

V.17 An ITU modem for transmitting fax and data signals at rates up to 14.4 Kbps over the PSTN.

V.27ter An ITU modem for transmitting fax and data signals at rates up to 4.8 Kbps over the PSTN.

V.29 An ITU modem for transmitting fax and data signals at rates up to 9.6 Kbps over the PSTN.

V.34 An ITU modem for transmitting fax and data signals at up to 33.6 Kbps over the PSTN.

Xerographic recording A recording method in which an electrically charged insulating surface is exposed by a signal-controlled light beam to selectively discharge it, forming a latent electrostatic image that is subsequently developed with a toner and printed on paper.

About the Authors

Kenneth R. McConnell is a fax consultant drawing on his many years of experience in facsimile engineering, starting at the U.S. Army Signal Corps Laboratories. At Times Facsimile Corporation (later Westrex and then Datalog), he designed military and commercial fax equipment and later became manager of the engineering department. Equipment designed and built there included facsimile systems for business documents, newsphotos, satellite photos, weather maps, fingerprints, x-ray films, and full-page newspaper masters. In addition, he designed fax modems and equipment for transmission of weather maps by fax on experimental HF radio station KE2XER. He moved on to Visual Sciences and later was vice-president of engineering at Panafax, USA.

Mr. McConnell, as chair of the IRE (later the IEEE) Facsimile Committee, was instrumental in the building of the first fax test chart issued as an engineering standard. That chart was widely used throughout the world for performance testing of fax equipment and in many other imaging applications. As first chair of EIA TR-29 Committee on Facsimile in 1962, he was a primary mover for compatibility between fax machines made by different manufacturers and in the development of fax standards that would make that compatibility possible. In 1992, TR-29 gave him an award for "recognition of outstanding service and achievement for his distinguished leadership and contributions to the development of facsimile and facsimile standards."

He organized and served as chair of facsimile conferences presented by the Institute for Graphic Communications both in the United States and Europe for many years in the 1970s and 1980s. At International Electronic Imaging Expositions and Conferences, he chaired sessions and presented minicourses on digital facsimile.

Besides the previous editions of *FAX* (published by Artech House), Mr. McConnell has authored numerous technical papers and magazine articles on facsimile. He wrote *Facsimile and Its Role in Electronic Imaging*, a technical publication for the Association for Information and Image Management (AIIM), and the facsimile sections in the *McGraw Hill Encyclopedia of Science and Industry*, and the *Reinhold Encyclopedia of Electronics*.

Mr. McConnell received his B.S. in electrical engineering cum laude from Michigan State University and is a member of Tau Beta Pi. He is a Life Member of IEEE and holds 12 patents.

Dr. Dennis Bodson received his B.E.E. and M.E.E. degrees in electrical engineering from The Catholic University of America and a master's degree in public administration from the University of Southern California Washington Center for Public Affairs. He completed his Ph.D. in electrical engineering at California Western University. He has over 35 years of telecommunications experience in government and industry and has played a prominent role in the development of international, national, industry, and federal standards relating to facsimile.

Dr. Bodson has served for many years as the head of the U.S. delegation to the ITU-T SG8, which is responsible for the development of ITU facsimile recommendations. Dr. Bodson is currently the rapporteur for Q2/SG8, which deals with facsimile test charts and images. He has participated in the development of facsimile standards in the TR-29 Committee and in his capacity as chief of the National Communications System Office of Technology and Standards. Dr. Bodson has been very active in the IEEE Standards Board and Standards Association activities. He is responsible for the recent development of a series of IEEE standards on facsimile test charts for black and white, gray scale, and color.

Dr. Bodson has been recognized for his standardization activities throughout his career. He has received the Federal Engineer of the Year award as the Outstanding Engineer from the National Communications System, given by the National Society of Professional Engineers, and the George S. Wham Leadership Medal from the American National Standards Institute. Dr. Bodson was recently awarded the Charles Proteus Steinmetz award from the IEEE for his work in facsimile, and the IEEE grade of Fellow was based on his achievements in facsimile. He has published approximately 50 technical publications, most of them facsimile related, and has received five patents in this field.

Stephen J. Urban is currently Vice President of Network Systems at Delta Information Systems, Incorporated. He holds a B.S. (cum laude) and an M.S.

degree in electrical engineering from Lehigh University and graduate credits from the University of Pennsylvania. From 1964 to 1967 he served as an engineer in the Advanced Radar Lab and in Advanced Engineering and Research for Philco-Ford Corporation. In 1967, Mr. Urban became manager of the Computer-Aided Design Technology Lab, responsible for the development of computer-aided systems used for design, fabrication, and testing of electronic equipment. From 1977 to 1981 he served as the Software and Computer Facility Manager for Fairchild Camera.

In 1981, Mr. Urban joined Delta Information Systems, Inc., and took responsibility for projects dealing with image communication including compression of binary, grayscale, and color images, and telematic protocols for transmitting these images. He was the principal investigator on various bi-level compression algorithm studies. He has been responsible for similar projects related to grayscale and color image compression using differential PCM, transform coding, vector quantization, and other forms of entropy coding. He is currently responsible for the development of protocol analyzers and test systems for video teleconferencing operating over a variety of networks.

Mr. Urban, a member of IEEE, ACM, Tau Beta Pi, and Eta Kappa Nu, is also the author of several papers. He is chair of the Telecommunication Industries Association TR-29 committee on Facsimile and Facsimile Systems and is active in international telematic standards as a U.S. Delegate to the ITU-T, especially in the area of standards development for facsimile over the Internet. Mr. Urban has been issued a patent on Group 3 Facsimile Error Reduction.

Index

Recent Titles in the Artech House Telecommunications Library

Vinton G. Cerf, Senior Series Editor

Access Networks: Technology and V5 Interfacing, Alex Gillespie

Advanced High-Frequency Radio Communications, Eric E. Johnson, Robert I. Desourdis, Jr., et al.

Advances in Telecommunications Networks, William S. Lee and Derrick C. Brown

Advances in Transport Network Technologies: Photonics Networks, ATM, and SDH, Ken-ichi Sato

Asynchronous Transfer Mode Networks: Performance Issues, Second Edition, Raif O. Onvural

ATM Switches, Edwin R. Coover

ATM Switching Systems, Thomas M. Chen and Stephen S. Liu

Broadband Network Analysis and Design, Daniel Minoli

Broadband Networking: ATM, SDH, and SONET, Mike Sexton and Andy Reid

Broadband Telecommunications Technology, Second Edition, Byeong Lee, Minho Kang, and Jonghee Lee

Client/Server Computing: Architecture, Applications, and Distributed Systems Management, Bruce Elbert and Bobby Martyna

Communication and Computing for Distributed Multimedia Systems, Guojun Lu

Communications Technology Guide for Business, Richard Downey, Seán Boland, and Phillip Walsh

Community Networks: Lessons from Blacksburg, Virginia, Andrew Cohill and Andrea Kavanaugh, editors

Computer Mediated Communications: Multimedia Applications, Rob Walters

Computer Telephony Integration, Second Edition, Rob Walters

Convolutional Coding: Fundamentals and Applications, Charles Lee

Desktop Encyclopedia of the Internet, Nathan J. Muller

Distributed Multimedia Through Broadband Communications Services, Daniel Minoli and Robert Keinath

Electronic Mail, Jacob Palme

Enterprise Networking: Fractional T1 to SONET, Frame Relay to BISDN, Daniel Minoli

FAX: Facsimile Technology and Systems, Third Edition, Kenneth R. McConnell, Dennis Bodson, and Stephen Urban

Guide to ATM Systems and Technology, Mohammad A. Rahman

Guide to Telecommunications Transmission Systems, Anton A. Huurdeman

A Guide to the TCP/IP Protocol Suite, Floyd Wilder

Information Superhighways Revisited: The Economics of Multimedia, Bruce Egan

International Telecommunications Management, Bruce R. Elbert

Internet E-mail: Protocols, Standards, and Implementation, Lawrence Hughes

Internetworking LANs: Operation, Design, and Management, Robert Davidson and Nathan Muller

Introduction to Satellite Communication, Second Edition, Bruce R. Elbert

Introduction to Telecommunications Network Engineering, Tarmo Anttalainen

Introduction to Telephones and Telephone Systems, Third Edition, A. Michael Noll

LAN, ATM, and LAN Emulation Technologies, Daniel Minoli and Anthony Alles

The Law and Regulation of Telecommunications Carriers, Henk Brands and Evan T. Leo

Marketing Telecommunications Services: New Approaches for a Changing Environment, Karen G. Strouse

Mutlimedia Communications Networks: Technologies and Services, Mallikarjun Tatipamula and Bhumip Khashnabish, Editors

Networking Strategies for Information Technology, Bruce Elbert

Packet Switching Evolution from Narrowband to Broadband ISDN, M. Smouts

Packet Video: Modeling and Signal Processing, Naohisa Ohta

Performance Evaluation of Communication Networks, Gary N. Higginbottom

Practical Computer Network Security, Mike Hendry

Practical Multiservice LANs: ATM and RF Broadband, Ernest O. Tunmann

Principles of Secure Communication Systems, Second Edition, Don J. Torrieri

Principles of Signaling for Cell Relay and Frame Relay, Daniel Minoli and George Dobrowski

Pulse Code Modulation Systems Design, William N. Waggener

Signaling in ATM Networks, Raif O. Onvural and Rao Cherukuri

Smart Cards, José Manuel Otón and José Luis Zoreda

Smart Card Security and Applications, Mike Hendry

SNMP-Based ATM Network Management, Heng Pan

Successful Business Strategies Using Telecommunications Services, Martin F. Bartholomew

Super-High-Definition Images: Beyond HDTV, Naohisa Ohta

Telecommunications Deregulation, James Shaw

Telemetry Systems Design, Frank Carden

Teletraffic Technologies in ATM Networks, Hiroshi Saito

Understanding Modern Telecommunications and the Information Superhighway, John G. Nellist and Elliott M. Gilbert

Understanding Networking Technology: Concepts, Terms, and Trends, Second Edition, Mark Norris

Understanding Token Ring: Protocols and Standards, James T. Carlo, Robert D. Love, Michael S. Siegel, and Kenneth T. Wilson

Videoconferencing and Videotelephony: Technology and Standards, Second Edition, Richard Schaphorst

Visual Telephony, Edward A. Daly and Kathleen J. Hansell

World-Class Telecommunications Service Development, Ellen P. Ward

For further information on these and other Artech House titles, including previously considered out-of-print books now available through our In-Print-Forever® (IPF®) program, contact:

Artech House
685 Canton Street
Norwood, MA 02062
Phone: 781-769-9750
Fax: 781-769-6334
e-mail: artech@artechhouse.com

Artech House
46 Gillingham Street
London SW1V 1AH UK
Phone: +44 (0)20 7596-8750
Fax: +44 (0)20 7630-0166
e-mail: artech-uk@artechhouse.com

Find us on the World Wide Web at:
www.artechhouse.com